SCHAUM'S OUTLINE OF

THEORY AND PROBLEMS

of Plane and Solid

ANALYTIC GEOMETRY

●

BY

JOSEPH H. KINDLE, Ph.D.

Professor of Mathematics
University of Cincinnati

●

SCHAUM'S OUTLINE SERIES

McGraw-Hill, Inc.

New York St. Louis San Francisco Auckland Bogotá
Caracas Lisbon London Madrid Mexico City Milan
Montreal New Delhi San Juan Singapore
Sydney Tokyo Toronto

ISBN 07-034575-9

26 27 28 29 30 31 32 33 34 35 36 37 38 39 40 BAW BAW 9 9 8 7 6 5 4 3

Cover design by Amy E. Becker.

Foreword to the Student

This problem book is designed to supplement courses in analytic geometry which are usually given in colleges and technical schools. It follows closely the order used in most textbooks of analytic geometry. A total of 345 carefully solved representative problems and 910 supplementary problems for practice, of varying degrees of difficulty, is included. The problems are so arranged as to present a natural development of each topic. Since analytic geometry is primarily a problem solving course and since one of the principal causes of poor work in any course in mathematics is disorderly ways of attempting to solve problems, we are convinced that this book, properly used, will prove very helpful. It is intended also to be of service to students who feel the need for a review of the fundamental theory and problem work of analytic geometry.

Proper use of the book necessitates a clear understanding of what it is not as well as what it is. It is definitely not a formal textbook and you should not attempt to use it as a means of avoiding a careful study of your regular text. Each chapter of the book has a brief summary of the necessary definitions, principles and theorems, followed by graded sets of solved and supplementary problems.

It cannot be emphasized too strongly that one learns mathematics by doing mathematics, by solving problems. A perfunctory reading of a textbook, the memorizing of a few likely-looking formulas, and a casual study of the problems solved in this book will do no more than give an illusory feeling of well-being; actually, you have gained little more than a vague impression of the material. To make effective use of this book you should reproduce the solutions on paper, pausing to see the why as well as the how of each step. There is something to be learned from each of the solved problems; and when you have learned it, you should have very little difficulty with most of the supplementary practice problems and with the problems of your text.

J. H. K.

Cincinnati, Ohio

November, 1950

Contents

CHAPTER 1

Rectangular Coordinates

RECTANGULAR COORDINATES. In earlier courses in Algebra and Trigonometry, use was made of rectangular coordinates. In this system the plane is divided into four quadrants by two perpendicular lines intersecting at a point O. The horizontal line $X'OX$ is called the x-axis, the vertical line $Y'OY$ the y-axis, and the two together are called the *coordinate axes*. The point O is the origin.

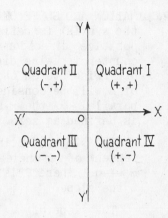

The distance from the y-axis is called the x-coordinate or *abscissa* of the point. The distance from the x-axis is called the y-coordinate or *ordinate* of the point. The two distances taken together are called the *coordinates* of the point and are represented by the symbol (x,y). Abscissas are positive when measured to the right of the y-axis, negative when measured to the left. Ordinates are positive when measured above the x-axis, negative if below.

When a set of points, whose coordinates are given, are to be plotted, a suitable scale is chosen and marked on the coordinate axes. The points can then be easily plotted.

DISTANCE BETWEEN TWO POINTS. The distance d between two points $P_1(x_1,y_1)$ and $P_2(x_2,y_2)$ is readily seen to be

$$d = \sqrt{(x_2 - x_1)^2 + (y_2 - y_1)^2}.$$

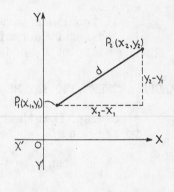

Thus, the distance between points $(4,-1)$ and $(7,3)$ is

$$d = \sqrt{(7-4)^2 + (3+1)^2}$$

$$= 5 \text{ units.}$$

POINT OF DIVISION is the point which divides a line segment in a given ratio. Consider that $P_1(x_1,y_1)$ and $P_2(x_2,y_2)$ are two points on a line directed from P_1 through P_2. Let $P(x,y)$ be a third point which divides the line so that $\dfrac{P_1P}{PP_2} = r$. Since P_1P and PP_2 are read in the same direction on the line, the ratio will be positive. If the point of division $P(x,y)$ were on the segment extended in either direction, then the ratio $\dfrac{P_1P}{PP_2} = r$ would be negative because P_1P and PP_2 would then have opposite directions.

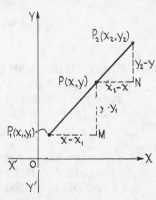

By similar triangles, $\dfrac{P_1M}{PN} = \dfrac{x - x_1}{x_2 - x} = \dfrac{P_1P}{PP_2} = r.$

1

Solving for x, $x = \dfrac{x_1 + rx_2}{1 + r}$.   Similarly, $y = \dfrac{y_1 + ry_2}{1 + r}$.

If $P(x,y)$ is the midpoint of line P_1P_2, $r = 1$ and $x = \dfrac{x_1 + x_2}{2}$, $y = \dfrac{y_1 + y_2}{2}$.

INCLINATION AND SLOPE OF A LINE. The *inclination* of a line L (not parallel to the x-axis) is defined as the smallest positive angle measured from the positive direction of the x-axis in a counterclockwise direction to L. Unless otherwise stated the positive direction of L will be considered upward. If L is parallel to the x-axis its inclination is defined as zero.

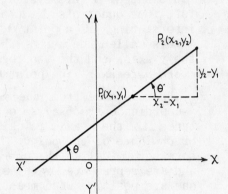

The *slope* of a line is defined as the tangent of the angle of inclination. Thus $m = \tan \theta$ where θ is the inclination and m the slope.

The slope of a line passing through two points $P_1(x_1,y_1)$ and $P_2(x_2,y_2)$ is

$$m = \tan \theta = \frac{y_2 - y_1}{x_2 - x_1}$$

regardless of the quadrants in which P_1 and P_2 lie.

PARALLEL AND PERPENDICULAR LINES. If two lines are parallel, their slopes are equal.

If two lines L_1 and L_2 are perpendicular, the slope of one of the lines is the negative reciprocal of the slope of the other line. Thus if m_1 is the slope of L_1 and m_2 is the slope of L_2, then $m_1 = -1/m_2$, or $m_1m_2 = -1$.

ANGLE BETWEEN TWO INTERSECTING LINES. The angle α measured in a positive direction, counterclockwise, from the line L_1 whose slope is m_1, to the line L_2 whose slope is m_2 is

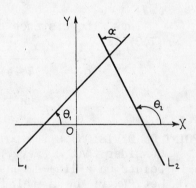

$$\tan \alpha = \frac{m_2 - m_1}{1 + m_2m_1}.$$

Proof: $\theta_2 = \alpha + \theta_1$, or $\alpha = \theta_2 - \theta_1$.

$$\tan \alpha = \tan(\theta_2 - \theta_1)$$

$$= \frac{\tan \theta_2 - \tan \theta_1}{1 + \tan \theta_2 \tan \theta_1} = \frac{m_2 - m_1}{1 + m_2m_1}.$$

AREA OF ANY POLYGON WHOSE VERTICES ARE GIVEN. Let $P_1(x_1,y_1)$, $P_2(x_2,y_2)$, $P_3(x_3,y_3)$ be the vertices of a triangle. The area A of the triangle in terms of the vertices is

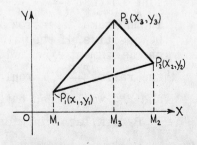

$$A = \tfrac{1}{2}(x_1y_2 + x_2y_3 + x_3y_1 - x_3y_2 - x_2y_1 - x_1y_3).$$

Proof: The area of the triangle = area of trapezoid $M_1P_1P_3M_3$ + area of trapezoid $M_3P_3P_2M_2$ − area of trapezoid $M_1P_1P_2M_2$. Then

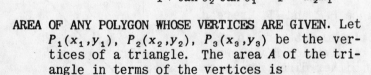

$$A = \tfrac{1}{2}(y_1 + y_3)(x_3 - x_1) + \tfrac{1}{2}(y_3 + y_2)(x_2 - x_3) - \tfrac{1}{2}(y_1 + y_2)(x_2 - x_1)$$
$$= \tfrac{1}{2}(x_1 y_2 + x_2 y_3 + x_3 y_1 - x_1 y_3 - x_2 y_1 - x_3 y_2).$$

This can be expressed also in the form of a determinant:

$$A = \tfrac{1}{2} \begin{vmatrix} x_1 & y_1 & 1 \\ x_2 & y_2 & 1 \\ x_3 & y_3 & 1 \end{vmatrix}$$

Another form which is convenient to use for the area of a triangle, but which is especially useful when the areas of polygons of more than three sides are to be determined, is in the form of an array.

$$A = \tfrac{1}{2}(x_1 y_2 + x_2 y_3 + x_3 y_1 - x_1 y_3 - x_3 y_2 - x_2 y_1). \qquad A = \tfrac{1}{2} \begin{vmatrix} x_1 & y_1 \\ x_2 & y_2 \\ x_3 & y_3 \\ x_1 & y_1 \end{vmatrix}$$

Note that the first row of the array is repeated.

SOLVED PROBLEMS

DISTANCE BETWEEN TWO POINTS.

1. Determine the distance between (a) $(-2, 3)$ and $(5, 1)$, (b) $(6, -1)$ and $(-4, -3)$.

a) $\quad d = \sqrt{(x_2 - x_1)^2 + (y_2 - y_1)^2} = \sqrt{(5 + 2)^2 + (1 - 3)^2} = \sqrt{49 + 4} = \sqrt{53}$

b) $\quad d = \sqrt{(x_2 - x_1)^2 + (y_2 - y_1)^2} = \sqrt{(-4 - 6)^2 + (-3 + 1)^2} = \sqrt{104} = 2\sqrt{26}$

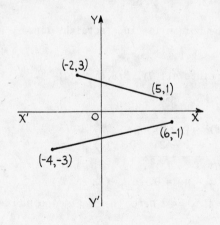

Problem 1.

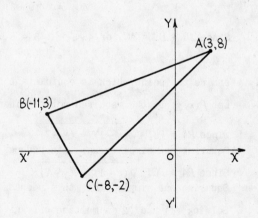

Problem 2.

2. Show that points $A(3, 8)$, $B(-11, 3)$, $C(-8, -2)$ are the vertices of an isosceles triangle.

$$AB = \sqrt{(3 + 11)^2 + (8 - 3)^2} = \sqrt{221}$$

$$BC = \sqrt{(-11 + 8)^2 + (3 + 2)^2} = \sqrt{34}$$

$$AC = \sqrt{(3 + 8)^2 + (8 + 2)^2} = \sqrt{221}. \qquad \text{Since } AB = AC, \text{ the triangle is isosceles.}$$

3. (a) Show that the points $A(7,5)$, $B(2,3)$, $C(6,-7)$ are the vertices of a right triangle.
(b) Find the area of the right triangle.

a) $AB = \sqrt{(7-2)^2 + (5-3)^2} = \sqrt{29}$ $BC = \sqrt{(2-6)^2 + (3+7)^2} = \sqrt{116}$

$$AC = \sqrt{(7-6)^2 + (5+7)^2} = \sqrt{145}$$

Since $(AB)^2 + (BC)^2 = (AC)^2$, or $29 + 116 = 145$, ABC is a right triangle.

b) Area $= \frac{1}{2}(AB)(BC) = \frac{1}{2}\sqrt{29}\sqrt{116} = 29$ square units.

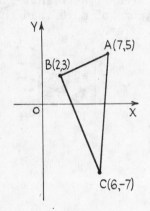

Problem 3.

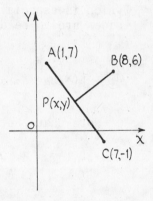

Problem 4.   *Problem 5.*

4. Show that the following points lie in a straight line: $A(-3,-2)$, $B(5,2)$, $C(9,4)$.

$$AB = \sqrt{(5+3)^2 + (2+2)^2} = 4\sqrt{5} \qquad BC = \sqrt{(9-5)^2 + (4-2)^2} = 2\sqrt{5}$$

$$AC = \sqrt{(9+3)^2 + (4+2)^2} = 6\sqrt{5}$$

Since $AB + BC = AC$, or $4\sqrt{5} + 2\sqrt{5} = 6\sqrt{5}$, the points lie in a straight line.

5. Determine the point which is equidistant from points $A(1,7)$, $B(8,6)$, $C(7,-1)$.

Let $P(x,y)$ be the required point.   Then $PA = PB = PC$.

Since $PA = PB$, $\sqrt{(x-1)^2 + (y-7)^2} = \sqrt{(x-8)^2 + (y-6)^2}$.
Squaring and simplifying, this reduces to $7x - y - 25 = 0$.   (1)

Since $PA = PC$, $\sqrt{(x-1)^2 + (y-7)^2} = \sqrt{(x-7)^2 + (y+1)^2}$.
Squaring and simplifying, this reduces to $3x - 4y = 0$.   (2)

Solving (1) and (2) simultaneously gives $x = 4$, $y = 3$. Hence the required point is $(4,3)$.

POINT WHICH DIVIDES A LINE SEGMENT IN A GIVEN RATIO.

6. Find the coordinates of the point $P(x,y)$ which divides the line
segment from $P_1(1,7)$ to $P_2(6,-3)$ in the ratio $r = 2/3$.

Since the ratio is positive, P_1P and PP_2 must have the same
direction and $P(x,y)$ must be on the segment P_1P_2 (internal division).

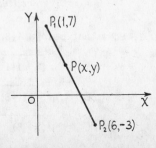

$$r = \frac{P_1P}{PP_2} = \frac{2}{3}$$

$$x = \frac{x_1 + rx_2}{1 + r} = \frac{1 + \frac{2}{3}(6)}{1 + \frac{2}{3}} = 3 \qquad y = \frac{y_1 + ry_2}{1 + r} = \frac{7 + \frac{2}{3}(-3)}{1 + \frac{2}{3}} = 3$$

The required point is (3,3).

7. Determine the coordinates of the point $P(x,y)$ which divides the line segment from $P_1(-2,1)$ to $P_2(3,-4)$ externally in the ratio $r = -8/3$.

Since the ratio is negative, P_1P and PP_2 have opposite directions and $P(x,y)$ must be outside segment P_1P_2 (external division). $\quad r = \dfrac{P_1P}{PP_2} = -\dfrac{8}{3}$.

$$x = \frac{x_1 + rx_2}{1 + r} = \frac{-2 + (-\frac{8}{3})(3)}{1 + (-\frac{8}{3})} = 6 \qquad y = \frac{y_1 + ry_2}{1 + r} = \frac{1 + (-\frac{8}{3})(-4)}{1 + (-\frac{8}{3})} = -7$$

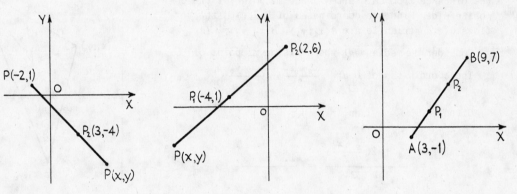

Problem 7. Problem 8. Problem 9.

8. A circle with center at $P_1(-4,1)$ has one end of a diameter at $P_2(2,6)$. Determine the coordinates $P(x,y)$ of the other end.

$$r = \frac{P_1P}{PP_2} = -\frac{1}{2}$$

Since P_1P and PP_2 are read in opposite directions along the line, the ratio r is negative.

$$x = \frac{x_1 + rx_2}{1 + r} = \frac{-4 + (-\frac{1}{2})(2)}{1 + (-\frac{1}{2})} = -10 \qquad y = \frac{y_1 + ry_2}{1 + r} = \frac{1 + (-\frac{1}{2})(6)}{1 + (-\frac{1}{2})} = -4$$

9. Determine the two points of trisection, $P_1(x_1,y_1)$ and $P_2(x_2,y_2)$, of the line segment joining $A(3,-1)$ and $B(9,7)$.

To find $P_1(x_1,y_1)$: $\quad r_1 = \dfrac{AP_1}{P_1B} = \dfrac{1}{2}, \qquad x_1 = \dfrac{3 + \frac{1}{2}(9)}{1 + \frac{1}{2}} = 5, \qquad y_1 = \dfrac{-1 + \frac{1}{2}(7)}{1 + \frac{1}{2}} = \dfrac{5}{3}.$

To find $P_2(x_2,y_2)$: $\quad r_2 = \dfrac{AP_2}{P_2B} = \dfrac{2}{1}, \qquad x_2 = \dfrac{3 + 2(9)}{1 + 2} = 7, \qquad y_2 = \dfrac{-1 + 2(7)}{1 + 2} = \dfrac{13}{3}.$

10. The point $B(-4,1)$ is three fifths of the distance from one end $A(2,-2)$ of a line segment to the other end $C(x,y)$. Find end point $C(x,y)$.

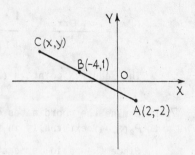

$$\frac{AB}{BC} = \frac{3}{2} \qquad r = \frac{AC}{CB} = -\frac{5}{2}$$

Since AC and CB are read in opposite directions, the ratio r is negative.

$$x = \frac{2 + \left(-\frac{5}{2}\right)(-4)}{1 + \left(-\frac{5}{2}\right)} = -8 \qquad\qquad y = \frac{-2 + \left(-\frac{5}{2}\right)(1)}{1 + \left(-\frac{5}{2}\right)} = 3$$

11. The medians of a triangle intersect in a point $P(x,y)$, called the centroid of the triangle, which is 2/3 of the distance from any vertex to the midpoint of the opposite side. Find the coordinates of $P(x,y)$ if the vertices of the triangle are $A(x_1,y_1)$, $B(x_2,y_2)$, $C(x_3,y_3)$.

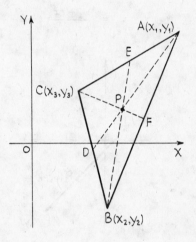

Consider the median APD, where D is the midpoint of BC. The coordinates of D are $\dfrac{x_2+x_3}{2}$, $\dfrac{y_2+y_3}{2}$.

Given $\dfrac{AP}{AD} = \dfrac{2}{3}$. Then $r = \dfrac{AP}{PD} = \dfrac{2}{1} = 2$.

$$x = \frac{x_1 + 2\left(\frac{x_2+x_3}{2}\right)}{1 + 2} = \frac{x_1 + x_2 + x_3}{3}$$

$$y = \frac{y_1 + 2\left(\frac{y_2+y_3}{2}\right)}{1 + 2} = \frac{y_1 + y_2 + y_3}{3}$$

The required point is $\frac{1}{3}(x_1 + x_2 + x_3)$, $\frac{1}{3}(y_1 + y_2 + y_3)$. The same result is obtained when

median BPE or median CPF is employed, and $r = \dfrac{AP}{PD} = \dfrac{BP}{PE} = \dfrac{CP}{PF} = \dfrac{2}{1} = 2$.

INCLINATION AND SLOPE OF A LINE.

12. Find the slope m and the angle of inclination θ of the lines through each of the following pairs of points:

a) $(-8,-4)$, $(5,9)$.   c) $(-11,4)$, $(-11,10)$.

b) $(10,-3)$, $(14,-7)$.   d) $(8,6)$, $(14,6)$.

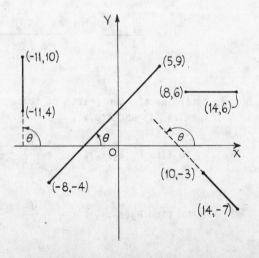

$$m = \tan\theta = \frac{y_2 - y_1}{x_2 - x_1}$$

a) $m = \dfrac{9+4}{5+8} = 1 \qquad\qquad \theta = \tan^{-1} 1 = 45°$

b) $m = \dfrac{-7+3}{14-10} = -1 \qquad\qquad \theta = \tan^{-1} -1 = 135°$

c) $m = \dfrac{10-4}{-11+11} = \dfrac{6}{0} = \infty \qquad \theta = \tan^{-1}\infty = 90°$

d) $m = \dfrac{6-6}{14-8} = \dfrac{0}{6} = 0 \qquad\qquad \theta = \tan^{-1} 0 = 0°$

13. Show that the three points $A(-3,4)$, $B(3,2)$ and $C(6,1)$ lie on the same straight line.

Slope of $AB = \dfrac{2-4}{3+3} = -\dfrac{1}{3}$. Slope of $AC = \dfrac{1-4}{6+3} = -\dfrac{1}{3}$.

Since the slope of AB equals the slope of AC, the three given points lie on the same straight line.

14. By means of slopes, show that the points $A(8,6)$, $B(4,8)$ and $C(2,4)$ are the vertices of a right triangle.

Slope of $AB = \dfrac{8-6}{4-8} = -\dfrac{1}{2}$. Slope of $BC = \dfrac{4-8}{2-4} = 2$.

Since the slope of AB is the negative reciprocal of the slope of BC, these two sides of the triangle are perpendicular.

ANGLE BETWEEN TWO INTERSECTING LINES.

15. The angle between two lines, L_1 and L_2 is $45°$. If the slope m_1 of L_1 is $2/3$, determine the slope m_2 of L_2.

$$\tan 45° = \frac{m_2 - m_1}{1 + m_2 m_1}, \quad \text{or} \quad 1 = \frac{m_2 - \dfrac{2}{3}}{1 + \dfrac{2}{3}m_2}. \quad \text{Solving, } m_2 = 5.$$

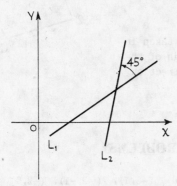

Problem 15. Problem 16.

16. Find the interior angles of the triangle whose vertices are $A(-3,-2)$, $B(2,5)$, $C(4,2)$.

$$m_{AB} = \frac{5+2}{2+3} = \frac{7}{5} \qquad m_{BC} = \frac{2-5}{4-2} = -\frac{3}{2} \qquad m_{CA} = \frac{2+2}{4+3} = \frac{4}{7}$$

$$\tan A = \frac{m_{AB} - m_{CA}}{1 + m_{AB}m_{CA}} = \frac{\dfrac{7}{5} - \dfrac{4}{7}}{1 + \dfrac{7}{5}\left(\dfrac{4}{7}\right)} = \frac{29}{63}, \qquad A = 24°43.1'.$$

$$\tan B = \frac{m_{BC} - m_{AB}}{1 + m_{BC}m_{AB}} = \frac{-\dfrac{3}{2} - \dfrac{7}{5}}{1 + \left(-\dfrac{3}{2}\right)\left(\dfrac{7}{5}\right)} = \frac{29}{11}, \quad B = 69°13.6'.$$

$$\tan C = \frac{m_{CA} - m_{BC}}{1 + m_{CA}m_{BC}} = \frac{\dfrac{4}{7} - \left(-\dfrac{3}{2}\right)}{1 + \dfrac{4}{7}\left(-\dfrac{3}{2}\right)} = \frac{29}{2}, \quad C = 86°3.3'. \quad \text{Check: } A + B + C = 180°.$$

AREA OF ANY POLYGON WHOSE VERTICES ARE GIVEN.

17. Find the area A of the triangle whose vertices are (2,3), (5,7), (-3,4).

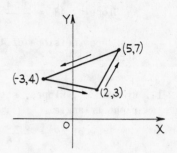

$$A = \frac{1}{2} \begin{vmatrix} 2 & 3 \\ 5 & 7 \\ -3 & 4 \\ 2 & 3 \end{vmatrix}$$

$$= \frac{1}{2}[2 \cdot 7 + 5 \cdot 4 + (-3)(3) - 2 \cdot 4 - (-3)(7) - 5 \cdot 3]$$

$$= \frac{1}{2}(14 + 20 - 9 - 8 + 21 - 15) = 11.5 \text{ square units.}$$

18. Find the area A of the pentagon whose vertices are (-5,-2), (-2,5), (2,7), (5,1), (2,-4).

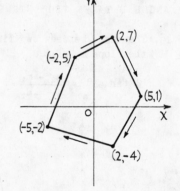

$$A = \frac{1}{2} \begin{vmatrix} -5 & -2 \\ -2 & 5 \\ 2 & 7 \\ 5 & 1 \\ 2 & -4 \\ -5 & -2 \end{vmatrix}$$

$$= \frac{1}{2}[(-5)(5) + (-2)(7) + 2 \cdot 1 + 5(-4) + 2(-2) \\ - (-5)(-4) - 2 \cdot 1 - 5 \cdot 7 - 2 \cdot 5 - (-2)(-2)]$$

$$= \frac{1}{2}(-132) = -66.$$

Answer: 66 square units. If the vertices are taken in order counterclockwise around the polygon, the sign of the area will be positive; if taken in a clockwise direction, the sign will be negative.

SUPPLEMENTARY PROBLEMS

1. Plot the points whose coordinates are: (2,3), (4,0), (-3,1), ($\sqrt{2}$,-1), (-2,0), (-2,$\sqrt{3}$), (0,1), (-2,$\sqrt{8}$), ($\sqrt{7}$,0), (0,0), (4.5,-2), ($\sqrt{10}$,-$\sqrt{2}$), (0,$\sqrt{3}$), (2.3,-6).

2. Draw the triangle whose vertices are: (a) (0,0), (-1,5), (4,2);
$\qquad$ (b) ($\sqrt{2}$, 0), (4,5), (-3,2);
$\qquad$ (c) (2 + $\sqrt{2}$, -3), ($\sqrt{3}$,3), (-2, 1+$\sqrt{8}$).

3. Draw the polygon whose vertices are: (a) (-3,2), (1,5), (5,3), (1,-2);
$\qquad$ (b) (-5,0), (-3,-4), (3,-3), (7,2), (1,6).

4. Find the distance between the pairs of points whose coordinates are:
$\quad$ (a) (4,1), (3,-2); $\qquad$ (c) (0,3), (-4,1); $\qquad$ (e) (2,-6), (2,-2);
$\quad$ (b) (-7,4), (1,-11); $\qquad$ (d) (-1,-5), (2,-3); $\qquad$ (f) (-3,1), (3,-1).
Ans. (a) $\sqrt{10}$, (b) 17, (c) $2\sqrt{5}$, (d) $\sqrt{13}$, (e) 4, (f) $2\sqrt{10}$.

5. Find the perimeters of the triangles whose vertices are:
$\quad$ (a) (-2,5), (4,3), (7,-2); $\qquad$ (c) (2,-5), (-3,4), (0,-3);
$\quad$ (b) (0,4), (-4,1), (3,-3); $\qquad$ (d) (-1,-2), (4,2), (-3,5).
Ans. (a) 23.56, (b) 20.67, (c) 20.74, (d) 21.30.

6. Show that the triangles whose vertices are given are isosceles:
$\quad$ (a) (2,-2), (-3,-1), (1,6); $\qquad$ (c) (2,4), (5,1), (6,5);
$\quad$ (b) (-2,2), (6,6), (2,-2); $\qquad$ (d) (6,7), (-8,-1), (-2,-7).

7. Show that the triangles whose vertices are given are right triangles. Find their areas.
   (a) (0,9), (-4,-1), (3,2);   (c) (3,-2), (-2,3), (0,4);
   (b) (10,5), (3,2), (6,-5);   (d) (-2,8), (-6,1), (0,4).
   Ans. Areas: (a) 29, (b) 29, (c) 7.5, (d) 15 square units.

8. Prove that the following points are the vertices of a parallelogram:
   (a) (-1,-2), (0,1), (-3,2), (-4,-1);
   (b) (-1,-5), (2,1), (1,5), (-2,-1);   (c) (2,4), (6,2), (8,6), (4,8).

9. Find the coordinates of the point which is equidistant from:
   (a) (3,3), (6,2), (8,-2);   (b) (4,3), (2,7), (-3,-8);   (c) (2,3), (4,-1), (5,2).
   Ans. (a) (3,-2), (b) (-5,1), (c) (3,1).

10. Show that the following points lie in a straight line. Use the distance method.
    (a) (0,4), (3,-2), (-2,8);   (c) (1,2), (-3,10), (4,-4);
    (b) (-2,3), (-6,1), (-10,-1);   (d) (1,3), (-2,-3), (3,7).

11. Prove that the sum of the squares of the distances of any point $P(x,y)$ from two opposite vertices of any rectangle is equal to the sum of the squares of its distances from the other two vertices. Choose the vertices (0,0), (0,b), (a,b) and (a,0).

12. Determine the point 10 units distance from (-3,6) and with the abscissa 3.
    Ans. (3,-2), (3,14).

13. Find the coordinates of the point $P(x, y)$ which divides the segment of the line from $P_1(x_1,y_1)$ to $P_2(x_2,y_2)$ such that $\dfrac{P_1P}{PP_2} = r$.

   (a) $P_1(4,-3)$, $P_2(1,4)$, $r = \dfrac{2}{1}$.

   (b) $P_1(5,3)$, $P_2(-3,-3)$, $r = \dfrac{1}{3}$.

   (c) $P_1(-2,3)$, $P_2(3,-2)$, $r = \dfrac{2}{5}$.

   (d) $P_1(0,3)$, $P_2(7,4)$, $r = -\dfrac{2}{7}$.

   (e) $P_1(-5,2)$, $P_2(1,4)$, $r = -\dfrac{5}{3}$.

   (f) $P_1(2,-5)$, $P_2(6,3)$, $r = \dfrac{3}{4}$.

   Ans. (a) $(2, \dfrac{5}{3})$, (b) $(3, \dfrac{3}{2})$, (c) $(-\dfrac{4}{7}, \dfrac{11}{7})$, (d) $(-\dfrac{14}{5}, \dfrac{13}{5})$, (e) (10,7), (f) $(\dfrac{26}{7}, -\dfrac{11}{7})$.

14. Find the coordinates of the centroid of each of the triangles whose vertices are:
    (a) (5,7), (1,-3), (-5,1);   (c) (3,6), (-5,2), (7,-6);   (e) (-3,1), (2,4), (6,-2).
    (b) (2,-1), (6,7), (-4,-3);   (d) (7,4), (3,-6), (-5,2);
    Ans. (a) $(\dfrac{1}{3}, \dfrac{5}{3})$, (b) $(\dfrac{4}{3}, 1)$, (c) $(\dfrac{5}{3}, \dfrac{2}{3})$, (d) $(\dfrac{5}{3}, 0)$, (e) $(\dfrac{5}{3}, 1)$.

15. If the point (9,2) divides the segment of the line from $P_1(6,8)$ to $P_2(x_2,y_2)$ in the ratio $r = 3/7$, find the coordinates of P_2. *Ans*. (16,-12).

16. Determine the coordinates of the vertices of a triangle if the midpoints of its sides are (-2,1), (5,2) and (2,-3). *Ans*. (1,6), (9,-2), (-5,-4).

17. Determine the coordinates of the vertices of a triangle if the midpoints of its sides are (3,2), (-1,-2) and (5,-4). *Ans*. (-3,4), (9,0), (1,-8).

18. Show analytically that the lines joining the midpoints of the adjacent sides of the quadrilateral $A(-3,2)$, $B(5,4)$, $C(7,-6)$, $D(-5,-4)$ form a second quadrilateral whose perimeter is equal to the sum of the diagonals of the first.

19. Show that the line joining the midpoints of the sides in each triangle in Problem 14 is parallel to the third side and one half as long.

20. Given the quadrilateral $A(-2,6)$, $B(4,4)$, $C(6,-6)$ and $D(2,-8)$.
  (a) Show that the line segment joining the midpoints of AD and BC bisects the segment joining the midpoints of AB and CD.
  (b) Show that the line segments joining the midpoints of adjacent sides of the quadrilateral form a parallelogram.

21. The line segment joining $A(-2,-1)$ and $B(3,3)$ is extended to C. If $BC = 3AB$ find the coordinates of C.   *Ans.* $(18,15)$

22. Show that the midpoint of the hypotenuse of a right triangle is equidistant from the vertices of the triangle. Hint: Take the vertex of the right angle at $(0,0)$ and the other vertices at $(a,0)$ and $(0,b)$.

23. In each of the isosceles triangles in Problem 6, show that two of the medians are equal in length.

24. Find the slopes of the lines passing through the following pairs of points:
  (a) $(3,4)$, $(1,-2)$;   (c) $(6,0)$, $(6,\sqrt{3})$;   (e) $(2,4)$, $(-2,4)$;
  (b) $(-5,3)$, $(2,-3)$;   (d) $(1,3)$, $(7,1)$;   (f) $(3,-2)$, $(3,5)$.

  Ans. (a) 3,   (b) $-\dfrac{6}{7}$,   (c) ∞,   (d) $-\dfrac{1}{3}$,   (e) 0,   (f) ∞

25. Find the inclination of the lines passing through the following pairs of points:
  (a) $(4,6)$ and $(1,3)$;   (c) $(2,3)$ and $(1,4)$;   (e) $(\sqrt{3},2)$ and $(0,1)$;
  (b) $(2,\sqrt{3})$ and $(1,0)$;   (d) $(3,-2)$ and $(3,5)$;   (f) $(2,4)$ and $(-2,4)$.

  Ans. (a) $\theta = \tan^{-1} 1 = 45°$.   (c) $\theta = \tan^{-1} -1 = 135°$.   (e) $\theta = \tan^{-1} 1/\sqrt{3} = 30°$.

  (b) $\theta = \tan^{-1} \sqrt{3} = 60°$.   (d) $\theta = \tan^{-1} \infty = 90°$.   (f) $\theta = \tan^{-1} 0 = 0°$.

26. Which of the following sets of points lie on a straight line? Use the method of slopes.
  (a) $(2,3)$, $(-4,7)$ and $(5,8)$.   (d) $(0,5)$, $(5,0)$ and $(6,-1)$.
  (b) $(4,1)$, $(5,-2)$ and $(6,-5)$.   (e) $(a,0)$, $(2a,-b)$ and $(-a,2b)$.
  (c) $(-1,-4)$, $(2,5)$ and $(7,-2)$.   (f) $(-2,1)$, $(3,2)$ and $(6,3)$.

  Ans. (a) No,   (b) Yes,   (c) No,   (d) Yes,   (e) Yes,   (f) No.

27. Prove that $(1,-2)$ is on the line joining the points $(-5,1)$ and $(7,-5)$ and is equidistant from them.

28. By using slopes show that the following sets of points are the vertices of a right triangle:
  (a) $(6,5)$, $(1,3)$ and $(5,-7)$;   (c) $(2,4)$, $(4,8)$ and $(6,2)$;
  (b) $(3,2)$, $(5,-4)$ and $(1,-2)$;   (d) $(3,4)$, $(-2,-1)$ and $(4,1)$.

29. Find the interior angles of the triangle whose vertices are:
  (a) $(3,2)$, $(5,-4)$ and $(1,-2)$.   *Ans.* $45°$, $45°$, $90°$.
  (b) $(4,2)$, $(0,1)$ and $(6,-1)$.   *Ans.* $109°39.2'$, $32°28.3'$, $37°52.5'$.
  (c) $(-3,-1)$, $(4,4)$ and $(-2,3)$.   *Ans.* $113°29.9'$, $40°25.6'$, $26°4.5'$.

30. By determining the interior angles show that the following triangles are isosceles. Check by finding the lengths of the sides.
  (a) $(2,4)$, $(5,1)$ and $(6,5)$.   *Ans.* $59°2.2'$, $61°55.6'$, $59°2.2'$.
  (b) $(8,2)$, $(3,8)$ and $(-2,2)$.   *Ans.* $50°11.7'$, $79°36.6'$, $50°11.7'$.

    (c) (3,2), (5,−4) and (1,−2).   *Ans.* 45°, 45°, 90°.
    (d) (1,5), (5,−1) and (9,6).   *Ans.* 63°26′, 63°26′, 53°8′.

31. The slope of a line through $A(3,2)$ is 3/4. Locate two points on this line that are 5 units from A.   *Ans.* (7,5), (−1,−1).

32. The angle from the line through (−4,5) and (3,y) to the line through (−2,4) and (9,1) is 135°. Find the value of y.   *Ans.* $y = 9$.

33. The line L_2 makes an angle of 60° with the line L_1. If the slope of L_1 is 1, find the slope of L_2.   *Ans.* $-(2+\sqrt{3})$

34. Find the slope of a line which makes an angle of 45° with the line through (2,−1) and (5,3).   *Ans.* $m_2 = -7$

35. Find the equation of the line through the point (2,5) which makes an angle of 45° with the line $x - 3y + 6 = 0$.   *Ans.* $2x - y + 1 = 0$

36. Find the areas of the triangles whose vertices are:
    (a) (2,−3), (4,2) and (−5,−2).   *Ans.* 18.5 square units
    (b) (−3,4), (6,2) and (4,−3).   *Ans.* 24.5
    (c) (−8,−2), (−4,−6) and (−1,5).   *Ans.* 28
    (d) (0,4), (−8,0) and (−1,−4).   *Ans.* 30
    (e) ($\sqrt{2}$,2), (−4,6) and (4, $-2\sqrt{2}$).   *Ans.* $7\sqrt{2} - 2 = 7.899$
    (f) (−7,5), (1,1) and (−3,3).   *Ans.* 0. Explain your answer.
    (g) $(a, b+c)$, $(b, c+a)$ and $(c, a+b)$.   *Ans.* 0

37. Find the areas of the polygons whose vertices are:
    (a) (2,5), (7,1), (3,−4) and (−2,3).   *Ans.* 39.5 square units
    (b) (0,4), (1,−6), (−2,−3) and (−4,2).   *Ans.* 25.5
    (c) (1,5), (−2,4), (−3,−1), (2,−3) and (5,1).   *Ans.* 40

38. Show that the lines joining the midpoints of the sides of each triangle in Problem 36 divide the triangle into four triangles whose areas are equal.

CHAPTER 2

Equation and Locus

THE TWO FUNDAMENTAL PROBLEMS OF ANALYTIC GEOMETRY are:

1. Given an equation, to find the corresponding locus.
2. Given a locus defined by some geometrical condition, to find the corresponding equation.

THE LOCUS, OR GRAPH, of an equation in two variables is the curve or straight line containing all the points, and only the points, whose coordinates satisfy the equation.

Before plotting the graph of an equation, it is most often very helpful to determine from the form of the equation certain properties of the curve. Such properties are: intercepts, symmetry, extent.

INTERCEPTS. The intercepts of a curve are the directed (positive or negative) distances from the origin to the points where the curve intersects the coordinate axes.

To determine the x-intercept, let $y = 0$ in the equation and solve for x. Similarly, to find the y-intercept, let $x = 0$ and solve for y.

Thus in the equation $y^2 + 2x = 16$, when $y = 0$, $x = 8$; when $x = 0$, $y = \pm 4$. Hence the x-intercept of the curve is 8, and the intercepts on the y axis are ± 4.

SYMMETRY. Two points are symmetric with respect to a line if that line is the perpendicular bisector of the line connecting the two points. Two points are symmetric about a point if that point is the midpoint of the line connecting the two given points. It follows that:

1. If an equation remains unchanged when x is replaced by $-x$, the graph is symmetric with respect to the y-axis. For every value of y in such an equation there are two numerically equal values of x with opposite signs.

   Example: $x^2 - 6y + 12 = 0$, or $x = \pm\sqrt{6y - 12}$.

2. If an equation remains unchanged when y is replaced by $-y$, the graph is symmetric with respect to the x-axis. For every value of x in such an equation there are two numerically equal values of y with opposite signs.

   Example: $y^2 - 4x - 7 = 0$, or $y = \pm\sqrt{4x + 7}$.

3. If an equation remains unchanged when x is replaced by $-x$ and y is replaced by $-y$, the graph is symmetric with respect to the origin.

   Example: $x^3 + x + y^3 = 0$.

EXTENT. If certain values of one variable cause the other variable to become imaginary, such values must be excluded.

Consider the equation $y^2 = 2x - 3$, or $y = \pm\sqrt{2x - 3}$. If x is less than 1.5, then $2x - 3$ is negative and y is imaginary. Therefore no value of x less than 1.5 may be used, and the curve lies entirely to the right of the line $x = 1.5$.

Solving for x, $x = \frac{1}{2}(y^2 + 3)$. Since any value of y gives a real value of x, no value of y is excluded and the locus extends to infinity, y increasing numerically as x increases from $x = 1.5$.

SOLVED PROBLEMS

LOCUS OF A GIVEN EQUATION.

1. Discuss and plot the locus of the ellipse $9x^2 + 16y^2 = 144$.

Intercepts. When $y = 0$, $x = \pm 4$. When $x = 0$, $y = \pm 3$. Hence the x-intercepts are ± 4, and the y-intercepts are ± 3.

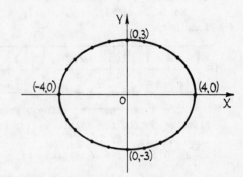

Symmetry. Since the equation contains only even powers of x and y, the curve is symmetric about both axes and therefore about the origin. Hence it is sufficient to plot only that part of the curve which lies in the first quadrant and determine the rest of the curve by symmetry.

Extent. Solving for y and for x,

$$y = \pm \frac{3}{4}\sqrt{16 - x^2}, \qquad x = \pm \frac{4}{3}\sqrt{9 - y^2}.$$

If x is *numerically* greater than 4, $16 - x^2$ is negative and y is imaginary. Hence x can have no value greater than 4 nor less than -4, or $4 \geqq x \geqq -4$. Similarly, y can have no value greater than 3 nor less than -3, or $3 \geqq y \geqq -3$.

x	0	± 1	± 2	± 3	± 3.5	± 4
y	± 3	± 2.9	± 2.6	± 2.0	± 1.5	0

2. Discuss and plot the graph of the parabola $y^2 - 2y - 4x + 9 = 0$.

Solving for y by the quadratic formula,

$$y = \frac{-b \pm \sqrt{b^2 - 4ac}}{2a}, \text{ where } a = 1,\ b = -2,\ c = -4x + 9:$$

$$y = 1 \pm 2\sqrt{x - 2}. \qquad (1)$$

Solving for x, $\quad x = \dfrac{y^2 - 2y + 9}{4}. \qquad (2)$

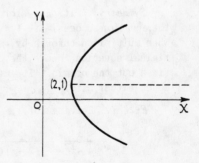

Intercepts. When $y = 0$, $x = 9/4$. When $x = 0$, y is imaginary $(1 \pm 2\sqrt{-2})$. Hence the x-intercept is $9/4$, and there is no y-intercept.

Symmetry. The curve is not symmetric about either of the coordinate axes nor about the origin.

It is symmetric about the line $y = 1$, as each value of x gives two values of y, one value being as much greater than 1 as the other is less than 1.

Extent. From (1): If x is less than 2, $x - 2$ is negative and y is imaginary. Hence x can have no value less than 2.

From (2): Since any value of y gives a real value of x, no value of y is excluded.

x	2	9/4	3	4	5	6
y	1	0, 2	3, -1	3.8, -1.8	4.5, -2.5	5, -3

3. Discuss and plot the graph of the hyperbola $xy - 2y - x = 0$.

Intercepts. When $x = 0$, $y = 0$; when $y = 0$, $x = 0$.

Symmetry. The curve is not symmetric about either of the coordinate axes or about the origin.

Extent. Solving for y, $y = \dfrac{x}{x-2}$. If $x = 2$, the denominator $x - 2$ becomes zero and y becomes infinite.

Solving for x, $x = \dfrac{2y}{y-1}$. If $y = 1$, the denominator $y - 1$ becomes zero and x becomes infinite.

No value of either variable will make the other imaginary.

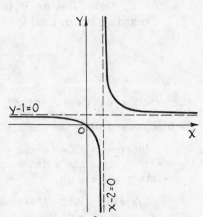

x	0	1	$1\frac{1}{2}$	$1\frac{3}{4}$	2	$2\frac{1}{4}$	$2\frac{1}{2}$	3	4	5	-1	-2	-3	-4
y	0	-1	-3	-7	∞	9	5	3	2	1.7	.3	.5	.6	.7

As x approaches the value 2 from the left, y becomes negatively infinite.   As x approaches 2 from the right, y increases without limit. The two parts of the curve approach indefinitely near the line $x = 2$, becoming tangent to the line at $\pm$ infinity.   The line $x - 2 = 0$ is called a vertical asymptote to the curve.

Now what happens when x becomes infinite? Consider $y = \dfrac{x}{x-2} = \dfrac{1}{1 - \dfrac{2}{x}}$. As x becomes positively or negatively infinite, $\dfrac{2}{x}$ approaches zero and y approaches 1. The line $y - 1 = 0$ is a horizontal asymptote.

4. Discuss and plot the graph of

$x^2 y - 4y + x = 0$.

Intercepts. When $x = 0$, $y = 0$.
When $y = 0$, $x = 0$.

Symmetry. Writing $-x$ for x and $-y$ for y, the equation becomes $-x^2 y + 4y - x = 0$, which when multiplied through by -1 becomes the o-riginal equation. Hence the curve is symmetric about the origin.   It is not symmetric about either axis.

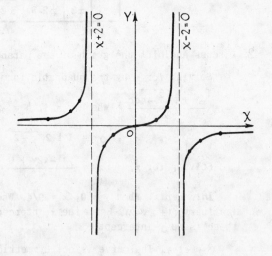

Extent.   Solving for y,

$$y = \frac{x}{4 - x^2} = \frac{x}{(2-x)(2+x)}.$$

The vertical asymptotes are $x - 2 = 0$, $x + 2 = 0$.

Solving for x by the quadratic formula, $x = \dfrac{-1 \pm \sqrt{1 + 16y^2}}{2y}$.   The horizontal asymptote is $y = 0$.

No value of either variable will make the other imaginary.

x	-4	-3	-2.5	-2	-1.5	-1	0	1	1.5	2	2.5	3	4
y	.3	.6	1.1	∞	-.9	-.3	0	.3	.9	∞	-1.1	-.6	-.3

5. Plot the locus of $x^2 - x + xy + y - 2y^2 = 0$.

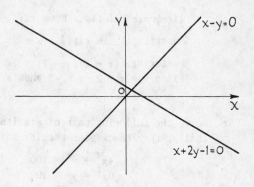

Sometimes an equation will factor, in which case the locus of the equation will consist of the loci of the several factors.

Since the equation factors into

$$(x - y)(x + 2y - 1) = 0,$$

its locus is the two intersecting lines

$$x - y = 0 \quad \text{and} \quad x + 2y - 1 = 0.$$

6. Determine the real points, if any, on the graphs of the following equations.

a. $(x + 4)^2 + (y - 2)^2 = -5$.

b. $x^2 + y^2 = 0$.

c. $x^2 + y^2 - 8x + 2y + 17 = 0$.

d. $x^2 + 2y^2 - 6x + 11 = 0$.

e. $(x^2 - 4y^2)^2 + (x + 3y - 10)^2 = 0$.

f. $x^2 + (2i - 1)x - (6i + 5)y - 1 = 0$.

a. Since the square of any real number is positive, both $(x + 4)^2$ and $(y - 2)^2$ are positive and the equation cannot be satisfied by any real values of x and y.

b. It is evident that the only real point which satisfies this equation is $(0,0)$.

c. Rearranging, $(x^2 - 8x + 16) + (y^2 + 2y + 1) = 0$, or $(x - 4)^2 + (y + 1)^2 = 0$. Then $x - 4 = 0$ and $y + 1 = 0$, or $x = 4$, $y = -1$. Hence the only real point on the graph of this equation is $(4,-1)$.

d. Rearranging, $x^2 - 6x + 9 + 2y^2 + 2 = 0$, or $(x - 3)^2 + 2y^2 + 2 = 0$. Since $(x - 3)^2$, $2y^2$ and 2 are each positive for any real values of x and y, the given equation cannot be satisfied by real values of x and y.

e. The equation is satisfied by values of x and y which make $x^2 - 4y^2 = 0$ and $x + 3y - 10 = 0$, simultaneously. Solving these two equations for x and y gives $(4, 2)$ and $(-20, 10)$ as the only real points on the locus.

f. Collecting real parts and imaginary parts, $(x^2 - x - 5y - 1) + 2i(x - 3y) = 0$. This equation is satisfied for values of x and y which make $x^2 - x - 5y - 1 = 0$ and $x - 3y = 0$, simultaneously. Solving these two equations for x and y gives $(3, 1)$ and $(-1/3, -1/9)$ as the only real points on the locus of the equation.

7. Solve the following pair of simultaneous equations graphically, then check results by solving them algebraically.

$$xy = 8 \qquad (1)$$
$$x - y + 2 = 0 \qquad (2)$$

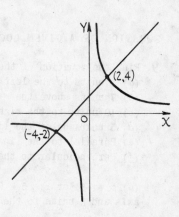

Solving (1) for y, $y = \dfrac{8}{x}$. When $x = 0$, y is infinite.

Solving (1) for x, $x = \dfrac{8}{y}$. When $y = 0$, x is infinite.

Then $y = 0$ is a horizontal asymptote, and $x = 0$ is a vertical asymptote.

x	0	1	2	3	4	-1	-2	-3	-4
y	∞	8	4	8/3	2	-8	-4	-8/3	-2

Equation (2) represents a straight line whose intercepts are $(-2,0)$ and $(0,2)$. From the graph the solutions read $(-4,-2)$ and $(2,4)$.

Algebraic Solution. From (2), $y = x + 2$.

Substituting in (1), $x(x + 2) = 8$, or $x^2 + 2x - 8 = 0$.

Factoring, $(x + 4)(x - 2) = 0$. Hence $x = -4$ and $x = 2$.
Since $y = x + 2$, $y = -2$ when $x = -4$, and $y = 4$ when $x = 2$.

8. Solve the following pair of simultaneous equations graphically, then check your results analytically.

$$4x^2 + y^2 = 100 \qquad (1)$$
$$9x^2 - y^2 = 108 \qquad (2)$$

Each of the curves is symmetric about both axes and
the origin.

Solving (1) for y, $y = \pm \sqrt{100 - 4x^2}$. Hence x cannot
have any value greater than 5 nor less than –5.

Solving (1) for x, $x = \frac{1}{2}\sqrt{100 - y^2}$. Hence y cannot
have any value greater than 10 or less than –10.

x	0	±1	±2	±3	±4	±5
y	±10	±9.8	±9.2	±8	±6	0

Solving (2) for y, $y = \pm 3\sqrt{x^2 - 12}$. Hence x can have
no value between $\sqrt{12}$ and $-\sqrt{12}$.

Solving (2) for x, $x = \pm \frac{1}{3}\sqrt{y^2 + 108}$. Hence y may have any value.

x	$\pm\sqrt{12}$	±4	±5	±6
y	0	±6	±10.8	±14.7

Reading from the graph the solutions are $(4, \pm 6)$, $(-4, \pm 6)$.

Algebraic Solution. $4x^2 + y^2 = 100$
$\underline{9x^2 - y^2 = 108}$

$\qquad 13x^2 = 208$, $x^2 = 16$, and $x = \pm 4$.

$y^2 = 9x^2 - 108 = 144 - 108 = 36$, and $y = \pm 6$.

EQUATION OF A GIVEN LOCUS.

9. Find the equation of the straight lines which are
   (a) 5 units to the left of the y-axis.
   (b) 7 units above the x-axis.
   (c) 10 units to the right of the line $x + 4 = 0$.
   (d) 5 units below the line $y = 2$.
   (e) Parallel to the line $y + 8 = 0$ and 6 units from $(2, 1)$.
   (f) Perpendicular to the line $y - 2 = 0$ and 4 units from $(-1, 7)$.

(a) $x = -5$, or $x + 5 = 0$. This is the equation of the line which is parallel to the y-axis and 5 units to the left of the y-axis.

(b) $y = 7$, or $y - 7 = 0$. This is the equation of the line which is parallel to the x-axis and 7 units above the x-axis.

(c) $x = -4 + 10$, or $x = 6$. This is the equation of the line which is 10 units to the right of the line $x + 4 = 0$. It is parallel to the y-axis and 6 units to the right of the y-axis.

(d) $y = 2 - 5$, or $y = -3$. This is the equation of the line which is 5 units below the line $y - 2 = 0$. It is parallel to the x-axis and 3 units below the x-axis.

(e) Since the line $y + 8 = 0$ is parallel to the x-axis, each required line is also parallel to the x-axis and 6 units above or below the line $y = 1$. Then $y = 1 \pm 6$, or $y = 7$ and $y = -5$.

(f) Since the line $y - 2 = 0$ is parallel to the x-axis, each required line is parallel to the y-axis and 4 units to the right or left of the line $x = -1$. Then $x = -1 \pm 4$, or $x = 3$ and $x = -5$.

10. Determine the equation of the line
    (a) parallel to the x-axis and 5 units from the point $(3, -4)$.
    (b) equidistant from the lines $x + 5 = 0$ and $x - 2 = 0$.
    (c) three times as far from the line $y - 9 = 0$ as from the line $y + 2 = 0$.

   Let (x, y) be any point on the required line.

   (a) $y = -4 \pm 5$, or $y = 1$ and $y = -9$.

   (b) $\dfrac{5 + x}{2 - x} = 1$, or $x = \dfrac{-5 + 2}{2} = -\dfrac{3}{2}$, or $2x + 3 = 0$.

   (c) $\dfrac{y + 2}{9 - y} = \pm \dfrac{1}{3}$. Simplifying, $4y - 3 = 0$ and $2y + 15 = 0$.

   For the line $4y - 3 = 0$, which is between the two given lines, the ratio is $+\dfrac{1}{3}$. For the line $2y + 15 = 0$, which is below the two given lines, the ratio is $-\dfrac{1}{3}$.

11. Derive the equation of the locus of a point $P(x, y)$ which moves in such a way that it is always equidistant from the points $A(-2, 3)$ and $B(3, -1)$.

   $$PA = PB, \quad \text{or} \quad \sqrt{(x + 2)^2 + (y - 3)^2} = \sqrt{(x - 3)^2 + (y + 1)^2}.$$

   Squaring and simplifying, we obtain $10x - 8y + 3 = 0$. This is the equation of the perpendicular bisector of the segment of the line connecting the two points.

12. Find the equation of a line
    (a) with slope 2/3 and passing through the point $(-4, 5)$.
    (b) passing through the two points $(3, -1)$ and $(0, 6)$.

   Let (x, y) be any point on the required line.
   The slope of the line through points (x_1, y_1) and (x_2, y_2) $= \dfrac{y_2 - y_1}{x_2 - x_1}$.

   (a) The slope of the line joining $(-4, 5)$ and (x, y) is $\dfrac{2}{3}$.

   Then $\dfrac{y - 5}{x + 4} = \dfrac{2}{3}$. Simplifying, $2x - 3y + 23 = 0$.

   (b) Slope of line joining $(3, -1)$ and $(0, 6)$ = slope of line joining $(0, 6)$ and (x, y).

   Then $\dfrac{6 + 1}{0 - 3} = \dfrac{y - 6}{x - 0}$. Simplifying, $7x + 3y - 18 = 0$.

13. Find the equation of the line
   (a) passing through (2,-1) and perpendicular to the line through (4,3) and (-2,5).
   (b) passing through (-4,1) and parallel to the line through (2,3) and (-5,0).

(a) If two lines are perpendicular, the slope of one of the lines is the negative reciprocal of the slope of the other line.

Slope of line through (4,3) and (-2,5) = $\dfrac{5-3}{-2-4}$ = $-\dfrac{1}{3}$.

Slope of the required line = negative reciprocal of $-\dfrac{1}{3}$ = 3.

Let (x,y) be any point on the required line. Then the slope of the required line through (x,y) and (2,-1) = $\dfrac{y+1}{x-2}$ = 3. Simplifying, $3x - y - 7 = 0$.

(b) If two lines are parallel, their slopes are equal.

Let (x,y) be any point on the required line.

Slope of line through (2,3) and (-5,0) = slope of line through (x,y) and (-4,1).

Then $\dfrac{3-0}{2+5}$ = $\dfrac{y-1}{x+4}$.   Simplifying, $3x - 7y + 19 = 0$.

14. A point $P(x,y)$ moves in such a way that its distance from $C(2,-1)$ is always 5. Find the equation of its locus.

The distance $PC = 5$,   or $\sqrt{(x-2)^2 + (y+1)^2}$ = 5.

Squaring and simplifying, the required equation is $x^2 + y^2 - 4x + 2y = 20$.
The locus is a circle with its center at (2,-1) and radius 5.

15. A point $P(x,y)$ moves so that the sum of the squares of its distances from points $A(0,0)$ and $B(2,-4)$ is always 20. Derive the equation of its locus.

$(PA)^2 + (PB)^2 = 20$,   or $x^2 + y^2 + [(x-2)^2 + (y+4)^2]$ = 20.

Simplifying, $x^2 + y^2 - 2x + 4y = 0$. This is the equation of a circle with diameter AB.

16. A point $P(x,y)$ moves so that the sum of its distances from the coordinate axes equals the square of its distance from the origin. Determine the equation of its locus.

Distance of $P(x,y)$ from y-axis + distance from x-axis = square of distance from (0,0).

Then $x + y = x^2 + y^2$,   or $x^2 + y^2 - x - y = 0$.   This is the equation of a circle with center $(\frac{1}{2},\frac{1}{2})$ and radius $\frac{1}{2}\sqrt{2}$.

17. The point $P(x,y)$ moves so that the ratio of its distance from the line $y - 4 = 0$ to its distance from the point (3,2) is 1. Determine the equation of its locus.

$\dfrac{\text{Distance of } P(x,y) \text{ from } y - 4 = 0}{\text{Distance of } P(x,y) \text{ from } (3,2)}$ = 1,   or $\dfrac{4-y}{\sqrt{(x-3)^2 + (y-2)^2}}$ = 1.

Squaring and simplifying, $(4-y)^2 = (x-3)^2 + (y-2)^2$,   or $x^2 - 6x + 4y - 3 = 0$.

This is the equation of a parabola.

18. Given two points $P_1(2,4)$ and $P_2(5,-3)$. Determine the equation of the locus of the point $P(x,y)$ if the slope of PP_1 is 1 more than the slope of PP_2.

Slope of PP_1 = slope of PP_2 + 1, or $\dfrac{y-4}{x-2} = \dfrac{y+3}{x-5} + 1$.

Simplifying, $x^2 + 3y - 16 = 0$, which is the equation of a parabola.

19. Derive the equation of the locus of a point $P(x,y)$ which moves in such a way that its distance from the point $F(3,2)$ is always equal to its distance from the y-axis.

$PF = x$, or $\sqrt{(x-3)^2 + (y-2)^2} = x$, or $x^2 - 6x + 9 + y^2 - 4y + 4 = x^2$.

Simplifying, $y^2 - 4y - 6x + 13 = 0$, which is the equation of a parabola.

20. A point $P(x,y)$ moves in such a way that the difference of its distances from $F_1(1,4)$ and $F_2(1,-4)$ is always equal to 6. Derive the equation of its locus.

$PF_1 - PF_2 = 6$, or $\sqrt{(x-1)^2 + (y-4)^2} - \sqrt{(x-1)^2 + (y+4)^2} = 6$.

Transfer one radical to the right of the equation.

$$\sqrt{(x-1)^2 + (y-4)^2} = 6 + \sqrt{(x-1)^2 + (y+4)^2}.$$

Squaring, $x^2 - 2x + 1 + y^2 - 8y + 16 = 36 + 12\sqrt{(x-1)^2 + (y+4)^2} + x^2 - 2x + 1 + y^2 + 8y + 16$.

Simplifying, $4y + 9 = -3\sqrt{(x-1)^2 + (y+4)^2}$.

Squaring, $16y^2 + 72y + 81 = 9x^2 - 18x + 9 + 9y^2 + 72y + 144$.

Simplifying, $9x^2 - 7y^2 - 18x + 72 = 0$, the equation of a hyperbola.

SUPPLEMENTARY PROBLEMS

LOCUS OF A GIVEN EQUATION.

Discuss and plot the graph of each of the following equations 1-18.

1. $x^2 + 2x - y + 3 = 0$

2. $4x^2 - 9y^2 + 36 = 0$

3. $x^2 + y^2 - 8x + 4y - 29 = 0$

4. $2x^2 + 3y^2 - 18 = 0$

5. $3x^2 + 5y^2 = 0$

6. $4y^2 - x^3 = 0$

7. $(xy-6)^2 + (x^2 + 3xy + y^2 + 5) = 0$

8. $8y - x^3 = 0$

9. $y^2 = x(x-2)(x+3)$

10. $y = x(x+2)(x-3)$

11. $(x^2 + 2xy - 24)^2 + (2x^2 + y^2 - 33)^2 = 0$

12. $x^2y + 4y - 8 = 0$

13. $x^2y^2 + 4x^2 - 9y^2 = 0$

14. $x^2 + y^2 + 4x - 6y + 17 = 0$

15. $2x^2 + y^2 - 2y^2i + x^2i - 54 - 17i = 0$

16. $y(x+2)(x-4) - 8 = 0$

17. $x^2 + xy - 2y^2 - 3x + 3y = 0$

18. $(x^2 - y) - yi = (5 - 2x) + 3(1 - x)i$

Plot the graphs of the following pairs of simultaneous equations and find the solutions of the equations from the graphs. Check results by solving the equations algebraically.

19. $y = x^2$, $x - y + 2 = 0$. *Ans.* $(2,4)$, $(-1,1)$.

20. $4y - x^2 = 0, \quad x^2y + 4y - 8 = 0.$ *Ans.*   (2,1), (-2,1), others imaginary.

21. $x^2 + y^2 - 20 = 0, \quad y^2 - 2x - 12 = 0.$ *Ans.*   (2, ±4), (-4, ±2).

22. $y^2 - 2x - 5 = 0, \quad 3x^2 - 2y^2 - 1 = 0.$ *Ans.*   (2.7, ±3.2), (-1.4, ±1.5).

23. $y^2 - 4x - 9 = 0, \quad x^2 + 2y - 6 = 0.$ *Ans.*   (-2,1), (-2,1), (4,-5), (0,3).

24. $2x^2 + y^2 - 6 = 0, \quad x^2 - y^2 - 4 = 0.$ *Ans.*   Imaginary.

25. $2x^2 - 5xy + 2y^2 = 0, \quad x^2 + y^2 - 5 = 0.$ *Ans.*   (2,1), (-2,-1), (1,2), (-1,-2).

26. $x^2 - y^2 + x - y = 0, \quad x^2 - 2xy - 3x + 6y = 0.$ *Ans.*   (3,-4), (-2/3,-1/3), (3,3), (0,0).

EQUATION OF A GIVEN LOCUS.

27. Determine the equations of the straight lines which are:

    (a) 3 units to the right of the y-axis.   *Ans.* $x - 3 = 0$

    (b) 5 units below the x-axis.   *Ans.* $y + 5 = 0$

    (c) Parallel to the y-axis and 7 units from (-2,2).   *Ans.* $x - 5 = 0, \quad x + 9 = 0.$

    (d) 8 units to the left of the line $x = -2$.   *Ans.* $x + 10 = 0$

    (e) Parallel to the x-axis and midway between the points (2,3) and (2,-7).

                                               Ans. $y + 2 = 0$

    (f) 4 times as far from the line $x = 3$ as from the line $x = -2$.

                                   Ans. $3x + 11 = 0, \quad x + 1 = 0.$

    (g) Through the point (-2,-3) and perpendicular to the line $x - 3 = 0$.   *Ans.* $y + 3 = 0$

    (h) Equidistant from the coordinate axes.   *Ans.* $y - x = 0, \quad y + x = 0.$

    (i) Through the point (3,-1) and parallel to the line $y + 3 = 0$.   *Ans.* $y + 1 = 0$

    (j) Equidistant from the lines $y - 7 = 0$ and $y + 2 = 0$.   *Ans.* $2y - 5 = 0$

28. A point $P(x,y)$ moves in such a way that its distance from (-2,3) is 4. Find the equation of its locus.   *Ans.* $x^2 + y^2 + 4x - 6y - 3 = 0$

29. Derive the equation of the locus of a point $P(x,y)$ which moves in such a way that it is always equidistant from the points (-3,1) and (7,5).   *Ans.* $5x + 2y - 16 = 0$

30. A point $P(x,y)$ moves in such a way that its distance from (3,2) is always one half of its distance from (-1,3). Find the equation of its locus.
    Ans. $3x^2 + 3y^2 - 26x - 10y + 42 = 0$

31. Derive the equation of the locus of a point $P(x,y)$ which moves so that its distance from (2,3) is always equal to its distance from the line $x + 2 = 0$.
    Ans. $y^2 - 8x - 6y + 9 = 0$

32. Derive the equation of the circle with its center at (3,5) and tangent to the line $y - 1 = 0$.   *Ans.* $x^2 + y^2 - 6x - 10y + 30 = 0$

33. Derive the equation of the locus of a point the sum of whose distances from $(c,0)$ and $(-c,0)$ is always equal to $2a$, $(2a > 2c)$.   *Ans.* $(a^2 - c^2)x^2 + a^2y^2 = a^4 - a^2c^2$

34. Derive the equation of the locus of a point $P(x,y)$ the sum of whose distances from (2,3) and (2,-3) is always 8.   *Ans.* $16x^2 + 7y^2 - 64x - 48 = 0$

35. Derive the equation of the locus of a point which moves so that the difference of its distances from (3, 2) and (−5, 2) is 6.   *Ans.* $7x^2 - 9y^2 + 14x + 36y - 92 = 0$

36. A point moves so that its distance from the line $y + 4 = 0$ is two-thirds of its distance from the point (3, 2). Find the equation of its locus.
Ans. $4x^2 - 5y^2 - 24x - 88y - 92 = 0$

37. Derive the equation of the locus of a point whose distance from (−2, 2) is always three times its distance from the line $x - 4 = 0$.   *Ans.* $8x^2 - y^2 - 76x + 4y + 136 = 0$

38. Derive the equation of the locus of a point, the sum of the squares of whose distances from the coordinate axes is 9.   *Ans.* $x^2 + y^2 = 9$

39. Derive the equation of the perpendicular bisector of the segment of the line connecting the points (−3, 2) and (5, −4).   *Ans.* $4x - 3y = 7$

40. Derive the equation of the locus of a point which moves in such a way that it is always 3 units from the origin of coordinates.   *Ans.* $x^2 + y^2 = 9$

41. Derive the equation of the circle with its center at (2, 3) and which passes through the point (5, −1).   *Ans.* $x^2 + y^2 - 4x - 6y - 12 = 0$

42. Given the points $A(0, -2)$, $B(0, 4)$, and $C(0, 0)$. Find the equation of the locus of the point $P(x, y)$, if the product of the slopes of PA and PB equals the slope of PC.
Ans. $y^2 - xy - 2y - 8 = 0$

43. A line segment 12 units in length moves so that its ends are always on the coordinate axes. Find the equation of the locus of its midpoint.   *Ans.* $x^2 + y^2 = 36$

44. Given the points $A(-2, 3)$ and $B(3, 1)$. Find the equation of the locus of $P(x, y)$, if the slope of PA is the negative reciprocal of the slope of PB.
Ans. $x^2 + y^2 - x - 4y - 3 = 0$

CHAPTER 3

The Straight Line

A STRAIGHT LINE is represented by an equation of the first degree in two variables. Conversely, the locus of an equation of the first degree in two variables is a straight line.

A straight line is completely determined if its direction is known and a point is given through which the line must pass.

POINT-SLOPE FORM. The equation of the straight line through point $P_1(x_1,y_1)$ whose slope is m is

$$y - y_1 = m(x - x_1).$$

SLOPE-INTERCEPT FORM. The equation of the straight line having slope m and y-intercept $(0,b)$ is

$$y = mx + b.$$

TWO-POINT FORM. The equation of the straight line through points $P_1(x_1,y_1)$ and $P_2(x_2,y_2)$ is

$$\frac{y - y_1}{x - x_1} = \frac{y_1 - y_2}{x_1 - x_2}.$$

INTERCEPT FORM. The equation of the straight line whose x and y intercepts are respectively $(a,0)$ and $(0,b)$ is

$$\frac{x}{a} + \frac{y}{b} = 1.$$

GENERAL FORM. Every equation of the first degree in x and y may be reduced to the form $Ax + By + C = 0$, where A, B and C are arbitrary constants. For an equation in this form, slope $m = -\dfrac{A}{B}$ and y-intercept $b = -\dfrac{C}{B}$.

NORMAL EQUATION OF A STRAIGHT LINE. A straight line is completely determined if the length of the perpendicular from the origin $(0,0)$ to the line is known and if the angle which this perpendicular makes with the x-axis is known.

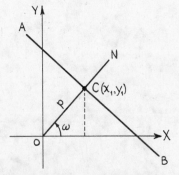

Let AB be the given line. Draw ON perpendicular to AB.

The perpendicular distance p from O to AB is taken as positive for all positions of AB, and ω is the angle from $0°$ to $360°$ which ON makes with the positive end of the x-axis.

Let the coordinates of point C be (x_1,y_1).

Then $x_1 = p\cos\omega$, $y_1 = p\sin\omega$, and slope of $AB = -\dfrac{1}{\tan\omega} = -\cot\omega = -\dfrac{\cos\omega}{\sin\omega}$.

If (x,y) is any other point on AB, then by the point-slope formula

$$y - y_1 = -\cot\omega\,(x - x_1), \quad \text{or} \quad y - p\sin\omega = -\frac{\cos\omega}{\sin\omega}(x - p\cos\omega).$$

Simplifying, $x \cos \omega + y \sin \omega - p = 0$, the normal form of the equation of a straight line.

REDUCTION TO NORMAL FORM. If $Ax + By + C = 0$ and $x \cos \omega + y \sin \omega - p = 0$ are respectively the general and normal forms of the same line, the coefficients of the two equations are equal or proportional.

Hence $\dfrac{\cos \omega}{A} = \dfrac{\sin \omega}{B} = \dfrac{-p}{C} = k$, where k is the constant ratio.

Then $\cos \omega = kA$, $\sin \omega = kB$, $-p = kC$. Squaring and adding the first two, $\cos^2 \omega + \sin^2 \omega = k^2(A^2 + B^2)$,   or $1 = k^2(A^2 + B^2)$,   and $k = \dfrac{1}{\pm \sqrt{A^2 + B^2}}$.

Substituting for k,

$$\cos \omega = \frac{A}{\pm \sqrt{A^2 + B^2}}, \quad \sin \omega = \frac{B}{\pm \sqrt{A^2 + B^2}}, \quad -p = \frac{C}{\pm \sqrt{A^2 + B^2}}.$$

Hence the normal form of $Ax + By + C = 0$ is

$$\frac{A}{\pm \sqrt{A^2 + B^2}} x + \frac{B}{\pm \sqrt{A^2 + B^2}} y + \frac{C}{\pm \sqrt{A^2 + B^2}} = 0$$

where the sign before the radical is chosen opposite to that of C. If $C = 0$, the sign before the radical is chosen the same as that of B.

DISTANCE FROM A LINE TO A POINT. To find the perpendicular distance d from line L to point (x_1, y_1), draw L_1 through (x_1, y_1) and parallel to L.

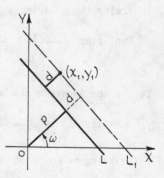

The equation of L is $x \cos \omega + y \sin \omega - p = 0$, and the equation of L_1 is $x \cos \omega + y \sin \omega - (p + d) = 0$, as the lines are parallel.

Since the coordinates of (x_1, y_1) satisfy the equation for L_1, $x_1 \cos \omega + y_1 \sin \omega - (p + d) = 0$. Solving for d,

$$d = x_1 \cos \omega + y_1 \sin \omega - p.$$

If (x_1, y_1) and the origin are on opposite sides of the line L, the distance d is positive; if they are on the same side of the line L, d is negative.

SOLVED PROBLEMS

1. Derive the equation of the line through point $P_1(x_1, y_1)$ whose slope is m. (See the adjacent figure.)

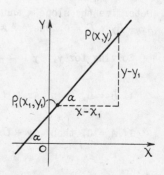

Let $P(x, y)$ be any other point on the line.
Slope m of the line through (x, y) and (x_1, y_1) is

$$m = \frac{y - y_1}{x - x_1}, \quad \text{or} \quad y - y_1 = m(x - x_1).$$

2. Derive the equation of the line whose slope is m and whose y-intercept is $(0, b)$.

Let $P(x,y)$ be any other point on the line.

The slope m of the line through (x,y) and $(0,b)$ is $m = \dfrac{y-b}{x-0}$. Then $y = mx + b$.

3. Write the equation of the line (a) passing through $(-4,3)$ with slope $\frac{1}{2}$, (b) passing through $(0,5)$ with slope -2, (c) passing through $(2,0)$ with slope $\frac{3}{4}$.

Let $P(x,y)$ be any other point on each line. Employ $y - y_1 = m(x - x_1)$.

a. $y - 3 = \frac{1}{2}(x + 4)$, or $2y - 6 = x + 4$, or $x - 2y + 10 = 0$.

b. $y - 5 = -2(x - 0)$, or $y - 5 = 2x$, or $2x + y - 5 = 0$.
   This equation can be found directly by substituting in $y = mx + b$.
   Then $y = -2x + 5$, or $2x + y - 5 = 0$.

c. $y - 0 = \frac{3}{4}(x - 2)$, or $4y = 3x - 6$, or $3x - 4y - 6 = 0$.

4. Derive the equation of the line passing through points (x_1, y_1) and (x_2, y_2).

Let (x,y) be any other point on the line passing through (x_1, y_1) and (x_2, y_2).

Slope of line through (x,y) and (x_1, y_1) = slope of line through (x_1, y_1) and (x_2, y_2).

Hence,
$$\frac{y - y_1}{x - x_1} = \frac{y_1 - y_2}{x_1 - x_2}.$$

5. Determine the equation of the line passing through points $(-2, -3)$ and $(4, 2)$.

Employing $\dfrac{y - y_1}{x - x_1} = \dfrac{y_1 - y_2}{x_1 - x_2}$, we have $\dfrac{y + 3}{x + 2} = \dfrac{-3 - 2}{-2 - 4}$ or $5x - 6y - 8 = 0$.

6. Derive the equation of the line whose intercepts are $(a,0)$ and $(0,b)$.

Substituting in $\dfrac{y - y_1}{x - x_1} = \dfrac{y_1 - y_2}{x_1 - x_2}$ gives $\dfrac{y - 0}{x - a} = \dfrac{0 - b}{a - 0}$ or $bx + ay = ab$.

Dividing $bx + ay = ab$ by ab gives $\dfrac{x}{a} + \dfrac{y}{b} = 1$. (Intercept form.)

7. Write the equation of the line whose intercepts on the x and y axes are respectively 5 and -3.

Employing $\dfrac{x}{a} + \dfrac{y}{b} = 1$, the required equation is $\dfrac{x}{5} + \dfrac{y}{-3} = 1$ or $3x - 5y - 15 = 0$.

8. Determine the slope m and the intercept b on the y-axis for the line whose equation is $Ax + By + C = 0$, where A, B and C are arbitrary constants.

Solving for y, $y = -\dfrac{A}{B}x - \dfrac{C}{B}$. Comparing this with $y = mx + b$, $m = -\dfrac{A}{B}$, $b = -\dfrac{C}{B}$.

If $B = 0$, then $Ax + C = 0$ or $x = -\dfrac{C}{A}$, a line parallel to the y-axis.

If $A = 0$, then $By + C = 0$ or $y = -\dfrac{C}{B}$, a line parallel to the x-axis.

9. Find the slope m and the intercept b on the y-axis for the line $2y + 3x = 7$.

Rearranging to the form $y = mx + b$, $y = -\frac{3}{2}x + \frac{7}{2}$. Hence its slope is $-3/2$ and its y-intercept is $7/2$.

Or, rearranging to the form $Ax + By + C = 0$, $3x + 2y - 7 = 0$. Hence its slope $m = -\frac{A}{B} = -\frac{3}{2}$, and its y-intercept $b = -\frac{C}{B} = -\frac{-7}{2} = \frac{7}{2}$.

10. Show that if lines $Ax + By + C = 0$ and $A'x + B'y + C' = 0$ are parallel, then $A/A' = B/B'$; if perpendicular, then $AA' + BB' = 0$.

If parallel, $m = m'$, or $-\frac{A}{B} = -\frac{A'}{B'}$, or $\frac{A}{A'} = \frac{B}{B'}$.

If perpendicular, $m = -\frac{1}{m'}$, or $-\frac{A}{B} = \frac{B'}{A'}$, or $AA' + BB' = 0$.

11. Determine the equation of the line passing through point $(2, -3)$ and parallel to the line passing through points $(4, 1)$ and $(-2, 2)$.

Parallel lines have equal slopes.
Let (x, y) be any other point on the required line through $(2, -3)$.

Slope of line through (x, y) and $(2, -3)$ = slope of line through $(4, 1)$ and $(-2, 2)$.

Hence, $\frac{y + 3}{x - 2} = \frac{1 - 2}{4 + 2}$. Simplifying, $x + 6y + 16 = 0$.

12. Find the equation of the line passing through point $(-2, 3)$ and perpendicular to the line $2x - 3y + 6 = 0$.

If two lines are perpendicular, the slope of one of the lines is the negative reciprocal of the slope of the other line.

Slope of $2x - 3y + 6 = 0$, which is in the form $Ax + By + C = 0$, is $-\frac{A}{B} = \frac{2}{3}$. Hence the slope of the required line is $-\frac{3}{2}$.

Let (x, y) be any other point on the required line through $(-2, 3)$ with slope $-\frac{3}{2}$.
Then, $y - 3 = -\frac{3}{2}(x + 2)$. Simplifying, $3x + 2y = 0$.

13. Find the equation of the line which is the perpendicular bisector of the segment connecting points $(7, 4)$ and $(-1, -2)$.

Midpoint (x_0, y_0) of the segment is

$$x_0 = \frac{x_1 + x_2}{2} = \frac{7 - 1}{2} = 3, \qquad y_0 = \frac{y_1 + y_2}{2} = \frac{4 - 2}{2} = 1.$$

Slope of segment $= \frac{4 + 2}{7 + 1} = \frac{3}{4}$. Hence, slope of required line $= -\frac{4}{3}$.

Let (x, y) be any other point on the required line through $(3, 1)$ with slope $-\frac{4}{3}$.

Then, $y - 1 = -\frac{4}{3}(x - 3)$. Simplifying, $4x + 3y - 15 = 0$.

14. Write the equation of the line through $(2,-3)$ with inclination $60°$.

Let (x,y) be any other point on the required line whose slope is $\tan 60° = \sqrt{3}$.

Then, $y + 3 = \sqrt{3}(x - 2)$. Simplifying, $\sqrt{3}x - y - 3 - 2\sqrt{3} = 0$.

15. Determine the particular value of the parameter k so that:
a) $3kx + 5y + k - 2 = 0$ passes through point $(-1,4)$;
b) $4x - ky - 7 = 0$ has the slope 3;
c) $kx - y = 3k - 6$ has the x-intercept 5.

$a.$ Substituting $x = -1$, $y = 4$: $3k(-1) + 5(4) + k - 2 = 0$, $2k = 18$, $k = 9$.

$b.$ Employing the given form $Ax + By + C = 0$, slope $= -\dfrac{A}{B} = -\dfrac{4}{-k} = 3$, $k = \dfrac{4}{3}$.

Or, reducing $4x - ky - 7 = 0$ to the form $y = mx + b$, $y = \dfrac{4}{k}x - \dfrac{7}{k}$.

Then, slope $= \dfrac{4}{k} = 3$, $3k = 4$, $k = \dfrac{4}{3}$.

$c.$ When $y = 0$, $x = \dfrac{3k - 6}{k} = 5$. Then, $3k - 6 = 5k$, $k = -3$.

16. Find the equations of the lines which have the slope $-3/4$ and form with the coordinate axes a triangle of area 24 square units.

A line with slope $-\dfrac{3}{4}$ and y-intercept b is given by $y = -\dfrac{3}{4}x + b$.

When $x = 0$, $y = b$; when $y = 0$, $x = \dfrac{4}{3}b$.

Area of triangle $= \dfrac{1}{2}$(product of intercepts) $= \dfrac{1}{2}(b \cdot \dfrac{4}{3}b) = \dfrac{2}{3}b^2 = 24$.

Then $2b^2 = 3(24)$, $b^2 = 36$, $b = \pm 6$, and the required equations are

$$y = -\frac{3}{4}x \pm 6, \quad \text{or} \quad 3x + 4y - 24 = 0 \text{ and } 3x + 4y + 24 = 0.$$

17. Determine the locus of each of the following equations:
a) $x^2 + 8xy - 9y^2 = 0$;
b) $x^3 - 4x^2 - x + 4 = 0$.

$a.$ Since the equation factors into $(x - y)(x + 9y) = 0$, its locus is the two straight lines $x - y = 0$, $x + 9y = 0$.

$b.$ Factoring, $(x - 1)(x^2 - 3x - 4) = (x - 1)(x + 1)(x - 4) = 0$.
Hence its locus is the three straight lines $x - 1 = 0$, $x + 1 = 0$, $x - 4 = 0$.

18. A point (x,y) moves so that it is numerically twice as far from the line $x = 5$ as from the line $y = 8$. Find the equation of its locus.

Distance of (x,y) from line $x = 5$ $= \pm 2$[distance of (x,y) from line $y = 8$],

or $x - 5 = \pm 2(y - 8)$.

Hence the locus of the point is the two straight lines

$$x - 2y + 11 = 0 \text{ and } x + 2y - 21 = 0, \text{ or } (x - 2y + 11)(x + 2y - 21) = 0.$$

NORMAL EQUATION OF A STRAIGHT LINE.

19. Construct each of the following lines AB using the given values of p and ω, and write their equations.

a) $p = 5$, $\omega = \pi/6 = 30°$.
b) $p = 6$, $\omega = 2\pi/3 = 120°$.

c) $p = 4$, $\omega = 4\pi/3 = 240°$.
d) $p = 5$, $\omega = 7\pi/4 = 315°$.

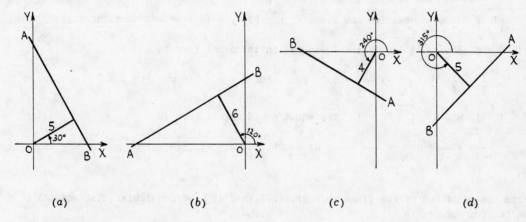

(a)   (b)   (c)   (d)

a. $x \cos 30° + y \sin 30° - 5 = 0$, or $\frac{1}{2}\sqrt{3}x + \frac{1}{2}y - 5 = 0$, or $\sqrt{3}x + y - 10 = 0$.

b. $x \cos 120° + y \sin 120° - 6 = 0$, or $-\frac{1}{2}x + \frac{1}{2}\sqrt{3}y - 6 = 0$, or $x - \sqrt{3}y + 12 = 0$.

c. $x \cos 240° + y \sin 240° - 4 = 0$, or $-\frac{1}{2}x - \frac{1}{2}\sqrt{3}y - 4 = 0$, or $x + \sqrt{3}y + 8 = 0$.

d. $x \cos 315° + y \sin 315° - 5 = 0$, or $\frac{1}{\sqrt{2}}x - \frac{1}{\sqrt{2}}y - 5 = 0$, or $x - y - 5\sqrt{2} = 0$.

20. Reduce each equation to the normal form and find p and ω.

a) $\sqrt{3}x + y - 9 = 0$.
b) $3x - 4y - 6 = 0$.

c) $x + y + 8 = 0$.
d) $12x - 5y = 0$.

e) $4y - 7 = 0$.
f) $x + 5 = 0$.

The normal form of $Ax + By + C = 0$ is $\dfrac{A}{\pm\sqrt{A^2+B^2}}x + \dfrac{B}{\pm\sqrt{A^2+B^2}}y + \dfrac{C}{\pm\sqrt{A^2+B^2}} = 0$.

a. $A = \sqrt{3}$, $B = 1$, $\sqrt{A^2+B^2} = \sqrt{3+1} = 2$. Since C $(= -9)$ is negative, $\sqrt{A^2+B^2}$ is given the positive sign. The equation in normal form is

$$\frac{\sqrt{3}}{2}x + \frac{1}{2}y - \frac{9}{2} = 0, \quad \text{and} \quad \cos\omega = \frac{\sqrt{3}}{2}, \quad \sin\omega = \frac{1}{2}, \quad p = \frac{9}{2}, \quad \omega = 30°.$$

Since $\sin\omega$ and $\cos\omega$ are both positive, ω is a first quadrant angle.

b. $A = 3$, $B = -4$, $\sqrt{A^2+B^2} = \sqrt{9+16} = 5$. The equation in normal form is

$$\frac{3}{5}x - \frac{4}{5}y - \frac{6}{5} = 0, \quad \text{and} \quad \cos\omega = \frac{3}{5}, \quad \sin\omega = -\frac{4}{5}, \quad p = \frac{6}{5}, \quad \omega = 306°52'.$$

Since $\cos\omega$ is positive and $\sin\omega$ is negative, ω is a fourth quadrant angle.

c. $A = 1$, $B = 1$, $\sqrt{A^2+B^2} = \sqrt{2}$. Since C $(= +8)$ is positive, the radical is given the negative sign. The equation in normal form is

$$-\frac{1}{\sqrt{2}}x - \frac{1}{\sqrt{2}}y - 4\sqrt{2} = 0, \quad \text{and} \quad \cos\omega = \sin\omega = -\frac{1}{\sqrt{2}}, \quad p = 4\sqrt{2}, \quad \omega = 225°.$$

Since cos ω and sin ω are both negative, ω is a third quadrant angle.

d. $\sqrt{A^2 + B^2} = \sqrt{144 + 25} = 13$. Since $C = 0$, the radical is given the same sign as B (= −5); this will make sin ω positive and ω < 180°. The equation in normal form is

$$-\frac{12}{13}x + \frac{5}{13}y = 0, \quad \text{and} \quad \cos ω = -\frac{12}{13}, \quad \sin ω = \frac{5}{13}, \quad p = 0, \quad ω = 157°23'.$$

Since cos ω is negative and sin ω is positive, ω is a second quadrant angle.

e. $A = 0$, $B = 4$, $\sqrt{A^2 + B^2} = 4$. The equation in normal form is

$$\frac{4}{4}y - \frac{7}{4} = 0, \quad \text{or} \quad y - \frac{7}{4} = 0, \quad \text{and} \quad \cos ω = 0, \quad \sin ω = 1, \quad p = \frac{7}{4}, \quad ω = 90°.$$

f. $A = 1$, $B = 0$, $\sqrt{A^2 + B^2} = 1$. The equation in normal form is

$$\frac{1}{-1}x + \frac{5}{-1} = 0, \quad \text{or} \quad -x - 5 = 0, \quad \text{and} \quad \cos ω = -1, \sin ω = 0, p = 5, ω = 180°.$$

21. Find the equations of the lines through (4,−2) and at a perpendicular distance (p) of 2 units from the origin.

The equation of the system of lines through (4,−2) and with slope m is

$$y + 2 = m(x - 4), \quad \text{or} \quad mx - y - (4m + 2) = 0.$$

The normal form of $mx - y - (4m + 2) = 0$ is $\dfrac{mx - y - (4m + 2)}{\pm\sqrt{m^2 + 1}} = 0.$

Then, $p = \dfrac{4m + 2}{\pm\sqrt{m^2 + 1}} = 2$, or $(4m + 2)^2 = 4(m^2 + 1)$. Solving, $m = 0, -\dfrac{4}{3}$.

The required equations are $y + 2 = 0$, and $y + 2 = -\dfrac{4}{3}(x - 4)$ or $4x + 3y - 10 = 0$.

22. Calculate the distance d from (a) line $8x + 15y - 24 = 0$ to point (−2,−3),
                              (b) line $6x - 8y + 5 = 0$ to point (−1,7).

a. Normal form of the equation is $\dfrac{8x + 15y - 24}{+\sqrt{8^2 + (15)^2}} = 0$, or $\dfrac{8x + 15y - 24}{17} = 0.$

$d = \dfrac{8(-2) + 15(-3) - 24}{17} = \dfrac{-85}{17} = -5$. Since d is negative, the point (−2,−3) and the origin are on the same side of the line.

b. Normal form of the equation is $\dfrac{6x - 8y + 5}{-\sqrt{6^2 + (-8)^2}} = 0$, or $\dfrac{6x - 8y + 5}{-10} = 0.$

$d = \dfrac{6(-1) - 8(7) + 5}{-10} = \dfrac{-57}{-10} = 5.7$. Since d is positive, the point (−1,7) and the origin are on opposite sides of the line.

23. Determine the equations of the bisectors of the angles between the lines
      (L₁) $3x - 4y + 8 = 0$
and (L₂) $5x + 12y - 15 = 0$.

Let $P'(x',y')$ be any point on bisector L_3.

Then, $d_1 = \dfrac{3x' - 4y' + 8}{-5}$, $d_2 = \dfrac{5x' + 12y' - 15}{13}$.

For every point on L_3, d_1 and d_2 are equal numerically.

P' and the origin are on the same side of L_1 but on opposite sides of L_2. Hence d_1 is negative and d_2 is positive, and $d_1 = -d_2$. Then the locus of P' is

$$\frac{3x' - 4y' + 8}{-5} = -\frac{5x' + 12y' - 15}{13}.$$

Simplifying and dropping primes, the equation of L_3 is $14x - 112y + 179 = 0$.

Similarly, let $P''(x'',y'')$ be any point on bisector L_4. Since P'' and the origin are on opposite sides of L_1 and L_2, d_3 and d_4 are both positive, and $d_3 = d_4$.

Then the locus of P'' is $\dfrac{3x'' - 4y'' + 8}{-5} = \dfrac{5x'' + 12y'' - 15}{13}$.

Simplifying and dropping primes, the equation of L_4 is $64x + 8y + 29 = 0$.

Note that L_3 and L_4 are perpendicular lines, their slopes being negative reciprocals of each other.

24. Find the equations of the lines parallel to the line $12x - 5y - 15 = 0$ and at a perpendicular distance from it numerically equal to 4.

Let $P'(x',y')$ be any point on the required line. Then, $\dfrac{12x' - 5y' - 15}{13} = \pm 4$.

Simplifying and dropping primes, the required equations are
$$12x - 5y - 67 = 0 \quad \text{and} \quad 12x - 5y + 37 = 0.$$

25. Determine the value of k so that the distance d from the line $8x + 15y + k = 0$ to the point $(2,3)$ is numerically 5 units.

$$d = \frac{8(2) + 15(3) + k}{\pm 17} = \pm 5. \qquad \text{Solving, } k = -146, 24.$$

26. Find the point of intersection of the bisectors of the interior angles of the triangle whose sides are: (L_1) $7x - y + 11 = 0$,
$\qquad (L_2)$ $x + y - 15 = 0$,
$\qquad (L_3)$ $7x + 17y + 65 = 0$.

The point of intersection (h,k) is the center of the circle inscribed in the triangle.
Then, the distance

of (h,k) from L_1 is $d_1 = \dfrac{7h - k + 11}{-\sqrt{50}}$,

of (h,k) from L_2 is $d_2 = \dfrac{h + k - 15}{\sqrt{2}}$,

of (h,k) from L_3 is $d_3 = \dfrac{7h + 17k + 65}{-\sqrt{338}}$.

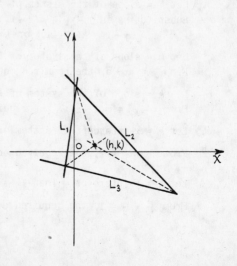

The distances are all negative, since the point and the origin are on the same side of each line. Then, $d_1 = d_2 = d_3$.

Since $d_1 = d_2$,　$\dfrac{7h - k + 11}{-5\sqrt{2}} = \dfrac{h + k - 15}{\sqrt{2}}$.　Simplifying,　$3h + k = 16$.

Since $d_1 = d_3$,　$\dfrac{7h - k + 11}{-5\sqrt{2}} = \dfrac{7h + 17k + 65}{-13\sqrt{2}}$.　Simplifying,　$4h - 7k = 13$.

Solving $3h + k = 16$　and　$4h - 7k = 13$ simultaneously, $h = 5$, $k = 1$.

27. Given the triangle $A(-2, 1)$, $B(5, 4)$, $C(2, -3)$, determine the length of the altitude through A and the area of the triangle.

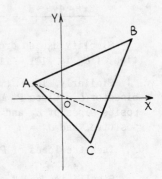

Equation of BC:　$\dfrac{y + 3}{x - 2} = \dfrac{4 + 3}{5 - 2}$,　or　$7x - 3y - 23 = 0$.

Distance from BC to A $= \dfrac{7(-2) - 3(1) - 23}{\sqrt{49 + 9}} = \dfrac{-40}{\sqrt{58}}$.

Length of $BC = \sqrt{(5 - 2)^2 + (4 + 3)^2} = \sqrt{58}$.

Area of the triangle $= \frac{1}{2}(\sqrt{58} \cdot \dfrac{40}{\sqrt{58}}) = 20$ square units.

SYSTEMS OF LINES.

28. Determine the equation of the system of lines
   (a) whose slope is -4,
   (b) passing through $(4, 1)$,
   (c) whose y-intercept is 7,
   (d) whose x-intercept is 5,
   (e) the sum of whose intercepts is 8,
   (f) whose y-intercept is twice the x-intercept,
   (g) having one intercept numerically twice the other intercept.

   In each case, let k be the arbitrary constant, or parameter, of the system of lines.

a. Let $k = y$-intercept of the system of lines whose slope is -4.
   From $y = mx + b$, the required equation is $y = -4x + k$, or $4x + y - k = 0$.

b. Let $k =$ slope of the system of lines passing through $(4, 1)$.
   Substituting in $y - y_1 = m(x - x_1)$, the required equation is
   　　$y - 1 = k(x - 4)$, or $kx - y + 1 - 4k = 0$.

c. Let $k =$ slope of the system of lines whose y-intercept is 7.
   From $y = mx + b$, the required equation is $y = kx + 7$, or $kx - y + 7 = 0$.

d. Let $k =$ slope of the system of lines whose x-intercept is 5.
   From $y - y_1 = m(x - x_1)$, the required equation is $y - 0 = k(x - 5)$, or $kx - y - 5k = 0$.

e. Let $k = x$-intercept of system of lines. Then $(8 - k) = y$-intercept of system.
   From $\dfrac{x}{a} + \dfrac{y}{b} = 1$, the required equation is $\dfrac{x}{k} + \dfrac{y}{8 - k} = 1$, or $(8 - k)x + ky - 8k + k^2 = 0$.

f. Let $k = y$-intercept. Then $\frac{1}{2}k = x$-intercept.
   From $\dfrac{x}{a} + \dfrac{y}{b} = 1$, the required equation is $\dfrac{x}{\frac{1}{2}k} + \dfrac{y}{k} = 1$, or $2x + y - k = 0$.

g. Slope of a line $= -\dfrac{y\text{-intercept}}{x\text{-intercept}}$. When the x-intercept is numerically ($\pm$) twice the y-intercept, slope of line is $\mp\frac{1}{2}$; when the y-intercept is numerically twice the x-intercept, slope of line is ∓ 2. Let $k = y$-intercept. Then, from $y = mx + b$, the required systems of lines are $y = \pm\frac{1}{2}x + k$ and $y = \pm 2x + k$.

29. Determine the equation of the line which passes through point $(-2,-4)$ and has the sum of its intercepts equal to 3.

The equation of the system of lines through $(-2,-4)$ is $y + 4 = m(x + 2)$.

When $x = 0$, $y = 2m - 4$; when $y = 0$, $x = \dfrac{4 - 2m}{m}$.

The sum of the intercepts is 3. Hence, $2m - 4 + \dfrac{4 - 2m}{m} = 3$.

Simplifying, $2m^2 - 9m + 4 = 0$. Solving, $(2m - 1)(m - 4) = 0$, $m = \frac{1}{2}$, 4.

Substituting these m values in $y + 4 = m(x + 2)$, the required equations are
$y + 4 = \frac{1}{2}(x + 2)$ and $y + 4 = 4(x + 2)$, or $x - 2y - 6 = 0$ and $4x - y + 4 = 0$.

30. Find the equation of the line passing through the point of intersection of the lines $3x - 2y + 10 = 0$ and $4x + 3y - 7 = 0$ and through point $(2, 1)$.

$3x - 2y + 10 + k(4x + 3y - 7) = 0$ is the equation of the system of lines passing through the intersection of the two given lines.
Since the required line passes through $(2, 1)$, $3 \cdot 2 - 2 \cdot 1 + 10 + k(4 \cdot 2 + 3 \cdot 1 - 7) = 0$.
Solving, $k = -7/2$. The required line is
$$3x - 2y + 10 - \frac{7}{2}(4x + 3y - 7) = 0, \text{ or } 22x + 25y - 69 = 0.$$

31. Find the equation of the line perpendicular to $4x + y - 1 = 0$ and passing through the point of intersection of $2x - 5y + 3 = 0$ and $x - 3y - 7 = 0$.

Slope of $4x + y - 1 = 0$ is -4. Then, slope of required line is $\frac{1}{4}$.
The equation of the system of lines passing through the point of intersection of $2x - 5y + 3 = 0$ and $x - 3y - 7 = 0$ is
$$2x - 5y + 3 + k(x - 3y - 7) = 0, \text{ or } (2 + k)x - (5 + 3k)y + (3 - 7k) = 0. \quad (1)$$
The slope of every line of this system is $\dfrac{2 + k}{5 + 3k}$, and slope of required line is $\frac{1}{4}$.

Then, $\dfrac{2 + k}{5 + 3k} = \dfrac{1}{4}$. Solving, $k = -3$.

Substituting $k = -3$ in (1), the required equation is $x - 4y - 24 = 0$.

SUPPLEMENTARY PROBLEMS

1. Write the equations of the lines satisfying the following conditions:
   (a) Through $(0, 2)$, $m = 3$.   Ans. $y - 3x - 2 = 0$
   (b) Through $(0, -3)$, $m = -2$.   Ans. $y + 2x + 3 = 0$
   (c) Through $(0, 4)$, $m = 1/3$.   Ans. $x - 3y + 12 = 0$
   (d) Through $(0, -1)$, $m = 0$.   Ans. $y + 1 = 0$
   (e) Through $(0, 3)$, $m = -4/3$.   Ans. $4x + 3y - 9 = 0$

2. Write the equation of the line which passes through:

      (a) $(2,-3)$ and $(4,2)$.   Ans. $5x - 2y - 16 = 0$

      (b) $(-4,1)$ and $(3,-5)$.   Ans. $6x + 7y + 17 = 0$

      (c) $(7,0)$ and $(0,4)$.   Ans. $4x + 7y - 28 = 0$

      (d) $(0,0)$ and $(5,-3)$.   Ans. $3x + 5y = 0$

      (e) $(5,-3)$ and $(5,2)$.   Ans. $x - 5 = 0$

      (f) $(-5,2)$ and $(3,2)$.   Ans. $y - 2 = 0$

3. Given the triangle whose vertices are $A(-5,6)$, $B(-1,-4)$ and $C(3,2)$.

   a. Derive the equations of the three medians.

      Ans. $7x + 6y - 1 = 0$, $x + 1 = 0$, $x - 6y + 9 = 0$

   b. Solve algebraically for their point of intersection.   Ans. $(-1, 4/3)$

4. a. Derive the equations of the three altitudes of the triangle in Problem 3.

      Ans. $2x + 3y - 8 = 0$, $2x - y - 2 = 0$, $2x - 5y + 4 = 0$

   b. Solve for the point of intersection of these altitudes.   Ans. $\left(\frac{7}{4}, \frac{3}{2}\right)$

5. a. Determine the equations of the perpendicular bisectors of the sides of the triangle in Problem 3.   Ans. $2x - 5y + 11 = 0$, $2x - y + 6 = 0$, $2x + 3y + 1 = 0$

   b. Solve for the point of intersection of these perpendicular bisectors.

      Ans. $(-19/8, 5/4)$. This is the center of the circle circumscribed about the triangle.

6. Prove that the points of intersection of the medians, of the altitudes, and of the perpendicular bisectors of the sides for the triangle in Problem 3, lie in a straight line. Ans. $2x - 33y + 46 = 0$

7. Find the equation of the line through the point $(2,3)$ so that its x-intercept will be twice its y-intercept.   Ans. $x + 2y - 8 = 0$

8. Determine the value of K in the equation $2x + 3y + K = 0$ so that this line will form a triangle with the coordinate axes whose area is 27 square units.   Ans. $K = \pm 18$

9. Determine the value of K so that the line whose equation is $2x + 3Ky - 13 = 0$ shall pass through the point $(-2,4)$.   Ans. $K = 17/12$

10. Determine the value of K so that the line whose equation is $3x - Ky - 8 = 0$ shall make an angle of $45°$ with the line $2x + 5y - 17 = 0$.   Ans. $K = 7, -9/7$

11. Find the point on the line $3x + y + 4 = 0$ that is equidistant from $(-5,6)$ and $(3,2)$. Ans. $(-2,2)$

12. Find the equations of the lines through $(1,-6)$ if the product of the intercepts for each line is 1.   Ans. $9x + y - 3 = 0$, $4x + y + 2 = 0$

13. Find the equation of the line whose x-intercept is $-3/7$, and which is perpendicular to the line $3x + 4y - 10 = 0$.   Ans. $28x - 21y + 12 = 0$

14. Find the equation of the line which is perpendicular to the line $2x + 7y - 3 = 0$ at its intersection with the line $3x - 2y + 8 = 0$.   Ans. $7x - 2y + 16 = 0$

15. Construct each of the following lines using the given values of p and ω, and write their equations.

      (a) $p = 6$, $\omega = 30°$.   Ans. $\sqrt{3}x + y - 12 = 0$

      (b) $p = \sqrt{2}$, $\omega = \pi/4$.   Ans. $x + y - 2 = 0$

      (c) $p = 3$, $\omega = 2\pi/3$.   Ans. $x - \sqrt{3}y + 6 = 0$

(d) $p = 4$, $\omega = 7\pi/4$.   Ans. $x - y - 4\sqrt{2} = 0$

(e) $p = 3$, $\omega = 0°$.   Ans. $x - 3 = 0$

(f) $p = 4$, $\omega = 3\pi/2$.   Ans. $y + 4 = 0$

16. Reduce each of the following straight line equations to the normal form. Find p and ω.

(a) $x - 3y + 6 = 0$.   Ans. $-\dfrac{x}{\sqrt{10}} + \dfrac{3}{\sqrt{10}}y - \dfrac{6}{\sqrt{10}} = 0$, $p = \dfrac{3\sqrt{10}}{5}$, $\omega = 108°26'$

(b) $2x + 3y - 10 = 0$ Ans. $\dfrac{2}{\sqrt{13}}x + \dfrac{3}{\sqrt{13}}y - \dfrac{10}{\sqrt{13}} = 0$, $p = \dfrac{10\sqrt{13}}{13}$, $\omega = 56°19'$

(c) $3x + 4y - 5 = 0$ Ans. $\dfrac{3}{5}x + \dfrac{4}{5}y - 1 = 0$, $p = 1$, $\omega = 53°8'$

(d) $5x + 12y = 0$ Ans. $\dfrac{5}{13}x + \dfrac{12}{13}y = 0$, $p = 0$, $\omega = 67°23'$

(e) $x + y - \sqrt{2} = 0$ Ans. $\dfrac{x}{\sqrt{2}} + \dfrac{y}{\sqrt{2}} - 1 = 0$, $p = 1$, $\omega = \pi/4$

17. Find the equations and the point of intersection of the bisectors of the interior angles of the triangle formed by the lines $4x - 3y - 65 = 0$, $7x - 24y + 55 = 0$, and $3x + 4y - 5 = 0$. *Ans.* $9x - 13y - 90 = 0$, $2x + 11y - 20 = 0$, $7x + y - 70 = 0$. Point $(10,0)$.

18. Find the equations and the point of intersection of the bisectors of the interior angles of the triangle formed by the lines $7x + 6y - 11 = 0$, $9x - 2y + 7 = 0$, and $6x - 7y - 16 = 0$. *Ans.* $x + 13y + 5 = 0$, $5x - 3y - 3 = 0$, $4x + y - 1 = 0$. Point $(6/17, -7/17)$.

19. Find the equations and the point of intersection of the bisectors of the interior angles of the triangle formed by the lines $y = 0$, $3x - 4y = 0$, and $4x + 3y - 50 = 0$. *Ans.* $x - 3y = 0$, $2x + 4y - 25 = 0$, $7x - y - 50 = 0$. Point $(15/2, 5/2)$.

20. Find the point of intersection of the bisectors of the interior angles of the triangle whose vertices are $(-1,3)$, $(3,6)$, and $(31/5, 0)$. *Ans.* $(17/7, 24/7)$.

21. Find the coordinates of the center and the radius of the circle inscribed in the triangle formed by the lines $15x - 8y + 25 = 0$, $3x - 4y - 10 = 0$, and $5x + 12y - 30 = 0$. *Ans.* $(4/7, 1/4)$. Radius $= 13/7$.

22. Find the value of K so that the line $y + 5 = K(x - 3)$ is at a distance from the origin whose numerical value is 3. *Ans.* $K = -8/15$, ∞

23. A point moves so that its distance from the line $5x + 12y - 20 = 0$ is always three times its distance from the line $4x - 3y + 12 = 0$. Find the equation of its locus. *Ans.* $181x - 57y + 368 = 0$, $131x - 177y + 568 = 0$

24. A point moves so that the square of its distance from $(3,-2)$ is numerically equal to its distance from the line $5x - 12y - 13 = 0$. Find the equation of its locus. *Ans.* $13x^2 + 13y^2 - 73x + 40y + 156 = 0$, $13x^2 + 13y^2 - 83x + 64y + 182 = 0$

25. Find the two points on the line $5x - 12y + 15 = 0$ that lie at a distance from the line $3x + 4y - 12 = 0$ numerically equal to 3. *Ans.* $(\dfrac{33}{7}, \dfrac{45}{14})$, $(-\dfrac{12}{7}, \dfrac{15}{28})$

26. Find two lines parallel to $8x - 15y + 34 = 0$ that lie at a distance from $(-2,3)$ numerically equal to 3. *Ans.* $8x - 15y + 112 = 0$, $8x - 15y + 10 = 0$

27. A point moves so that its distance from $3x - 4y - 2 = 0$ is always equal to its distance from the point $(-1, 2)$. Derive the equation of its locus.
   Ans. $16x^2 + 24xy + 9y^2 + 62x - 116y + 121 = 0$

28. Find the length of the altitude from A to side BC, also find the area, of the triangle whose vertices are:

   (a) $A(-3, 3)$, $B(5, 5)$, $C(2, -4)$.   *Ans.* Altitude = $\dfrac{11\sqrt{10}}{5}$,   Area = 33 square units

   (b) $A(5, 6)$, $B(1, -4)$, $C(-4, 0)$.   *Ans.* Altitude = $\dfrac{66\sqrt{41}}{41}$,   Area = 33 square units

   (c) $A(-1, 4)$, $B(1, -4)$, $C(5, 4)$.   *Ans.* Altitude = $\dfrac{12\sqrt{5}}{5}$,   Area = 24 square units

   (d) $A(0, 4)$, $B(5, 1)$, $C(1, -3)$.   *Ans.* Altitude = $4\sqrt{2}$,   Area = 16 square units

29. In each of the following equations of a line, find the value of K so that the corresponding condition is satisfied.
   (a) $(2 + K)x - (3 - K)y + 4K + 14 = 0$, passes through the point $(2, 3)$.   *Ans.* $K = -1$
   (b) $Kx + (3 - K)y + 7 = 0$, the slope of the line being 7.   *Ans.* $K = 7/2$
   (c) $5x - 12y + 3 + K = 0$, this line being at a distance from $(-3, 2)$ whose numerical value is 4.   *Ans.* $K = -16$, $K = 88$

30. Find the equation of the line which passes through the point of intersection of the lines $3x - 5y + 9 = 0$ and $4x + 7y - 28 = 0$ and which satisfies the following condition:
   (a) Passes through $(-3, -5)$.   *Ans.* $13x - 8y - 1 = 0$
   (b) Passes through $(4, 2)$.   *Ans.* $38x + 87y - 326 = 0$
   (c) Is parallel to $2x + 3y - 5 = 0$.   *Ans.* $82x + 123y - 514 = 0$
   (d) Is perpendicular to $4x + 5y - 20 = 0$.   *Ans.* $205x - 164y + 95 = 0$
   (e) Whose intercepts are equal.   *Ans.* $41x + 41y - 197 = 0$, $120x - 77y = 0$

31. Find the equation of the line through the point of intersection of the lines $x - 3y + 1 = 0$ and $2x + 5y - 9 = 0$ and whose distance from the origin is (a) 2, (b) $\sqrt{5}$.
   Ans. (a) $x - 2 = 0$, $3x + 4y - 10 = 0$; (b) $2x + y - 5 = 0$.

CHAPTER 4

The Circle

A CIRCLE is represented by an equation of the second degree in two variables. The converse is true only for certain forms of equations of the second degree.

A circle is completely determined if its center and radius are known.

THE EQUATION OF THE CIRCLE with its center at (h,k) and radius r is

$$(x - h)^2 + (y - k)^2 = r^2 .$$

If the center is at the origin the equation becomes $x^2 + y^2 = r^2$.

Every equation of the circle can be reduced to the form

$$x^2 + y^2 + Dx + Ey + F = 0.$$

If we write this equation in the form

$$x^2 + Dx + y^2 + Ey + F = 0$$

and complete squares, we have

$$x^2 + Dx + \frac{D^2}{4} + y^2 + Ey + \frac{E^2}{4} = \frac{D^2}{4} + \frac{E^2}{4} - F$$

or

$$(x + \frac{D}{2})^2 + (y + \frac{E}{2})^2 = \frac{D^2 + E^2 - 4F}{4} .$$

The center is at the point $(-\frac{D}{2}, -\frac{E}{2})$, and the radius $r = \frac{1}{2}\sqrt{D^2 + E^2 - 4F}$.

If $D^2 + E^2 - 4F > 0$, the circle is real.

If $D^2 + E^2 - 4F < 0$, the circle is imaginary.

If $D^2 + E^2 - 4F = 0$, the radius is zero and the equation will represent the point $(-\frac{D}{2}, -\frac{E}{2})$.

SOLVED PROBLEMS

1. Write the equation of the circle with its center at point $(-2,3)$ and radius 4.

$$(x + 2)^2 + (y - 3)^2 = 16 \quad \text{or} \quad x^2 + y^2 + 4x - 6y = 3.$$

2. For the circle whose equation is $x^2 + y^2 - 3x + 5y - 14 = 0$, find the coordinates of the center and the radius by (a) completing the square, (b) the formula.

a. $x^2 - 3x + \frac{9}{4} + y^2 + 5y + \frac{25}{4} = 14 + \frac{9}{4} + \frac{25}{4}$ or $(x - \frac{3}{2})^2 + (y + \frac{5}{2})^2 = \frac{90}{4}$.

Hence the center is at point $(\frac{3}{2}, -\frac{5}{2})$ and the radius $r = \frac{3\sqrt{10}}{2}$.

b. $h = -\frac{D}{2} = \frac{3}{2}$, $k = -\frac{E}{2} = -\frac{5}{2}$, and $r = \frac{1}{2}\sqrt{D^2 + E^2 - 4F} = \frac{1}{2}\sqrt{9 + 25 + 56} = \frac{3\sqrt{10}}{2}$.

3. Determine the value of k so that $x^2 + y^2 - 8x + 10y + k = 0$ is the equation of a circle whose radius is 7.

Since $r = \frac{1}{2}\sqrt{D^2 + E^2 - 4F}$, then $7 = \frac{1}{2}\sqrt{64 + 100 - 4k}$. Squaring and solving, $k = -8$.

4. Derive the equation of the circle whose center is $(5,-2)$ and which passes through point $(-1,5)$.

The radius of the circle is $r = \sqrt{(5+1)^2 + (-2-5)^2} = \sqrt{36 + 49} = \sqrt{85}$.

Then $(x-5)^2 + (y+2)^2 = 85$, or $x^2 + y^2 - 10x + 4y = 56$.

5. Find the equation of the circle which has for a diameter the segment joining the points $(5,-1)$ and $(-3,7)$.

The coordinates of the center are $h = \dfrac{5-3}{2} = 1$, $k = \dfrac{-1+7}{2} = 3$.

The radius is $r = \sqrt{(5-1)^2 + (-1-3)^2} = \sqrt{16 + 16} = 4\sqrt{2}$.

Then $(x-1)^2 + (y-3)^2 = 32$, or $x^2 + y^2 - 2x - 6y = 22$.

6. Find the equation of the circle which passes through $(0,0)$, $r = 13$, and the abscissa of its center is -12.

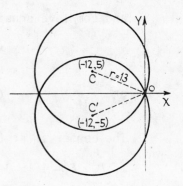

Since the circle passes through the origin,

$$h^2 + k^2 = r^2, \quad \text{or} \quad 144 + k^2 = 169$$

Solving, $k^2 = 169 - 144 = 25$, $k = \pm 5$.

Then, $(x+12)^2 + (y-5)^2 = 169$

and $(x+12)^2 + (y+5)^2 = 169$.

Expanding, $x^2 + y^2 + 24x - 10y = 0$

and $x^2 + y^2 + 24x + 10y = 0$.

7. Determine the equation of the circle passing through the three points $(5,3)$, $(6,2)$, and $(3,-1)$.

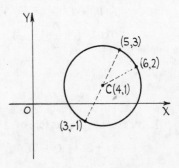

Since each of the standard forms

$$(x - h)^2 + (y - k)^2 = r^2$$

or $$x^2 + y^2 + Dx + Ey + F = 0$$

contains three undetermined constants, three conditions are necessary to determine these coefficients. As the circle must pass through these points, the coefficients may be determined by substituting the coordinates of the points for x and y and solving the three linear equations for D, E, and F. Then,

$$25 + 9 + 5D + 3E + F = 0,$$
$$36 + 4 + 6D + 2E + F = 0,$$
$$9 + 1 + 3D - E + F = 0.$$

Solving these three equations, $D = -8$, $E = -2$, and $F = 12$.

Substituting for D, E, and F, the equation of the circle is $x^2 + y^2 - 8x - 2y + 12 = 0$.

8. Find the equation of the circle which passes through the points (2,3) and (-1,1) and has its center on the line $x - 3y - 11 = 0$.

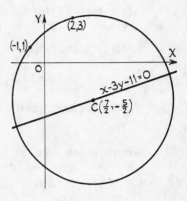

Let (h,k) be the coordinates of the center of the circle. Since (h,k) must be equidistant from (2,3) and (-1,1),

$$\sqrt{(h-2)^2 + (k-3)^2} = \sqrt{(h+1)^2 + (k-1)^2}.$$

Squaring and simplifying, $6h + 4k = 11$.

Since the center must lie on the line $x - 3y - 11 = 0$, then $h - 3k = 11$.

Solving these equations for h and k, $h = \frac{7}{2}$, $k = -\frac{5}{2}$.

Then $r = \sqrt{(\frac{7}{2} + 1)^2 + (-\frac{5}{2} - 1)^2} = \frac{1}{2}\sqrt{130}$.

The required equation is $(x - \frac{7}{2})^2 + (y + \frac{5}{2})^2 = \frac{130}{4}$ or $x^2 + y^2 - 7x + 5y - 14 = 0$.

9. Find the equation of the circle inscribed in the triangle determined by the lines, L_1: $2x - 3y + 21 = 0$,
$\qquad L_2$: $3x - 2y - 6 = 0$,
$\qquad L_3$: $2x + 3y + 9 = 0$.

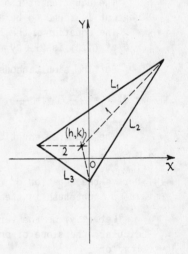

Since the center of the circle lies at the point of intersection of the bisectors of the interior angles of the triangle, it is necessary to find the equations of these bisectors. Let (h,k) be the coordinates of the center. To determine the bisector (1): (See the figure.)

$$\frac{2h - 3k + 21}{-\sqrt{13}} = \frac{3h - 2k - 6}{\sqrt{13}}, \quad \text{or} \quad h - k + 3 = 0.$$

To obtain bisector (2):

$$\frac{2h + 3k + 9}{-\sqrt{13}} = \frac{2h - 3k + 21}{-\sqrt{13}}, \quad \text{or} \quad 6k - 12 = 0.$$

Then $k = 2$, $h = -1$, and $r = \frac{2(-1) + 3(2) + 9}{\sqrt{13}} = \frac{13}{\sqrt{13}} = \sqrt{13}$.

Substituting in $(x - h)^2 + (y - k)^2 = r^2$, $(x + 1)^2 + (y - 2)^2 = 13$ or $x^2 + y^2 + 2x - 4y = 8$.

10. Derive the equation of the circle circumscribing the triangle determined by the lines, $x + y = 8$,
$\qquad 2x + y = 14$,
$\qquad 3x + y = 22$.

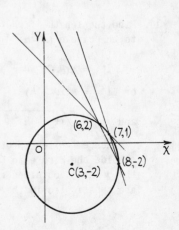

Solving these equations in pairs, the vertices of the triangle are (6,2), (7,1), and (8,-2).

Substituting the coordinates of these three points in the general equation $x^2 + y^2 + Dx + Ey + F = 0$, the following equations are obtained: $\qquad 6D + 2E + F = -40$,
$\qquad\qquad 7D + E + F = -50$,
$\qquad\qquad 8D - 2E + F = -68$.

Solving, $D = -6$, $E = 4$, and $F = -12$.

Substituting these values, $x^2 + y^2 - 6x + 4y - 12 = 0$.

11. Derive the equation of the circle with its center at $(-4, 2)$ and tangent to the line $3x + 4y - 16 = 0$.

The radius can be determined by finding the distance from $(-4, 2)$ to the line.

$$r = \left| \frac{3(-4) + 4(2) - 16}{5} \right| = \left| -\frac{20}{5} \right| = \left| -4 \right| \text{ or } 4.$$

The required equation is $(x + 4)^2 + (y - 2)^2 = 16$, or $x^2 + y^2 + 8x - 4y + 4 = 0$.

12. Derive the equation of the circle which passes through the point $(-2, 1)$ and is tangent to the line $3x - 2y - 6 = 0$ at the point $(4, 3)$.

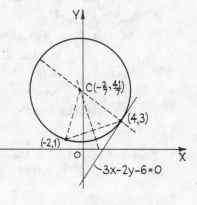

Since the circle must pass through the two points $(-2, 1)$ and $(4, 3)$, its center will lie on the perpendicular bisector of the segment connecting these points. The center must also lie on the line perpendicular to $3x - 2y - 6 = 0$ at the point $(4, 3)$.

The equation of the perpendicular bisector of the segment is found to be $3x + y - 5 = 0$.

The equation of the line perpendicular to $3x - 2y - 6 = 0$ at $(4, 3)$ is $2x + 3y - 17 = 0$.

Solving simultaneously $2x + 3y - 17 = 0$ and $3x + y - 5$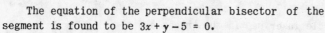

$= 0$, $x = -\frac{2}{7}$, $y = \frac{41}{7}$. Then $r = \sqrt{(4 + \frac{2}{7})^2 + (3 - \frac{41}{7})^2} = \frac{10}{7}\sqrt{13}$.

The required equation is $(x + \frac{2}{7})^2 + (y - \frac{41}{7})^2 = \frac{1300}{49}$ or $7x^2 + 7y^2 + 4x - 82y + 55 = 0$.

13. Derive the equation of the locus of the vertex of the right angle of the right triangle whose hypotenuse is the segment joining the points $(0, b)$ and (a, b).

Let (x, y) be the vertex of the right angle. Then, since the two legs must be perpendicular, the slope of one leg will be the negative reciprocal of the slope of the other leg, or

$$\frac{y - b}{x - 0} = -\frac{1}{\dfrac{y - b}{x - a}} = -\frac{x - a}{y - b}.$$

Simplifying, $(y - b)^2 = -x(x - a)$ or $x^2 + y^2 - ax - 2by + b^2 = 0$ (a circle).

14. Find the length of the tangent from the point $P_1(x_1, y_1)$ to the circle $(x - h)^2 + (y - k)^2 = r^2$.

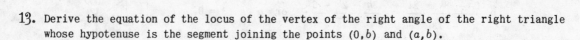

$$l^2 = (P_1C)^2 - r^2,$$

or $l^2 = (x_1 - h)^2 + (y_1 - k)^2 - r^2,$

and $l = \sqrt{(x_1 - h)^2 + (y_1 - k)^2 - r^2}.$

Hence the length of the tangent drawn from any point outside the circle to the circle is the square root of the quantity obtained when the coordinates of the point are substituted in the equation of the circle.

15. *Definition.* The locus of a point from which the lengths of the tangents to two circles are equal is a straight line called their *radical axis*.

Derive the equation of the radical axis for the two circles,

$$x^2 + y^2 + d_1 x + e_1 y + f_1 = 0$$

and

$$x^2 + y^2 + d_2 x + e_2 y + f_2 = 0.$$

Let $P'(x', y')$ be any point on the radical axis.

Then, $l_1 = \sqrt{x'^2 + y'^2 + d_1 x' + e_1 y' + f_1}$ and $l_2 = \sqrt{x'^2 + y'^2 + d_2 x' + e_2 y' + f_2}.$

Since $l_1 = l_2$, $\sqrt{x'^2 + y'^2 + d_1 x' + e_1 y' + f_1} = \sqrt{x'^2 + y'^2 + d_2 x' + e_2 y' + f_2}.$

Squaring, simplifying and dropping primes, $(d_1 - d_2)x + (e_1 - e_2)y + f_1 - f_2 = 0,$ a straight line.

16. Write the equation of the system of circles passing through the points of intersection of two given circles.

Let $x^2 + y^2 + d_1 x + e_1 y + f_1 = 0$ and $x^2 + y^2 + d_2 x + e_2 y + f_2 = 0$ be two intersecting circles.

Then $x^2 + y^2 + d_1 x + e_1 y + f_1 + K(x^2 + y^2 + d_2 x + e_2 y + f_2) = 0$ represents such a system, since the coordinates of the points of intersection must reduce the equation of each of the circles to zero.

For every value of K except $K = -1$, a circle is obtained. When $K = -1$, the equation reduces to a straight line which is the common chord of the two circles.

17. Find the equations of the circles passing through the points $A(1, 2)$, $B(3, 4)$ and tangent to the line $3x + y - 3 = 0$.

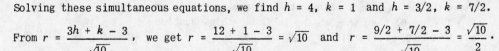

To determine the coordinates of the center $C(h, k)$, we know that $CA = CB$ and $CA = CN$, or

$$(h - 1)^2 + (k - 2)^2 = (h - 3)^2 + (k - 4)^2$$

and

$$(h - 1)^2 + (k - 2)^2 = (\frac{3h + k - 3}{\sqrt{10}})^2.$$

Expanding and simplifying these two equations, we have

$$h + k = 5$$

and

$$h^2 + 9k^2 - 6hk - 2h - 34k + 41 = 0.$$

Solving these simultaneous equations, we find $h = 4$, $k = 1$ and $h = 3/2$, $k = 7/2$.

From $r = \dfrac{3h + k - 3}{\sqrt{10}}$, we get $r = \dfrac{12 + 1 - 3}{\sqrt{10}} = \sqrt{10}$ and $r = \dfrac{9/2 + 7/2 - 3}{\sqrt{10}} = \dfrac{\sqrt{10}}{2}.$

Using $(x - h)^2 + (y - k)^2 = r^2$, we have

$$(x - 4)^2 + (y - 1)^2 = 10 \quad \text{and} \quad (x - \tfrac{3}{2})^2 + (y - \tfrac{7}{2})^2 = \frac{10}{4}.$$

Expanding these equations, we get $x^2 + y^2 - 8x - 2y + 7 = 0$ and $x^2 + y^2 - 3x - 7y + 12 = 0.$

18. Derive the equation of the circle of radius 5 and tangent to $3x + 4y - 16 = 0$ at $(4, 1)$.

Use (h,k) for the coordinates of the center.

Then $\dfrac{3h + 4k - 16}{5} = \pm 5$, or $3h + 4k - 16 = \pm 25$.

Also, $(h - 4)^2 + (k - 1)^2 = 25$, or $h^2 + k^2 - 8h - 2k = 8$.

Solving simultaneously, the two solutions are $(7,5)$ and $(1,-3)$.

The equations of the circles are $(x - 7)^2 + (y - 5)^2 = 25$ and $(x - 1)^2 + (y + 3)^2 = 25$.

19. Derive the equations of the two circles tangent to the lines $3x - 4y + 1 = 0$ and $4x + 3y - 7 = 0$ and passing through the point $(2,3)$.

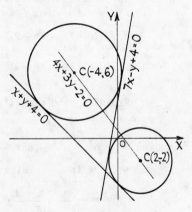

Use (h,k) for the coordinates of the center. Then

$$\frac{3h - 4k + 1}{-5} = \frac{4h + 3k - 7}{5} \quad \text{or} \quad 7h - k - 6 = 0. \quad (a)$$

Also, since $r = \dfrac{3h - 4k + 1}{-5}$,

$$(h - 2)^2 + (k - 3)^2 = \left(\frac{3h - 4k + 1}{-5}\right)^2$$

or $16h^2 + 9k^2 - 106h - 142k + 24hk + 324 = 0.$ $\qquad (b)$

Solving (a) and (b) simultaneously, the coordinates of the two centers are $(2,8)$ and $(6/5, 12/5)$.

Using $(2,8)$ as center, $r = \dfrac{3h - 4k + 1}{-5} = \dfrac{6 - 32 + 1}{-5} = 5$, and the equation of the circle is $(x - 2)^2 + (y - 8)^2 = 25$.

Using $(\frac{6}{5}, \frac{12}{5})$ as center, $r = 1$, and the circle is $(x - \frac{6}{5})^2 + (y - \frac{12}{5})^2 = 1$.

20. Write the equation of the circle tangent to the lines $x + y + 4 = 0$ and $7x - y + 4 = 0$ and having its center on the line $4x + 3y - 2 = 0$.

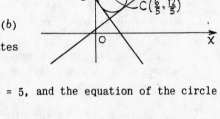

Use (h,k) for the coordinates of the center. Then

$$\frac{h + k + 4}{\sqrt{2}} = \pm \frac{7h - k + 4}{5\sqrt{2}}$$

or $h - 3k - 8 = 0$ and $3h + k + 6 = 0$,

which are the equations of the bisectors of the angles between the two lines. Since the center must lie on $4x + 3y - 2 = 0$, then $4h + 3k - 2 = 0$. Using this equation and $h - 3k - 8 = 0$, we find $h = 2$, $k = -2$.

Then $r = \dfrac{2 - 2 + 4}{\sqrt{2}} = 2\sqrt{2}$, and the equation of this circle is $(x - 2)^2 + (y + 2)^2 = 8$.

Using the equations $4h + 3k - 2 = 0$ and $3h + k + 6 = 0$, $h = -4$, $k = 6$, and $r = 3\sqrt{2}$.

Then the equation of this circle is $(x + 4)^2 + (y - 6)^2 = 18$.

21. Determine the locus of the point (x',y') which moves so that the sum of the squares of its distances from the lines $5x + 12y - 4 = 0$ and $12x - 5y + 10 = 0$ is 5.

The distance of (x', y') from $5x + 12y - 4 = 0$ is $\dfrac{5x' + 12y' - 4}{13}$, and from $12x - 5y + 10$

$= 0$ is $\dfrac{12x' - 5y' + 10}{-13}$. Then, $(\dfrac{5x' + 12y' - 4}{13})^2 + (\dfrac{12x' - 5y' + 10}{-13})^2 = 5$.

Simplifying and dropping primes, $169x^2 + 169y^2 + 200x - 196y = 729$, a circle.

22. Determine the locus of a point (x, y) the sum of the squares of whose distances from $(2, 3)$ and $(-1, -2)$ is 34.

$(x - 2)^2 + (y - 3)^2 + (x + 1)^2 + (y + 2)^2 = 34$. Simplifying, $x^2 + y^2 - x - y = 8$, a circle.

23. Derive the equation of the locus of a point (x, y) the ratio of whose distance from $(-1, 3)$ to its distance from $(3, -2)$ is a/b.

$$\frac{\sqrt{(x + 1)^2 + (y - 3)^2}}{\sqrt{(x - 3)^2 + (y + 2)^2}} = \frac{a}{b}.$$ Squaring and simplifying, the required equation is

$(b^2 - a^2)x^2 + (b^2 - a^2)y^2 + 2(b^2 + 3a^2)x - 2(3b^2 + 2a^2)y = 13a^2 - 10b^2$, a circle.

24. Derive the equation of the locus of a point (x, y) the square of whose distance from $(-5, 2)$ is equal to its distance from $5x + 12y - 26 = 0$.

$(x + 5)^2 + (y - 2)^2 = \pm(\dfrac{5x + 12y - 26}{13})$. Expanding and simplifying,

$13x^2 + 13y^2 + 125x - 64y + 403 = 0$ and $13x^2 + 13y^2 + 135x - 40y + 351 = 0$, circles.

25. Write the equation of a circle concentric with the circle $x^2 + y^2 - 4x + 6y - 17 = 0$, and tangent to the line $3x - 4y + 7 = 0$.

The center of the given circle is $(2, -3)$. The radius of the required circle is the distance from $(2, -3)$ to the line $3x - 4y + 7 = 0$, or $r = \dfrac{6 + 12 + 7}{5} = 5$.

Hence the required circle is $(x - 2)^2 + (y + 3)^2 = 25$.

26. Write the equations of the circles with radius 15 and tangent to the circle $x^2 + y^2 = 100$ at point $(6, -8)$.

The center of these circles must lie on the line through $(0, 0)$ and $(6, -8)$. The equation of this line is $y = -\dfrac{4}{3}x$.

Use (h, k) for the center. Then $k = -\dfrac{4}{3}h$ and $(h - 6)^2 + (k + 8)^2 = 225$.

Solving these equations for h and k, the coordinates of the two centers are $(-3, 4)$ and $(15, -20)$.

The equations of the two circles are $(x + 3)^2 + (y - 4)^2 = 225$ and
$$(x - 15)^2 + (y + 20)^2 = 225.$$

SUPPLEMENTARY PROBLEMS

1. Find the equation of the circle

   (a) whose center is (3,-1) and radius is 5.   *Ans.* $x^2 + y^2 - 6x + 2y - 15 = 0.$

   (b) whose center is (0,5) and radius is 5.   *Ans.* $x^2 + y^2 - 10y = 0.$

   (c) whose center is (-4,2) and diameter is 8. *Ans.* $x^2 + y^2 + 8x - 4y + 4 = 0.$

   (d) whose center is (4,-1) and which passes through the point (-1,3).
    Ans. $x^2 + y^2 - 8x + 2y - 24 = 0.$

   (e) which has as diameter the segment joining the points (-3,5) and (7,-3).
    Ans. $x^2 + y^2 - 4x - 2y - 36 = 0.$

   (f) whose center is (-4,3) and which is tangent to the y-axis.
    Ans. $x^2 + y^2 + 8x - 6y + 9 = 0.$

   (g) whose center is (3,-4) and which passes through the origin.
    Ans. $x^2 + y^2 - 6x + 8y = 0.$

   (h) whose center is at the origin and which crosses the x-axis at 6.
    Ans. $x^2 + y^2 - 36 = 0.$

   (i) which is tangent to both axes, its center being in the first quadrant and r = 8.
    Ans. $x^2 + y^2 - 16x - 16y + 64 = 0.$

   (j) which passes through the origin, radius = 10, and abscissa of the center is -6.
    Ans. $x^2 + y^2 + 12x - 16y = 0,$ $x^2 + y^2 + 12x + 16y = 0.$

2. Find the center and the radius of each of the following circles. State whether the circle is a real circle, a point circle, or an imaginary circle. Use the formula and check by completing the square.

   (a) $x^2 + y^2 - 8x + 10y - 12 = 0.$ *Ans.* (4,-5), $r = \sqrt{53},$ real.

   (b) $3x^2 + 3y^2 - 4x + 2y + 6 = 0.$ *Ans.* $(\frac{2}{3}, -\frac{1}{3}),$ $r = \frac{1}{3}\sqrt{-13},$ imaginary.

   (c) $x^2 + y^2 - 8x - 7y = 0.$ *Ans.* $(4, \frac{7}{2}),$ $r = \frac{1}{2}\sqrt{113},$ real.

   (d) $x^2 + y^2 = 0.$ *Ans.* (0,0), $r = 0,$ point.

   (e) $2x^2 + 2y^2 - x = 0.$ *Ans.* $(\frac{1}{4}, 0),$ $r = \frac{1}{4},$ real.

3. Derive the equation of the circle passing through the three points

   (a) (4,5), (3,-2), and (1,-4).   *Ans.* $x^2 + y^2 + 7x - 5y - 44 = 0.$

   (b) (8,-2), (6,2), and (3,-7).   *Ans.* $x^2 + y^2 - 6x + 4y - 12 = 0.$

   (c) (1,1), (1,3), and (9,2).   *Ans.* $8x^2 + 8y^2 - 79x - 32y + 95 = 0.$

   (d) (-4,-3), (-1,-7), and (0,0).   *Ans.* $x^2 + y^2 + x + 7y = 0.$

   (e) (1,2), (3,1), and (-3,-1).   *Ans.* $x^2 + y^2 - x + 3y - 10 = 0.$

4. Derive the equation of the circle circumscribing the triangle formed by the lines

   (a) $x - y + 2 = 0,$ $2x + 3y - 1 = 0,$ and $4x + y - 17 = 0.$
    Ans. $5x^2 + 5y^2 - 32x - 8y - 34 = 0.$

   (b) $x + 2y - 5 = 0,$ $2x + y - 7 = 0,$ and $x - y + 1 = 0.$
    Ans. $3x^2 + 3y^2 - 13x - 11y + 20 = 0.$

(c) $3x + 2y - 13 = 0$, $x + 2y - 3 = 0$, and $x + y - 5 = 0$.
   Ans. $x^2 + y^2 - 17x - 7y + 52 = 0$.

(d) $2x + y - 8 = 0$, $x - y - 1 = 0$, and $x - 7y - 19 = 0$.
   Ans. $3x^2 + 3y^2 - 8x + 8y - 31 = 0$.

(e) $2x - y + 7 = 0$, $3x + 5y - 9 = 0$, and $x - 7y - 13 = 0$.
   Ans. $169x^2 + 169y^2 - 8x + 498y - 3707 = 0$.

5. Derive the equation of the circle inscribed in the triangle formed by the lines

   (a) $4x - 3y - 65 = 0$, $7x - 24y + 55 = 0$, and $3x + 4y - 5 = 0$.
      Ans. $x^2 + y^2 - 20x + 75 = 0$.

   (b) $7x + 6y - 11 = 0$, $9x - 2y + 7 = 0$, and $6x - 7y - 16 = 0$.
      Ans. $85x^2 + 85y^2 - 60x + 70y - 96 = 0$.

   (c) $y = 0$, $3x - 4y = 0$, and $4x + 3y - 50 = 0$.
      Ans. $4x^2 + 4y^2 - 60x - 20y + 225 = 0$.

   (d) $15x - 8y + 25 = 0$, $3x - 4y - 10 = 0$, and $5x + 12y - 30 = 0$.
      Ans. $784x^2 + 784y^2 - 896x - 392y - 2399 = 0$.

   (e) inscribed in the triangle whose vertices are $(-1, 3)$, $(3, 6)$ and $(\frac{31}{5}, 0)$.
      Ans. $7x^2 + 7y^2 - 34x - 48y + 103 = 0$.

6. Derive the equation of the circle with its center at $(-2, 3)$ and which is tangent to the line $20x - 21y - 42 = 0$.   Ans. $x^2 + y^2 + 4x - 6y - 12 = 0$.

7. Derive the equation of the circle with its center at the origin and which is tangent to the line $8x - 15y - 12 = 0$.   Ans. $289x^2 + 289y^2 = 144$.

8. Derive the equation of the circle with its center at $(-1, -3)$ and tangent to the line through the points $(-2, 4)$ and $(2, 1)$.   Ans. $x^2 + y^2 + 2x + 6y - 15 = 0$.

9. Derive the equation of the circle with its center on the x-axis and which passes through the two points $(-2, 3)$ and $(4, 5)$.   Ans. $3x^2 + 3y^2 - 14x - 67 = 0$.

10. Derive the equation of the circle which passes through the points $(1, -4)$ and $(5, 2)$ and has its center on the line $x - 2y + 9 = 0$.   Ans. $x^2 + y^2 + 6x - 6y - 47 = 0$.

11. Derive the equation of the circle through the points $(-3, 2)$ and $(4, 1)$ and tangent to the x-axis.   Ans. $x^2 + y^2 - 2x - 10y + 1 = 0$, $x^2 + y^2 - 42x - 290y + 441 = 0$.

12. Derive the equation of the circle through the points $(2, 3)$ and $(3, 6)$ and tangent to the line $2x + y - 2 = 0$.   Ans. $x^2 + y^2 - 26x - 2y + 45 = 0$, $x^2 + y^2 - 2x - 10y + 21 = 0$.

13. Derive the equation of the circle through the point $(11, 2)$ and tangent to the line $2x + 3y - 18 = 0$ at the point $(3, 4)$.   Ans. $5x^2 + 5y^2 - 98x - 142y + 737 = 0$.

14. Derive the equation of the circle tangent to the line $3x - 4y - 13 = 0$ at the point $(7, 2)$ and with radius 10.   Ans. $x^2 + y^2 - 26x + 12y + 105 = 0$, $x^2 + y^2 - 2x - 20y + 1 = 0$.

15. Derive the equation of the circle tangent to the two lines $x - 2y + 4 = 0$ and $2x - y - 8 = 0$ and which passes through the point $(4, -1)$.
   Ans. $x^2 + y^2 - 30x + 6y + 109 = 0$, $x^2 + y^2 - 70x + 46y + 309 = 0$.

16. Derive the equation of the circle tangent to the two lines $x - 3y + 9 = 0$ and $3x + y - 3 = 0$ and with its center on the line $7x + 12y - 32 = 0$.
   Ans. $x^2 + y^2 + 8x - 10y + 31 = 0$, $961x^2 + 961y^2 + 248x - 5270y + 7201 = 0$.

17. Derive the equation of the circle which is the locus of the vertex of the right angle of the right triangle with the hypotenuse joining points $(-4, 1)$ and $(3, 2)$.
    Ans. $x^2 + y^2 + x - 3y - 10 = 0$.

18. Derive the equation of a circle which is tangent to the two lines $4x + 3y - 50 = 0$ and $3x - 4y - 25 = 0$ and whose radius is 5.   Ans. $x^2 + y^2 - 20x + 10y + 100 = 0$,
    $x^2 + y^2 - 36x - 2y + 300 = 0$,
    $x^2 + y^2 - 24x - 18y + 200 = 0$,
    $x^2 + y^2 - 8x - 6y = 0$.

19. Write the equation of the locus of a point the sum of the squares of whose distances from the two perpendicular lines $2x + 3y - 6 = 0$ and $3x - 2y + 8 = 0$ is 10. If a circle, determine the center and the radius.
    Ans. $13x^2 + 13y^2 + 24x - 68y - 30 = 0$. Center $\left(-\dfrac{12}{13}, \dfrac{34}{13}\right)$, $r = \sqrt{10}$.

20. Show that the locus of a point, the sum of the squares of whose distances from two perpendicular lines $a_1x + b_1y + c_1 = 0$ and $b_1x - a_1y + c_2 = 0$ is equal to a constant K^2, is a circle.

21. Write the equation of the locus of a point, the sum of the squares of whose distances from $(-2, -5)$ and $(3, 4)$ is equal to 70. If a circle, determine the center and the radius.
    Ans. $x^2 + y^2 - x + y - 8 = 0$. Center $(\frac{1}{2}, -\frac{1}{2})$, $r = \frac{1}{2}\sqrt{34}$.

22. Write the equation of the locus of a point the ratio of whose distance from $(2, -1)$ to its distance from $(-3, 4)$ is $2/3$. If a circle, determine its center and radius.
    Ans. $x^2 + y^2 - 12x + 10y - 11 = 0$. Center $(6, -5)$, $r = 6\sqrt{2}$.

23. Show that the locus of a point the ratio of whose distance from any two points (a, b) and (c, d) is equal to K (a constant) will be a circle.

24. Write the equation of the locus of a point, the square of whose distance from $(-2, -5)$ is three times its distance from the line $8x + 15y - 34 = 0$.
    Ans. $17x^2 + 17y^2 + 44x + 125y + 595 = 0$, $17x^2 + 17y^2 + 92x + 215y + 391 = 0$.

25. Write the equation of a circle which is tangent to the line $3x - 4y + 17 = 0$ and is concentric with the circle $x^2 + y^2 - 4x + 6y - 11 = 0$.
    Ans. $x^2 + y^2 - 4x + 6y - 36 = 0$.

26. Write the equation of a circle which is tangent to the circle $x^2 + y^2 = 25$ at $(3, 4)$ and has a radius 10.
    Ans. $x^2 + y^2 - 18x - 24y + 125 = 0$, $x^2 + y^2 + 6x + 8y - 75 = 0$.

27. A rod 30 inches long moves with its end points on two perpendicular wires. Determine the locus of its midpoint.   Ans. A circle, $x^2 + y^2 = 225$.

28. Find the longest and the shortest distance from $(10, 7)$ to the circle $x^2 + y^2 - 4x - 2y - 20 = 0$.   Ans. 15 and 5.

29. Find the length of the tangent from point $(7, 8)$ to the circle $x^2 + y^2 = 9$.   Ans. $2\sqrt{26}$.

30. Find the length of the tangent from point $(6, 4)$ to the circle $x^2 + y^2 + 4x + 6y - 19 = 0$.
    Ans. 9.

31. Determine the value of K for which the length of the tangent from point $(5, 4)$ to the circle $x^2 + y^2 + 2Ky = 0$ is (a) 1, (b) 0.   Ans. (a) $K = -5$, (b) $K = -5.125$.

32. Find the equations of the three radical axes of the given circles taken in pairs, and show that the three axes meet in a point.
$$x^2 + y^2 + 3x - 2y - 4 = 0, \quad x^2 + y^2 - 2x - y - 6 = 0, \quad \text{and} \quad x^2 + y^2 - 1 = 0.$$
Ans. $5x - y + 2 = 0$, $3x - 2y - 3 = 0$, $2x + y + 5 = 0$. Point of intersection $(-1, -3)$. This point is called the radical center of the circles.

33. Find the equations of the three radical axes of the three given circles taken in pairs. Find the radical center.
$$x^2 + y^2 + x = 0, \quad x^2 + y^2 + 4y + 7 = 0, \quad \text{and} \quad 2x^2 + 2y^2 + 5x + 3y + 9 = 0.$$
Ans. $x - 4y - 7 = 0$, $x + y + 3 = 0$, $x - y - 1 = 0$. Center $(-1, -2)$.

34. Find the equations of the radical axes of the three given circles taken in pairs. Also find the radical center.
$$x^2 + y^2 + 12x + 11 = 0, \quad x^2 + y^2 - 4x - 21 = 0, \quad \text{and} \quad x^2 + y^2 - 4x + 16y + 43 = 0.$$
Ans. $x + 2 = 0$, $x - y - 2 = 0$, $y + 4 = 0$. Center $(-2, -4)$.

35. Find the equation of the circle containing the point $(-2, 2)$ and which passes through the points of intersection of the two circles
$$x^2 + y^2 + 3x - 2y - 4 = 0 \quad \text{and} \quad x^2 + y^2 - 2x - y - 6 = 0.$$
Ans. $5x^2 + 5y^2 - 7y - 26 = 0$.

36. Find the equation of the circle containing the point $(3, 1)$ and which passes through the points of intersection of the two circles
$$x^2 + y^2 - x - y - 2 = 0 \quad \text{and} \quad x^2 + y^2 + 4x - 4y - 8 = 0.$$
Ans. $3x^2 + 3y^2 - 13x + 3y + 6 = 0$.

37. Find the equation of the circle passing through the points of intersection of the circles $x^2 + y^2 - 6x + 2y + 4 = 0$ and $x^2 + y^2 + 2x - 4y - 6 = 0$, and with its center on the line $y = x$.
Ans. $7x^2 + 7y^2 - 10x - 10y - 12 = 0$.

CHAPTER 5

Conic Sections — The Parabola

DEFINITION. The path of a point which moves so that its distance from a fixed point is in constant ratio to its distance from a fixed line is called a *conic section*, or a *conic*.

The fixed point is the *focus* of the conic, the fixed line the *directrix*, and the constant ratio the *eccentricity*, usually represented by *e*.

The conic sections fall into three classes, which vary in form and in certain properties. These classes are distinguished by the value of the eccentricity *e*.

If $e < 1$, the conic is an *ellipse*.
If $e = 1$, the conic is a *parabola*.
If $e > 1$, the conic is a *hyperbola*.

THE PARABOLA. Let $L'L$ be the fixed line and F the fixed point. Through F draw the x-axis perpendicular to the fixed line. Let the distance from F to $L'L$ be $2a$. Then from the definition of the parabola the curve must cross the x-axis at a point O, midway between F and $L'L$. Through O draw the y-axis.

The coordinates of F are $(a,0)$, and the equation of the directrix is $x = -a$ or $x + a = 0$.

Choose any point $P(x,y)$ so that $\dfrac{PF}{PM} = e = 1$.

Then $\sqrt{(x-a)^2 + (y-0)^2} = x + a$.

Squaring, $x^2 - 2ax + a^2 + y^2 = x^2 + 2ax + a^2$,

or $y^2 = 4ax$.

From the form of the equation it is seen that the curve is symmetrical with respect to the x-axis. The point in which the curve cuts its axis of symmetry is called its *vertex*. The chord $C'C$ through the focus and perpendicular to the axis is called the *latus rectum*. The length of the latus rectum is $4a$, the coefficient of the first degree term.

If the focus were to the left of the directrix the equation would have the form

$$y^2 = -4ax.$$

If the focus were on the y-axis, the form of the equation would be

$$x^2 = \pm 4ay$$

the sign depending on the position of the focus above or below the directrix.

Consider now a parabola with its vertex at (h,k), axis parallel to the x-axis, and focus at a distance a to the right of the vertex. The directrix, parallel to the y-axis and at a distance $2a$ to the left of the focus, has the equation $x = h - a$ or $x - h + a = 0$.

Let $P(x,y)$ be any point on the parabola. Then, since $PF = PM$,

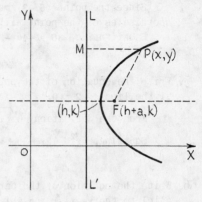

$$\sqrt{(x-h-a)^2 + (y-k)^2} = x - h + a,$$

or $\qquad y^2 - 2ky + k^2 = 4ax - 4ah,$

or $\qquad (y - k)^2 = 4a(x - h).$

Other standard forms are:

$$(y - k)^2 = -4a(x - h);$$
$$(x - h)^2 = 4a(y - k);$$
$$(x - h)^2 = -4a(y - k).$$

When expanded, the equations take the form $\quad x = ay^2 + by + c,$
$$y = ax^2 + bx + c.$$

SOLVED PROBLEMS

1. Locate the focus, find the equation of the directrix, and determine the length of the latus rectum for the parabola $3y^2 = 8x$, or $y^2 = \frac{8}{3}x$.

Here $4a = \frac{8}{3}$, or $a = \frac{2}{3}$. The focus is therefore at $(\frac{2}{3}, 0)$, and the equation of the directrix is $x = -\frac{2}{3}$.

To determine the length of the latus rectum, find y when $x = \frac{2}{3}$. When $x = \frac{2}{3}$, $y = \frac{4}{3}$ and the length of the latus rectum $= 2(\frac{4}{3}) = \frac{8}{3}$.

2. Find the equation of the parabola with its focus at $(0, -\frac{4}{3})$ and directrix $y - \frac{4}{3} = 0$. Find the length of the latus rectum.

Let $P(x,y)$ be any point on the parabola. Then, $\sqrt{(x - 0)^2 + (y + \frac{4}{3})^2} = y - \frac{4}{3}$.

Squaring and simplifying, $x^2 + \frac{16}{3}y = 0$. Latus rectum $= 4a = \frac{16}{3}$.

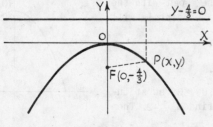

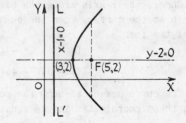

Problem 2.   Problem 3.

3. Derive the equation of the parabola with its vertex at $(3, 2)$ and its focus at $(5, 2)$.

Since the vertex is at $(3, 2)$ and the focus at $(5, 2)$, then $a = 2$ and the equation has the form $(y - k)^2 = 4a(x - h)$ or $(y - 2)^2 = 8(x - 3)$.
Simplifying, $y^2 - 4y - 8x + 28 = 0$.

4. Find the equation of the parabola with its vertex at the origin, axis on the y-axis, and which passes through the point (6,−3).

The standard form of the equation to use is $x^2 = -4ay$.

Since the point (6,−3) must lie on the curve, the value of a must be such that the coordinates of the point will satisfy the equation.

Substituting, $36 = -4a(-3)$, or $a = 3$. The required equation is $x^2 = -12y$.

5. Write the equation of the parabola with its focus at the point (6,−2) and whose directrix is the line $x - 2 = 0$.

From the definition, $\sqrt{(x-6)^2 + (y+2)^2} = x - 2$.

Squaring, $x^2 - 12x + 36 + y^2 + 4y + 4 = x^2 - 4x + 4$. Simplifying, $y^2 + 4y - 8x + 36 = 0$.

6. Write the equation of the parabola with its vertex at the point (2,3), with its axis parallel to the y-axis, and which passes through the point (4,5).

The standard form to use in this problem is $(x - h)^2 = 4a(y - k)$,

$$\text{or } (x - 2)^2 = 4a(y - 3).$$

Since point (4,5) is on the curve, $(4 - 2)^2 = 4a(5 - 3)$ and $a = \frac{1}{2}$.

The required equation is $(x - 2)^2 = 2(y - 3)$ or $x^2 - 4x - 2y + 10 = 0$.

7. Derive the equation of the parabola with its axis parallel to the x-axis, and which passes through the points (−2,1), (1,2), and (−1,3).

Use the standard form $y^2 + Dx + Ey + F = 0$.

Substituting for x and y the coordinates of the points, $1 - 2D + E + F = 0$,
$$4 + D + 2E + F = 0,$$
$$9 - D + 3E + F = 0.$$

Solving these three simultaneous equations, $D = \frac{2}{5}$, $E = -\frac{21}{5}$, $F = 4$.

Hence the required equation is $y^2 + \frac{2}{5}x - \frac{21}{5}y + 4 = 0$, or $5y^2 + 2x - 21y + 20 = 0$.

8. How high is a parabolic arch, of span 24 ft and height 18 ft, at a distance 8 ft from the center of the span?

Choose the x-axis along the base of the span with the origin at the center. Then the equation of the parabola will be of the form
$$(x - h)^2 = 4a(y - k)$$
or
$$(x - 0)^2 = 4a(y - 18).$$

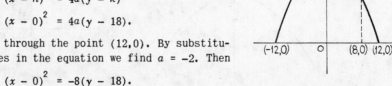

The curve passes through the point (12,0). By substituting these coordinates in the equation we find $a = -2$. Then
$$(x - 0)^2 = -8(y - 18).$$

To find the height of the arch 8 ft from the center, let $x = 8$ and solve for y. Then $8^2 = -8(y - 18)$, or $y = 10$ feet. The strongest simple arch is parabolic in shape.

9. Given the parabola whose equation is $y^2 + 8y - 6x + 4 = 0$, determine the coordinates of the vertex, the coordinates of the focus, and the equation of the directrix.

Completing the square, $y^2 + 8y + 16 = 6x - 4 + 16 = 6x + 12$, or $(y + 4)^2 = 6(x + 2)$.

The vertex is at $(-2, -4)$. Since $4a = 6$, then $a = 3/2$. Hence the focus is at point $(-\frac{1}{2}, -4)$, and the equation of the directrix is $x = -7/2$.

10. Derive the equation of the parabola with latus rectum joining points $(3,5)$ and $(3,-3)$.

Use $(y - k)^2 = \pm 4a(x - h)$.
Since the length of the latus rectum is 8, $4a = 8$ and $(y - k)^2 = \pm 8(x - h)$.

To determine (h,k), $(5 - k)^2 = \pm 8(3 - h)$ and $(-3 - k)^2 = \pm 8(3 - h)$, since the points $(3,5)$ and $(3,-3)$ lie on the curve. Solving this pair of equations, we find (h,k) to be $(1,1)$ and $(5,1)$.

The required equations are: (1) $(y - 1)^2 = 8(x - 1)$ or $y^2 - 2y - 8x + 9 = 0$

and (2) $(y - 1)^2 = -8(x - 5)$ or $y^2 - 2y + 8x - 39 = 0$.

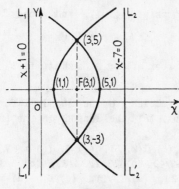

Problem 10.

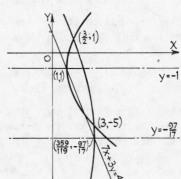

Problem 11.

11. Derive the equation of the parabola with its vertex on the line $7x + 3y - 4 = 0$ and containing the points $(3,-5)$ and $(3/2, 1)$, the axis being horizontal.

Use $(y - k)^2 = 4a(x - h)$. Substituting the coordinates of the points, we obtain

$$(-5 - k)^2 = 4a(3 - h) \quad \text{and} \quad (1 - k)^2 = 4a(3/2 - h).$$

Since (h,k) lies on $7x + 3y - 4 = 0$, then $7h + 3k - 4 = 0$. Solve simultaneously these three equations to obtain $h = 1$, $k = -1$, $4a = 8$; and $h = 359/119$, $k = -97/17$, $4a = -504/17$.
Hence the required equations are $(y + 1)^2 = 8(x - 1)$ and $(y + \frac{97}{17})^2 = -\frac{504}{17}(x - \frac{359}{119})$.

12. The path of a projectile thrown horizontally from a point y feet above the ground with a velocity v ft/sec is a parabola whose equation is

$$x^2 = -\frac{2v^2}{g}y,$$

where x is the distance measured horizontally from the point of projection and $g = 32$ ft/sec^2 approximately, the origin being at the starting point.

A stone was thrown horizontally from a point 10 feet above the ground. If the horizontal velocity of the stone was 160 ft/sec, how far will the stone travel horizontally before it hits the ground?

$$x^2 = -\frac{2v^2}{g}y = -\frac{2(160)^2}{32}(-10), \quad \text{and} \quad x = 40\sqrt{10} = 126.5 \text{ ft.}$$

SUPPLEMENTARY PROBLEMS

1. Find the coordinates of the focus, the length of the latus rectum, and the equation of the directrix of each of the following parabolas and plot the graph:
   (a) $y^2 = 6x$.   Ans. $(3/2, 0)$, 6, $x + 3/2 = 0$.
   (b) $x^2 = 8y$.   Ans. $(0, 2)$, 8, $y + 2 = 0$.
   (c) $3y^2 = -4x$.   Ans. $(-1/3, 0)$, $4/3$, $x - 1/3 = 0$.

2. Derive the equation of each of the following parabolas:
   (a) Focus at $(3, 0)$ and the directrix is $x + 3 = 0$.   Ans. $y^2 - 12x = 0$.
   (b) Focus at $(0, 6)$ and the directrix is the x-axis.   Ans. $x^2 - 12y + 36 = 0$.
   (c) Vertex at the origin, axis along x-axis, and passes through $(-3, 6)$. Ans. $y^2 = -12x$.

3. Derive the equation of the locus of a point which moves so that its distance from $(-2, 3)$ equals its distance from the line $x + 6 = 0$.   Ans. $y^2 - 6y - 8x - 23 = 0$.

4. Derive the equation of the parabola with its focus at $(-2, -1)$, and whose latus rectum joins the points $(-2, 2)$ and $(-2, -4)$.
   Ans. $y^2 + 2y - 6x - 20 = 0$, $y^2 + 2y + 6x + 4 = 0$.

5. Derive the equation of the parabola with its vertex at $(-2, 3)$ and its focus at $(1, 3)$.
   Ans. $y^2 - 6y - 12x - 15 = 0$.

6. Reduce the equation of each of the following parabolas to the standard form, and find the coordinates of (a) the vertex, (b) the focus, (c) the length of the latus rectum, and (d) the equation of the directrix.
   (1) $y^2 - 4y + 6x - 8 = 0$.   Ans. (a) $(2, 2)$, (b) $(1/2, 2)$, (c) 6, (d) $x - 7/2 = 0$.
   (2) $3x^2 - 9x - 5y - 2 = 0$.   Ans. (a) $(3/2, -7/4)$, (b) $(3/2, -4/3)$, (c) $5/3$.
   (3) $y^2 - 4y - 6x + 13 = 0$.   Ans. (a) $(3/2, 2)$, (b) $(3, 2)$, (c) 6, (d) $x = 0$.

7. Find the equation of a parabola with its axis parallel to the x-axis, and which passes through the points $(3, 3)$, $(6, 5)$ and $(6, -3)$.   Ans. $y^2 - 2y - 4x + 9 = 0$.

8. Find the equation of a parabola with its axis vertical, and which passes through points $(4, 5)$, $(-2, 11)$ and $(-4, 21)$.   Ans. $x^2 - 4x - 2y + 10 = 0$.

9. Find the equation of a parabola with its vertex on the line $2y - 3x = 0$, its axis parallel to the x-axis, and which passes through the two points $(3, 5)$ and $(6, -1)$.
   Ans. $y^2 - 6y - 4x + 17 = 0$, $11y^2 - 98y - 108x + 539 = 0$.

10. When the load is uniformly distributed, the cable of a suspension bridge hangs in the shape of an arc of a parabola. The supporting towers are 60 feet high and 500 feet apart, and the lowest point on the cable is 10 feet above the roadway. Using the floor of the bridge as the x-axis and the axis of symmetry of the parabola as the y-axis, find the equation of the parabola which the cable assumes. Find the length of a supporting rod 80 feet from the center of the bridge.   Ans. $x^2 - 1250y + 12500 = 0$; 15.12 ft.

11. A baseball is thrown horizontally from the top of the Washington Monument, 555 ft high, with an inital velocity of 40 ft/sec. How far from the foot of the monument will it strike the ground, supposed to be level?   Ans. 235.58 feet.

12. A plane 4000 feet high and flying south at 120 miles per hour releases a bomb. How far south will the bomb be when it hits the ground?   Ans. 2783 feet.

13. A parabolic arch has a height of 25 feet and a span of 40 feet. How high is the arch 8 feet each side of the center?   Ans. 21 feet.

CHAPTER 6

The Ellipse

DEFINITION. The path of a point which moves so that the sum of its distances from two fixed points is constant is an ellipse. The two fixed points are the *foci* of the ellipse.

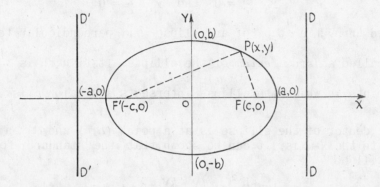

Let the two fixed points be $F(c,0)$ and $F'(-c,0)$, and let the constant sum be $2a$, $(a > c)$. Let $P(x,y)$ be any point on the locus. Then from the definition

$$F'P + PF = 2a,$$

or

$$\sqrt{(x+c)^2 + (y-0)^2} + \sqrt{(x-c)^2 + (y-0)^2} = 2a,$$

or

$$\sqrt{(x+c)^2 + (y-0)^2} = 2a - \sqrt{(x-c)^2 + (y-0)^2}.$$

Squaring and collecting terms, $cx - a^2 = -a\sqrt{(x-c)^2 + (y-0)^2}$.

Squaring and simplifying, $(a^2 - c^2)x^2 + a^2 y^2 = a^2(a^2 - c^2)$.

Dividing through by $a^2(a^2 - c^2)$, the equation becomes $\dfrac{x^2}{a^2} + \dfrac{y^2}{a^2 - c^2} = 1$.

Since $a > c$, $a^2 - c^2$ will be positive. Write $a^2 - c^2 = b^2$. We then have the standard form for the equation of the ellipse,

$$\frac{x^2}{a^2} + \frac{y^2}{b^2} = 1,$$

or

$$b^2 x^2 + a^2 y^2 = a^2 b^2.$$

Since this equation contains only even powers of x and y the curve is symmetric with respect to the x- and y-axis, and with the origin. The point O is the center of the ellipse, the long axis being called the *major axis* and the shorter axis the *minor axis*.

If the foci had been at $(0,c)$ and $(0,-c)$, the major axis would have been on the y-axis and the standard form would then be $\dfrac{x^2}{b^2} + \dfrac{y^2}{a^2} = 1$.

The *eccentricity* $e = \dfrac{c}{a} = \dfrac{\sqrt{a^2 - b^2}}{a}$, or $c = ae$.

Since the ellipse has two foci, it will have two directrices. The equations of the directrices $D'D'$ and DD are respectively

$$x + \frac{a}{e} = 0 \quad \text{and} \quad x - \frac{a}{e} = 0.$$

Had the foci been on the y-axis, the equations of the directrices would have been

$$y + \frac{a}{e} = 0 \quad \text{and} \quad y - \frac{a}{e} = 0.$$

A chord through a focus of an ellipse and perpendicular to the major axis is called a *latus rectum* of the ellipse. Its length is $\dfrac{2b^2}{a}$.

The points in which the ellipse intersects its major axis are called *vertices*.

If the center of the ellipse is at a point (h, k) and the major axis is parallel to the x-axis, it can be shown that the standard form of the ellipse will be

$$\frac{(x - h)^2}{a^2} + \frac{(y - k)^2}{b^2} = 1,$$

or

$$\frac{(x - h)^2}{b^2} + \frac{(y - k)^2}{a^2} = 1 \quad \text{if the major axis is par-}$$

allel to the y-axis. In either case, the general form of the equation of the ellipse is

$$Ax^2 + By^2 + Dx + Ey + F = 0$$

if A and B agree in sign.

SOLVED PROBLEMS

1. For the ellipse $9x^2 + 16y^2 = 576$, find the semi-major axis, the semi-minor axis, the eccentricity, the coordinates of the foci, the equations of the directrices, and the length of the latus rectum.

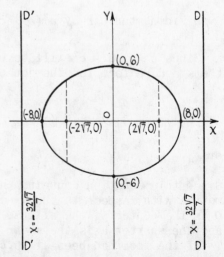

Dividing by 576, we have $\dfrac{x^2}{64} + \dfrac{y^2}{36} = 1$.

Hence, $a = 8$ and $b = 6$.

$$e = \frac{\sqrt{a^2 - b^2}}{a} = \frac{\sqrt{7}}{4}, \quad c = \sqrt{a^2 - b^2} = 2\sqrt{7}.$$

Coordinates of foci: $(2\sqrt{7}, 0)$ and $(-2\sqrt{7}, 0)$.

The equations of the directrices are

$$x = \pm \frac{a}{e} \quad \text{or} \quad x = \pm \frac{32\sqrt{7}}{7}.$$

The length of the latus rectum of the ellipse is $2b^2/a = 72/8 = 9$.

2. Derive the equation of the ellipse having its center at the origin, one focus at $(0,3)$, and length of semi-major axis 5.

Given: $c = 3$ and $a = 5$. Therefore $b = \sqrt{a^2 - c^2} = \sqrt{25-9} = 4$.

Using the standard form $\dfrac{x^2}{b^2} + \dfrac{y^2}{a^2} = 1$, the required equation is $\dfrac{x^2}{16} + \dfrac{y^2}{25} = 1$.

3. Find the equation of the ellipse having its center at the origin, major axis on the x-axis, and passing through the point $(4,3)$ and $(6,2)$.

The standard form is $\dfrac{x^2}{a^2} + \dfrac{y^2}{b^2} = 1$. Substituting for x and y the coordinates of the

points, we have $\dfrac{16}{a^2} + \dfrac{9}{b^2} = 1$ and $\dfrac{36}{a^2} + \dfrac{4}{b^2} = 1$. Solving simultaneously, $a^2 = 52$, $b^2 = 13$.

Hence the required equation is $\dfrac{x^2}{52} + \dfrac{y^2}{13} = 1$ or $x^2 + 4y^2 = 52$.

4. A point moves so that its distance from the point $(4,0)$ is always one half its distance from the line $x - 16 = 0$. Find the equation of its locus.

From the statement of the problem

$$\sqrt{(x-4)^2 + (y-0)^2} = \frac{1}{2}(x - 16), \quad \text{or} \quad x^2 - 8x + 16 + y^2 = \frac{x^2 - 32x + 256}{4}.$$

Simplifying, the required equation is $3x^2 + 4y^2 = 192$, an ellipse.

5. A line segment AB, 12 units long, containing an internal point $P(x,y)$ 8 units from A, is moved so that A is always on the y-axis and B on the x-axis. Find the equation of the locus of the point P.

By similar triangles, $\dfrac{MA}{AP} = \dfrac{y}{PB}$ or $\dfrac{\sqrt{64 - x^2}}{8} = \dfrac{y}{4}$.

Then, $64 - x^2 = 4y^2$ or $x^2 + 4y^2 = 64$. The locus is an ellipse with its center at the origin and major axis on the x-axis.

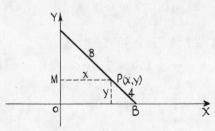

Problem 5.

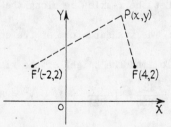

Problem 6.

6. A point $P(x,y)$ moves so that the sum of its distances from the points $(4,2)$ and $(-2,2)$ is 8. Derive the equation of its locus.

$$F'P + PF = 8, \quad \text{or} \quad \sqrt{(x+2)^2 + (y-2)^2} + \sqrt{(x-4)^2 + (y-2)^2} = 8.$$

Rearranging, $\quad \sqrt{(x+2)^2 + (y-2)^2} = 8 - \sqrt{(x-4)^2 + (y-2)^2}.$

Squaring and collecting terms, $3x - 19 = -4\sqrt{(x-4)^2 + (y-2)^2}$.

Squaring and collecting terms, the required equation is $7x^2 + 16y^2 - 14x - 64y - 41 = 0$, an ellipse.

7. Given the ellipse whose equation is $4x^2 + 9y^2 - 48x + 72y + 144 = 0$, find its center, semi-axes, vertices, and foci.

This equation can be put in standard form $\dfrac{(x-h)^2}{a^2} + \dfrac{(y-k)^2}{b^2} = 1$, by completing the square.

$$4(x^2 - 12x + 36) + 9(y^2 + 8y + 16) = 144,$$

$$4(x - 6)^2 + 9(y + 4)^2 = 144,$$

$$\frac{(x - 6)^2}{36} + \frac{(y + 4)^2}{16} = 1.$$

Hence the center of the ellipse is point $(6,-4)$; $a = 6$, $b = 4$; the vertices are $(0,-4)$, $(12,-4)$; and the foci are $(6 + 2\sqrt{5}, -4)$, $(6 - 2\sqrt{5}, -4)$.

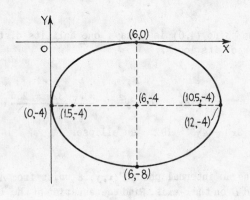

Problem 7.

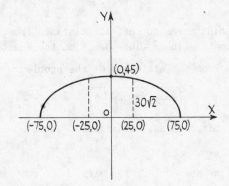

Problem 8.

8. An arch in the form of a semi-ellipse has a span of 150 feet and its greatest height is 45 feet. There are two vertical supports equidistant from each other and the ends of the arch. Find their height.

Let the x-axis lie along the base of the arch, with the origin at the middle of the base. The standard form of the equation will then be $\dfrac{x^2}{a^2} + \dfrac{y^2}{b^2} = 1$, and $a = 75$, $b = 45$.

To find the height of the right support, set $x = 25$ in the equation and solved for y.

Then $\dfrac{625}{5625} + \dfrac{y^2}{2025} = 1$, $y^2 = 8(225)$, and $y = 30\sqrt{2}$ feet.

9. The earth's orbit is an ellipse with the sun at one of the foci. If the semi-major axis of the ellipse is 93,000,000 miles and the eccentricity is 1/62 nearly, find the greatest and least distances of the earth from the sun.

Eccentricity $e = \dfrac{c}{a}$. Therefore, $\dfrac{1}{62} = \dfrac{c}{93,000,000}$ or $c = 1,500,000$.

The longest distance is $a + c = 94,500,000$ miles.
The shortest distance is $a - c = 91,500,000$ miles.

10. Find the equation of the ellipse with its center at $(1,2)$, focus at $(6,2)$ and containing the point $(4,6)$.

   Use the equation $\dfrac{(x-1)^2}{a^2} + \dfrac{(y-2)^2}{b^2} = 1$.

   Since $(4,6)$ lies on the curve, $\dfrac{(4-1)^2}{a^2} + \dfrac{(6-2)^2}{b^2} = 1$ or $\dfrac{9}{a^2} + \dfrac{16}{b^2} = 1$.

   Since $c = 5$, then $b^2 = a^2 - c^2 = a^2 - 25$ and $\dfrac{9}{a^2} + \dfrac{16}{a^2 - 25} = 1$.

   Solving, $a^2 = 45$ and $b^2 = 20$. Substituting, $\dfrac{(x-1)^2}{45} + \dfrac{(y-2)^2}{20} = 1$.

11. Find the equation of the ellipse with its center at $(-1,-1)$, vertex $(5,-1)$, and $e = \dfrac{2}{3}$.

   Since the center is at $(-1,-1)$ and the vertex at $(5,-1)$, then $a = 6$, $e = \dfrac{c}{a} = \dfrac{c}{6} = \dfrac{2}{3}$, and $c = 4$.   Also, $b^2 = a^2 - c^2 = 36 - 16 = 20$.

   The required equation is $\dfrac{(x+1)^2}{36} + \dfrac{(y+1)^2}{20} = 1$.

12. Find the equation of the ellipse with its focus $(4,-3)$, directrix $x = -1$, and eccentricity $2/3$.

   From the general definition of a conic section, if $\dfrac{PF}{PM} = e$, $e < 1$, the curve is an ellipse.

   Therefore, $\dfrac{\sqrt{(x-4)^2 + (y+3)^2}}{x+1} = \dfrac{2}{3}$.

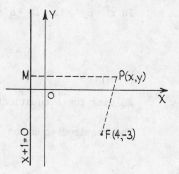

   Squaring both sides of this equation and simplifying, we have $5x^2 + 9y^2 - 80x + 54y = -221$.

   Completing the square, $5(x^2 - 16x + 64) + 9(y^2 + 6y + 9) = -221 + 320 + 81$,

   or $5(x-8)^2 + 9(y+3)^2 = 180$,

   or $\dfrac{(x-8)^2}{36} + \dfrac{(y+3)^2}{20} = 1$.

13. Determine the locus of a point $P(x,y)$ so that the product of the slopes of the lines joining $P(x,y)$ to $(3,-2)$ and $(-2,1)$ is -6.

   $\left(\dfrac{y+2}{x-3}\right)\left(\dfrac{y-1}{x+2}\right) = -6$, or $6x^2 + y^2 + y - 6x = 38$, an ellipse.

14. Determine the equation of the ellipse with foci $(0,\pm 4)$ and which passes through $(\dfrac{12}{5}, 3)$.

   Substitute $x = \dfrac{12}{5}$, $y = 3$ in $\dfrac{x^2}{b^2} + \dfrac{y^2}{a^2} = 1$ to get $\dfrac{144}{25b^2} + \dfrac{9}{a^2} = 1$.

   Since the foci are at $(0,\pm 4)$, then $c = 4$ and $a^2 - b^2 = 4^2 = 16$.

   Solving simultaneously these two equations, $a^2 = 25$, $b^2 = 9$. Hence, $\dfrac{x^2}{9} + \dfrac{y^2}{25} = 1$.

15. Determine the locus of the points which divide the ordinates of points on the circle $x^2 + y^2 = 25$ in the ratio $\frac{3}{5}$.

Let $y' = \frac{3}{5}y$ or $y = \frac{5}{3}y'$, and $x = x'$. Then $x'^2 + \frac{25}{9}y'^2 = 25$.

Dropping primes and simplifying, the equation is $9x^2 + 25y^2 = 225$, an ellipse.

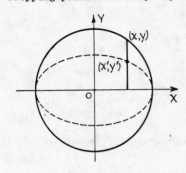   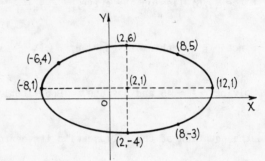

     Problem 15.   *Problem 16.*

16. Find the equation of the ellipse which passes through $(-6,4)$, $(-8,1)$, $(2,-4)$ and $(8,-3)$ and which has axes parallel to the coordinate axes.

In $x^2 + By^2 + Cx + Dy + E = 0$, substitute for x and y the coordinates of the four points.

$$16B - 6C + 4D + E = -36,$$
$$B - 8C + D + E = -64,$$
$$16B + 2C - 4D + E = -4,$$
$$9B + 8C - 3D + E = -64.$$

Solving these equations, $B = 4$, $C = -4$, $D = -8$, and $E = -92$.

The required equation is $x^2 + 4y^2 - 4x - 8y - 92 = 0$ or $\frac{(x-2)^2}{100} + \frac{(y-1)^2}{25} = 1$.

17. Find the equation of the locus of the center of a circle which is tangent to the circles $x^2 + y^2 = 1$ and $x^2 + y^2 - 4x - 21 = 0$.

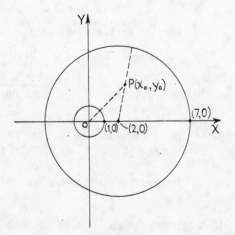

Use (x_0, y_0) for coordinates of the center. The given circles have radii 1 and 5 respectively.

(a) $5 - \sqrt{(x_0 - 2)^2 + (y_0 - 0)^2} = \sqrt{x_0^2 + y_0^2} - 1$.

Squaring, simplifying, and dropping subscripts, the equation becomes $8x^2 + 9y^2 - 16x - 64 = 0$, an ellipse. Writing this equation in the form

$$\frac{(x-1)^2}{9} + \frac{(y-0)^2}{8} = 1,$$

it is seen that the center of the ellipse is $(1,0)$.

(b) $\sqrt{x_0^2 + y_0^2} + 1 = 5 - \sqrt{(x_0 - 2)^2 + y_0^2}$. Squaring, simplifying, and dropping subscripts,

the equation becomes $3x^2 + 4y^2 - 6x - 9 = 0$ or $\frac{(x-1)^2}{4} + \frac{(y-0)^2}{3} = 1$.

The center of this ellipse is $(1,0)$.

18. Lines drawn from the foci to any point on an ellipse are called the focal radii of the ellipse. Find the equations of the focal radii drawn to the point $(2,3)$ on the ellipse $3x^2 + 4y^2 = 48$.

Write this equation as $\dfrac{x^2}{16} + \dfrac{y^2}{12} = 1$.   Then $c = \pm\sqrt{16 - 12} = \pm 2$.

The foci are at $(\pm 2, 0)$. The equation of the focal radius from $(2,0)$ to $(2,3)$ is $x - 2 = 0$, and from $(-2,0)$ to $(2,3)$ it is $y - 0 = \dfrac{3-0}{2+2}(x+2)$ or $3x - 4y + 6 = 0$.

SUPPLEMENTARY PROBLEMS

1. For each of the following ellipses find (a) the length of the semi-major axis, (b) the length of the semi-minor axis, (c) the coordinates of the foci, (d) the eccentricity.

(1) $\dfrac{x^2}{169} + \dfrac{y^2}{144} = 1$.   *Ans.* (a) 13, (b) 12, (c) $(\pm 5, 0)$, (d) $\dfrac{5}{13}$

(2) $\dfrac{x^2}{8} + \dfrac{y^2}{12} = 1$.   *Ans.* (a) $2\sqrt{3}$, (b) $2\sqrt{2}$, (c) $(0, \pm 2)$, (d) $\dfrac{\sqrt{3}}{3}$

(3) $225x^2 + 289y^2 = 65025$.   *Ans.* (a) 17, (b) 15, (c) $(\pm 8, 0)$, (d) $\dfrac{8}{17}$

2. Each of the following ellipses is in a standard position and has its center at the origin. Find its equation if it satisfies the additional conditions given.

(1) Foci $(\pm 4, 0)$, vertices $(\pm 5, 0)$.   *Ans.* $\dfrac{x^2}{25} + \dfrac{y^2}{9} = 1$

(2) Foci $(0, \pm 8)$, vertices $(0, \pm 17)$.   *Ans.* $\dfrac{x^2}{225} + \dfrac{y^2}{289} = 1$

(3) Length of latus rectum = 5, vertices $(\pm 10, 0)$.   *Ans.* $\dfrac{x^2}{100} + \dfrac{y^2}{25} = 1$

(4) Foci $(0, \pm 6)$, semi-minor axis = 8.   *Ans.* $\dfrac{x^2}{64} + \dfrac{y^2}{100} = 1$

(5) Foci $(\pm 5, 0)$, eccentricity $= \dfrac{5}{8}$.   *Ans.* $\dfrac{x^2}{64} + \dfrac{y^2}{39} = 1$

3. Write the equation of the ellipse with its center at the origin, its foci on the x-axis, and which passes through the points $(-3, 2\sqrt{3})$ and $(4, 4\sqrt{5}/3)$.   *Ans.* $4x^2 + 9y^2 = 144$.

4. Write the equation of the ellipse with its center at the origin, semi-major axis on the y-axis and 4 units long, and length of latus rectum = 9/2.   *Ans.* $16x^2 + 9y^2 = 144$.

5. A point $P(x,y)$ moves so that the sum of its distances from the two points $(3,1)$ and $(-5,1)$ is 10. Derive the equation of its locus. What curve is it?
Ans. $9x^2 + 25y^2 + 18x - 50y - 191 = 0$, an ellipse.

6. A point $P(x,y)$ moves so that the sum of its distances from $(2,-3)$ and $(2,7)$ is 12. Derive the equation of its locus.   *Ans.* $36x^2 + 11y^2 - 144x - 44y - 208 = 0$.

7. A point moves so that its distance from the point $(3,2)$ is one half of its distance from the line $x + 2 = 0$. Derive the equation of its locus. What curve is this?
Ans. $3x^2 + 4y^2 - 28x - 16y + 48 = 0$, an ellipse.

8. Given the ellipse whose equation is $9x^2 + 16y^2 - 36x + 96y + 36 = 0$. Find (a) the coordinates of its center, (b) the semi-major axis, (c) the semi-minor axis, (d) the foci, and (e) the length of a latus rectum.
   Ans. (a) (2,-3), (b) 4, (c) 3, (d) $(2 \pm \sqrt{7}, -3)$, (e) 4.5.

9. Find the equation of the ellipse with its center at (4,-1), focus at (1,-1), and passing through (8,0).   *Ans.* $\dfrac{(x-4)^2}{18} + \dfrac{(y+1)^2}{9} = 1$, or $x^2 + 2y^2 - 8x + 4y = 0$.

10. Find the equation of the ellipse with its center at (3,1), vertex (3,-2), and $e = 1/3$.
    Ans. $\dfrac{(x-3)^2}{8} + \dfrac{(y-1)^2}{9} = 1$, or $9x^2 + 8y^2 - 54x - 16y + 17 = 0$.

11. Find the equation of the ellipse with a focus at (-1,-1), directrix $x = 0$, and $e = \dfrac{\sqrt{2}}{2}$.
    Ans. $x^2 + 2y^2 + 4x + 4y + 4 = 0$.

12. A point $P(x,y)$ moves so that the product of the slopes of the two lines joining P to the two points (-2,1) and (6,5) is -4. Show that the locus is an ellipse and locate its center.   *Ans.* $4x^2 + y^2 - 16x - 6y - 43 = 0$.   Center (2,3).

13. A line segment AB, 18 units long, is moved so that A is always on the y-axis and B on the x-axis. Find the equation of the locus of a point $P(x,y)$, where P is on the line and 6 units from B.   *Ans.* $x^2 + 4y^2 = 144$, an ellipse.

14. An arch is in the form of a semi-ellipse, with the major axis as the span. If the span is 80 feet and the height is 30 feet, find the height of the arch at a point 15 feet from the minor axis.   *Ans.* $15\sqrt{55}/4$ feet.

15. The earth's orbit is an ellipse with the sun at one of the foci. If the semi-major axis of the ellipse is 92.9 million miles and the eccentricity is 0.017, find, to three significant figures the greatest and least distance of the earth from the sun.
    Ans. (94.5, 91.3) million miles.

16. Find the equation of the ellipse with foci at (± 8,0) and which passes through (8, 18/5).
    Ans. $\dfrac{x^2}{100} + \dfrac{y^2}{36} = 1$.

17. Determine the locus of the points which divide the ordinates of points on the circle $x^2 + y^2 = 16$ in the ratio $\frac{1}{2}$.   *Ans.* $x^2 + 4y^2 = 16$.

18. Find the equations of the focal radii drawn to the point (1,-1) on the ellipse
    $$x^2 + 5y^2 - 2x + 20y + 16 = 0.$$
    Ans. $x - 2y - 3 = 0$, $x + 2y + 1 = 0$.

19. Find the equation of the ellipse which passes through the points (0,1), (1,-1), (2,2), (4,0), and whose axes are parallel to the coordinate axes.
    Ans. $13x^2 + 23y^2 - 51x - 19y - 4 = 0$.

20. Find the equation of the locus of the center of a circle tangent to the circles
    $$x^2 + y^2 = 4 \quad \text{and} \quad x^2 + y^2 - 6x - 27 = 0.$$
    Ans. $220x^2 + 256y^2 - 660x - 3025 = 0$ and $28x^2 + 64y^2 - 84x - 49 = 0$.

CHAPTER 7

The Hyperbola

THE HYPERBOLA. A point moves so that the difference of its distances from two points $F(c,0)$ and $F'(-c,0)$ is $2a$, where a is constant and $a < c$. Determine the equation of its locus. See Figure (a).

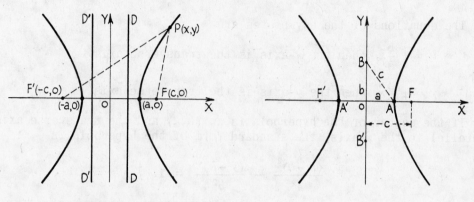

Figure (a).   *Figure (b).*

Let $P(x,y)$ be any point on the locus.

Then $F'P - PF = 2a$, or $\sqrt{(x+c)^2 + (y-0)^2} - \sqrt{(x-c)^2 + (y-0)^2} = 2a$.

Transposing one radical, $\sqrt{(x+c)^2 + (y-0)^2} = 2a + \sqrt{(x-c)^2 + (y-0)^2}$.

Squaring and collecting terms, $cx - a^2 = a\sqrt{(x-c)^2 + y^2}$.

Squaring and simplifying, $(c^2 - a^2)x^2 - a^2 y^2 = a^2(c^2 - a^2)$.

Dividing through by $a^2(c^2 - a^2)$, the equation becomes $\dfrac{x^2}{a^2} - \dfrac{y^2}{c^2 - a^2} = 1$.

Since $c > a$, $c^2 - a^2$ will be positive. Write $c^2 - a^2 = b^2$. We then have the standard form of the equation of a hyperbola with its center at the origin and its foci on the x-axis,

$$\frac{x^2}{a^2} - \frac{y^2}{b^2} = 1.$$

If the foci are at $(0,c)$ and $(0,-c)$ the standard form is $\dfrac{y^2}{a^2} - \dfrac{x^2}{b^2} = 1$.

The general form of the equation with the center at the origin and foci on the coordinate axes is $Ax^2 - By^2 = \pm 1$, the positive sign occurring when the foci are on the x-axis.

Since the equation contains only even powers of x and y, the curve is symmetric about the x and y axes and the origin.

The transverse axis is $A'A$, of length $2a$. The conjugate axis is $B'B$, of length $2b$. See Figure (b) above.

The eccentricity $e = \dfrac{c}{a} = \dfrac{\sqrt{a^2 + b^2}}{a}$. It is seen that $e > 1$, which agrees with the general definition of the conic sections. The equations of the directrices DD and $D'D'$ are $x = \pm \dfrac{a}{e}$ when the foci are on the x-axis, and $y = \pm \dfrac{a}{e}$ if the foci are on the y-axis.

The vertices of the hyperbola are the points in which the curve cuts its transverse axis.

The length of the latus rectum is $\dfrac{2b^2}{a}$.

The equations of the asymptotes are

$$y = \pm \frac{b}{a}x \quad \text{when the } x\text{-axis is the transverse axis,}$$

and $\quad y = \pm \dfrac{a}{b}x \quad$ when the y-axis is the transverse axis.

If the center of the hyperbola is at (h, k) and the transverse axis is parallel to the x-axis, the standard form of the hyperbola is

$$\frac{(x-h)^2}{a^2} - \frac{(y-k)^2}{b^2} = 1.$$

If the transverse axis is parallel to the y-axis, the equation is

$$\frac{(y-k)^2}{a^2} - \frac{(x-h)^2}{b^2} = 1.$$

The equations of the asymptotes are

$$y - k = \pm \frac{b}{a}(x - h) \quad \text{if the transverse axis is parallel to the } x\text{-axis,}$$

and $\quad y - k = \pm \dfrac{a}{b}(x - h) \quad$ if the transverse axis is parallel to the y-axis.

The general form of the equation of the hyperbola with axes parallel to the x- and y-axes is

$$Ax^2 - By^2 + Dx + Ey + F = 0,$$

where A and B agree in sign.

SOLVED PROBLEMS

1. Find the equation of the hyperbola with center at the origin, transverse axis on the y-axis, and passing through the points $(4, 6)$ and $(1, -3)$.

In the equation $\dfrac{y^2}{a^2} - \dfrac{x^2}{b^2} = 1$, substitute for x and y the coordinates of the given points. Then, $\dfrac{36}{a^2} - \dfrac{16}{b^2} = 1$ and $\dfrac{9}{a^2} - \dfrac{1}{b^2} = 1$.

Solving this pair of simultaneous equations, $a^2 = 36/5$ and $b^2 = 4$.

Substituting and simplifying, $\dfrac{5y^2}{36} - \dfrac{x^2}{4} = 1$ or $5y^2 - 9x^2 = 36$.

2. Find the coordinates of the vertices and the foci, the equations of the directrices, the equations of the asymptotes, the length of the latus rectum, the eccentricity, and plot the graph of the hyperbola $9x^2 - 16y^2 = 144$.

Write the equation in the form $\dfrac{x^2}{16} - \dfrac{y^2}{9} = 1$. Then $a = 4$, $b = 3$, $c = \sqrt{16 + 9} = 5$.

The intercepts are $(\pm 4, 0)$, and the foci are $(\pm 5, 0)$.

The eccentricity $e = \dfrac{c}{a} = \dfrac{5}{4}$, and the equations of the directrices are $x = \pm \dfrac{a}{e} = \pm \dfrac{16}{5}$.

The latus rectum $= \dfrac{2b^2}{a} = \dfrac{18}{4} = \dfrac{9}{2}$.

The equations of the asymptotes are $y = \pm \dfrac{b}{a} x = \pm \dfrac{3}{4} x$.

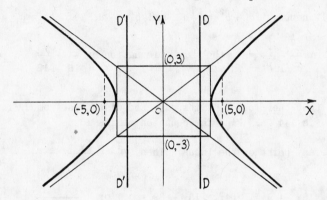

Problem 2.

3. Determine the equation of the hyperbola with its axes parallel to the coordinate axes and center at the origin, if the latus rectum is 18 and the distance between the foci is 12.

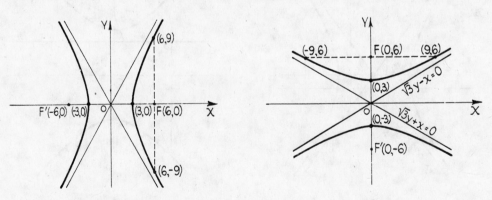

Problem 3(a).   *Problem 3(b).*

Latus rectum $= 2b^2/a = 18$, and $2c = 12$. Then $b^2 = 9a$ and $c = 6$.

Since $b^2 = c^2 - a^2 = 36 - a^2$, then $9a = 36 - a^2$ or $a^2 + 9a - 36 = 0$.
Solving, $(a - 3)(a + 12) = 0$ and $a = 3, -12$. Reject $a = -12$.

For $a^2 = 9$, $b^2 = 36 - 9 = 27$ and the two required equations are

(a) $\dfrac{x^2}{9} - \dfrac{y^2}{27} = 1$ or $3x^2 - y^2 = 27$, and (b) $\dfrac{y^2}{9} - \dfrac{x^2}{27} = 1$ or $3y^2 - x^2 = 27$.

4. Determine the equation of the hyperbola having foci at $(0, \pm 3)$ and conjugate axis 5.

Given: $c = 3$ and $b = \frac{5}{2}$.   Then $a^2 = c^2 - b^2 = 9 - \frac{25}{4} = \frac{11}{4}$.

Substituting in $\frac{y^2}{a^2} - \frac{x^2}{b^2} = 1$,   we obtain $\frac{y^2}{11/4} - \frac{x^2}{25/4} = 1$ or $100y^2 - 44x^2 = 275$.

5. Determine the equation of the hyperbola having its center at the origin, transverse axis on the x-axis, eccentricity $\frac{1}{2}\sqrt{7}$, and length of latus rectum 6.

Given: $e = \frac{\sqrt{a^2 + b^2}}{a} = \frac{\sqrt{7}}{2}$,   and latus rectum $= \frac{2b^2}{a} = 6$ or $b^2 = 3a$.

Solving simultaneously $a^2 + b^2 = \frac{7}{4}a^2$ and $b^2 = 3a$, we obtain $a^2 = 16$, $b^2 = 12$.

Substituting in $\frac{x^2}{a^2} - \frac{y^2}{b^2} = 1$,   the required equation is $\frac{x^2}{16} - \frac{y^2}{12} = 1$ or $3x^2 - 4y^2 = 48$.

6. A point moves so that the product of its directed distances from the lines $4x - 3y + 11 = 0$ and $4x + 3y + 5 = 0$ is $144/25$.   Find the equation of its locus.

Let $P(x, y)$ be any point on the locus.   Then $(\frac{4x - 3y + 11}{-5})(\frac{4x + 3y + 5}{-5}) = \frac{144}{25}$.

Simplifying, $16x^2 - 9y^2 + 64x + 18y - 89 = 0$ or $\frac{(x + 2)^2}{9} - \frac{(y - 1)^2}{16} = 1$,

which is the equation of a hyperbola having the given lines as asymptotes.

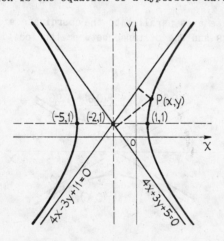

Problem 6.

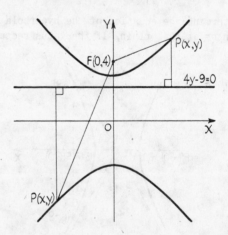

Problem 7.

7. A point (x, y) moves so that its distance from $(0, 4)$ is $4/3$ its distance from the line $4y - 9 = 0$.   Find the equation of its locus.

$$\sqrt{(x - 0)^2 + (y - 4)^2} = \frac{4}{3}(\frac{4y - 9}{4}).$$

Squaring and simplifying, $9x^2 - 7y^2 + 63 = 0$ or $\frac{y^2}{9} - \frac{x^2}{7} = 1$,   a hyperbola.

8. Find the equation of the hyperbola having its center at the origin, one vertex at (6,0), and the equation of one asymptote $4x - 3y = 0$.

Write the equation of the given asymptote in the form $y = \frac{4}{3}x$.

The asymptotes for $\frac{x^2}{a^2} - \frac{y^2}{b^2} = 1$ are $y = \pm\frac{b}{a}x$. Then $\frac{b}{a} = \frac{4}{3}$.

Since one vertex is at (6,0), $a = 6$ and $b = \frac{4a}{3} = 8$, and the equation is $\frac{x^2}{36} - \frac{y^2}{64} = 1$.

9. Determine the equation of the hyperbola with its center at (-4,1), vertex at (2,1), and semi-conjugate axis 4.

The distance between center and vertex is 6; hence $a = 6$.
The semi-conjugate axis is 4; hence $b = 4$.

Substituting in $\frac{(x-h)^2}{a^2} - \frac{(y-k)^2}{b^2} = 1$, we obtain $\frac{(x+4)^2}{36} - \frac{(y-1)^2}{16} = 1$.

10. Find (a) the center, (b) the vertices, (c) the foci, (d) the equations of the asymptotes, and (e) sketch the hyperbola whose equation is $9x^2 - 16y^2 - 18x - 64y - 199 = 0$.

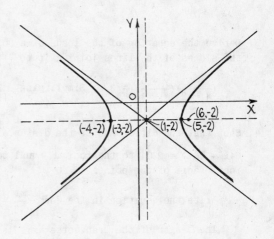

Complete the square and write the equation in the standard form

$$\frac{(x-h)^2}{a^2} - \frac{(y-k)^2}{b^2} = 1.$$

$$9(x^2 - 2x + 1) - 16(y^2 + 4y + 4) = 199 - 64 + 9,$$

$$9(x-1)^2 - 16(y+2)^2 = 144,$$

$$\frac{(x-1)^2}{16} - \frac{(y+2)^2}{9} = 1.$$

Ans. (a) (1,-2); (b) (-3,-2),(5,-2); (c) (-4,-2),(6,-2); (d) $y + 2 = \pm\frac{3}{4}(x - 1)$.

11. Write the equation of the hyperbola passing through the point (4,6) and whose asymptotes are $y = \pm\sqrt{3}\,x$.

The asymptotes of the hyperbola $\frac{x^2}{a^2} - \frac{y^2}{b^2} = 1$ are given by $y = \pm\frac{b}{a}x$.

Rearranging, $\frac{y}{b} = \pm\frac{x}{a}$, or $\frac{x}{a} - \frac{y}{b} = 0$ and $\frac{x}{a} + \frac{y}{b} = 0$.

Since the product $(\frac{x}{a} - \frac{y}{b})(\frac{x}{a} + \frac{y}{b}) = \frac{x^2}{a^2} - \frac{y^2}{b^2} = 0$, it follows that the equations of

the asymptotes for $\frac{x^2}{a^2} - \frac{y^2}{b^2} = 1$ can be determined by setting the constant term equal to zero and factoring.

Thus, in this problem the equation of the hyperbola has the form

$$(y - \sqrt{3}\,x)(y + \sqrt{3}\,x) = C \text{ (a constant)}.$$

Substituting the coordinates of point $(4,6)$, $(6 - 4\sqrt{3})(6 + 4\sqrt{3}) = C = -12$.

Hence the required equation is $(y - \sqrt{3}\,x)(y + \sqrt{3}\,x) = -12$ or $3x^2 - y^2 = 12$.

Definition. Two hyperbolas are called *conjugate* hyperbolas if the transverse and conjugate axes of one are respectively the conjugate and transverse axes of the other. If the equation of a hyperbola is written in the typical form, then the equation of the conjugate hyperbola is found by changing the signs of the coefficients of x^2 and y^2 in the given equation.

12. Write the equation of the hyperbola conjugate to the hyperbola $\dfrac{x^2}{9} - \dfrac{y^2}{16} = 1$. Write the equations of the asymptotes and find the coordinates of the foci for each hyperbola.

     The equation of the conjugate hyperbola is $-\dfrac{x^2}{9} + \dfrac{y^2}{16} = 1$.

     For each hyperbola, $c = \sqrt{9 + 16} = 5$. Then the coordinates of the foci are $(\pm 5, 0)$ for the given hyperbola and $(0, \pm 5)$ for the conjugate hyperbola.

     The equations of the asymptotes, $y = \pm \dfrac{4}{3}x$, are the same for both hyperbolas.

13. Derive the equation of the locus of a point $P(x,y)$ which moves so that the product of the slopes of the lines joining it to $(-2,1)$ and $(4,5)$ is 3.

     $\left(\dfrac{y - 1}{x + 2}\right)\left(\dfrac{y - 5}{x - 4}\right) = 3$.   Simplifying, $3x^2 - y^2 + 6y - 6x - 29 = 0$,   a hyperbola.

14. Show that the difference of the distances from the point $\left(8, \dfrac{8\sqrt{7}}{3}\right)$ on the hyperbola $64x^2 - 36y^2 = 2304$ to the foci is equal to the transverse axis. These distances are the focal radii of the point.

     Write the equation in the form $\dfrac{x^2}{36} - \dfrac{y^2}{64} = 1$.   Then $c = \pm\sqrt{36 + 64} = \pm 10$.

     The length of the transverse axis is $2a = 12$.

     The difference of the distances from $\left(8, \dfrac{8\sqrt{7}}{3}\right)$ to the foci $(\pm 10, 0)$ is

$$\sqrt{(8 + 10)^2 + \left(\dfrac{8\sqrt{7}}{3} - 0\right)^2} - \sqrt{(8 - 10)^2 + \left(\dfrac{8\sqrt{7}}{3} - 0\right)^2} = \dfrac{58}{3} - \dfrac{22}{3} = 12.$$

SUPPLEMENTARY PROBLEMS

1. Find (a) the vertices, (b) the foci, (c) the eccentricity, (d) the latus rectum, and (e) the equations of the asymptotes of each of the following hyperbolas:

     (1) $4x^2 - 45y^2 = 180$;   (2) $49y^2 - 16x^2 = 784$;   (3) $x^2 - y^2 = 25$.

Ans. (1):   (a) $(\pm 3\sqrt{5}, 0)$;   (b) $(\pm 7, 0)$;   (c) $\dfrac{7\sqrt{5}}{15}$;   (d) $\dfrac{8\sqrt{5}}{15}$;   (e) $y = \pm\dfrac{2\sqrt{5}}{15}x$.

     (2):   (a) $(0, \pm 4)$;   (b) $(0, \pm\sqrt{65})$;   (c) $\dfrac{\sqrt{65}}{4}$;   (d) $\dfrac{49}{2}$;   (e) $y = \pm\dfrac{4}{7}x$.

     (3):   (a) $(\pm 5, 0)$;   (b) $(\pm 5\sqrt{2}, 0)$;   (c) $\sqrt{2}$;   (d) 10;   (e) $y = \pm x$.

2. Write the equations of the hyperbolas for which the following conditions are given:
   (a) Transverse axis 8, foci at $(\pm5,0)$.   *Ans.* $9x^2 - 16y^2 = 144$.
   (b) Conjugate axis 24, foci at $(0, \pm13)$.   *Ans.* $144y^2 - 25x^2 = 3600$.
   (c) Center at $(0,0)$, a focus at $(8,0)$, a vertex at $(6,0)$.   *Ans.* $7x^2 - 9y^2 = 252$.

3. A point moves so that the difference of its distances from $(0,3)$ and $(0,-3)$ is 5. Derive the equation of its locus.   *Ans.* $44y^2 - 100x^2 = 275$.

4. Find the equation of the locus of a point which moves so that its distance from $(0,6)$ is 3/2 of its distance from the line $y - 8/3 = 0$.   *Ans.* $5y^2 - 4x^2 = 80$.

5. Write the equation of the hyperbola with its center at the origin, transverse axis on the y-axis, length of latus rectum 36, and distance between its foci 24.

   Ans. $3y^2 - x^2 = 108$.

6. Write the equation of the hyperbola with its center at the origin, transverse axis on the y-axis, eccentricity $2\sqrt{3}$, length of latus rectum 18.   *Ans.* $121y^2 - 11x^2 = 81$.

7. Write the equation of the hyperbola with its center at the origin, axes on the coordinate axes, and which passes through $(3,1)$ and $(9,5)$.   *Ans.* $x^2 - 3y^2 = 6$.

8. Derive the equation of the hyperbola with vertices at $(\pm6,0)$ and asymptotes $6y = \pm 7x$.

   Ans. $49x^2 - 36y^2 = 1764$.

9. Determine the equation of the locus of a point which moves so that the difference of its distances from $(-6,-4)$ and $(2,-4)$ is 6.   *Ans.* $\dfrac{(x+2)^2}{9} - \dfrac{(y+4)^2}{7} = 1$.

10. Find the coordinates of (a) the center, (b) the foci, (c) the vertices, and (d) the equations of the asymptotes for the hyperbola $9x^2 - 16y^2 - 36x - 32y - 124 = 0$.
    Ans. (a) $(2,-1)$; (b) $(7,-1),(-3,-1)$; (c) $(6,-1),(-2,-1)$; (d) $y + 1 = \pm \frac{3}{4}(x - 2)$.

11. A point moves so that the product of the slopes of the lines connecting it with $(-2,1)$ and $(3,2)$ is 4. Show that the locus is a hyperbola.   *Ans.* $4x^2 - y^2 - 4x + 3y - 26 = 0$.

12. A point moves so that the product of its directed distances from the lines $3x - 4y + 1 = 0$ and $3x + 4y - 7 = 0$ is 144/25. Find the equation of its locus. What curve is it?
    Ans. $9x^2 - 16y^2 - 18x + 32y - 151 = 0$.   Hyperbola.

13. Find the equation of the hyperbola with its center at $(0,0)$, a vertex at $(3,0)$, and the equation of one asymptote $2x - 3y = 0$.   *Ans.* $4x^2 - 9y^2 = 36$.

14. Write the equation of the hyperbola which is conjugate to the hyperbola of Problem 13.

    Ans. $9y^2 - 4x^2 = 36$.

15. Find the points of intersection of the hyperbolas, and trace the curves

$$x^2 - 2y^2 + x + 8y - 8 = 0,$$

$$3x^2 - 4y^2 + 3x + 16y - 18 = 0.$$

Ans. $(1,1)$, $(1,3)$, $(-2,1)$, $(-2,3)$.

16. Show that the difference of the distances from the point $(6, \dfrac{3\sqrt{5}}{2})$ on the hyperbola $9x^2 - 16y^2 = 144$ to the foci is equal to the length of the transverse axis. These distances are the focal radii of the point.

CHAPTER 8

Transformation of Coordinates

INTRODUCTION. It sometimes happens that the choice of axes at the beginning of the solution of a problem does not lead to the simplest form of the equation. By a proper transformation of axes an equation may be simplified. This may be accomplished in two steps, one called *translation* of axes, the other *rotation* of axes.

TRANSLATION OF AXES. Let OX and OY be the original axes, and let $O'X'$ and $O'Y'$ be the new axes, parallel respectively to the old ones. Also, let O' referred to the old axes be (h,k).

Let P be any point in the plane, and let its coordinates referred to the old axes be (x,y) and referred to the new axes be (x',y'). To determine x and y in terms of x', y', h and k:

$$x = MP = MM' + M'P = h + x' \quad \text{and}$$

$$y = NP = NN' + N'P = k + y'.$$

Hence the equations of translation are

$$x = x' + h, \quad y = y' + k.$$

ROTATION OF AXES. Let OX and OY be the original axes and OX' and OY' the new axes. O is the origin for each set of axes. Let the angle $X'OX$ through which the axes have been rotated be represented by θ. Let P be any point in the plane, and let its coordinates referred to the old axes be (x,y) and referred to the new axes be (x',y'). To determine x and y in terms of x', y' and θ:

$$x = OM = ON - MN$$
$$= x' \cos\theta - y' \sin\theta \quad \text{and}$$

$$y = MP = MM' + M'P = NN' + M'P$$
$$= x' \sin\theta + y' \cos\theta.$$

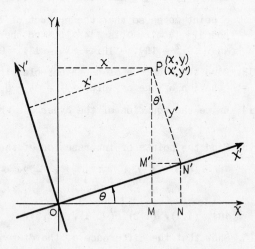

Hence the formulas for the rotation of the axes through an angle θ are

$$x = x' \cos\theta - y' \sin\theta,$$

$$y = x' \sin\theta + y' \cos\theta.$$

66

SOLVED PROBLEMS

1. Determine the equation of the curve $2x^2 + 3y^2 - 8x + 6y = 7$ when the origin is translated to the point $(2,-1)$.

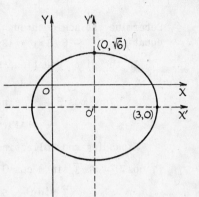

Substituting $x = x' + 2$, $y = y' - 1$, in the equation, we obtain

$$2(x' + 2)^2 + 3(y' - 1)^2 - 8(x' + 2) + 6(y' - 1) = 7.$$

Expanding and simplifying, the equation of the curve referred to the new axes is

$$2x'^2 + 3y'^2 = 18.$$

This is the standard equation of the ellipse with its center at the new origin and its major axis on the x'-axis, with semi-axes $a = 3$, $b = \sqrt{6}$.

2. Determine a translation of axes that will transform the equation $3x^2 - 4y^2 + 6x + 24y = 135$ into one in which the coefficients of the first degree terms are zero.

Substitute for x and y the values $x' + h$ and $y' + k$ respectively and collect the coefficients of the various powers of x' and y'.

$$3(x' + h)^2 - 4(y' + k)^2 + 6(x' + h) + 24(y' + k) = 135, \quad \text{or}$$

$$3x'^2 - 4y'^2 + (6h + 6)x' - (8k - 24)y' + 3h^2 - 4k^2 + 6h + 24k = 135.$$

From $6h + 6 = 0$ and $8k - 24 = 0$ we obtain $h = -1$ and $k = 3$, and the equation becomes

$$3x'^2 - 4y'^2 = 102.$$

This is the standard form for the hyperbola with its center at the origin, transverse axis on the x-axis, and semi-transverse axis $= \sqrt{34}$.

Another method. The following method is often used to eliminate first degree terms.

By completing the square, $\quad 3x^2 - 4y^2 + 6x + 24y = 135$

becomes $\qquad\qquad 3(x^2 + 2x + 1) - 4(y^2 - 6y + 9) = 102$

or $\qquad\qquad\qquad 3(x + 1)^2 - 4(y - 3)^2 = 102.$

For $x + 1$ substitute x', and for $y - 3$ substitute y'. Then the equation becomes

$$3x'^2 - 4y'^2 = 102.$$

3. Determine the equation of the parabola $x^2 - 2xy + y^2 + 2x - 4y + 3 = 0$ when the axes have been rotated $45°$.

$$x = x' \cos 45° - y' \sin 45° = \frac{x' - y'}{\sqrt{2}} \quad \text{and} \quad y = x' \sin 45° + y' \cos 45° = \frac{x' + y'}{\sqrt{2}}.$$

Substituting these values in the given equation,

$$(\frac{x' - y'}{\sqrt{2}})^2 - 2(\frac{x' - y'}{\sqrt{2}})(\frac{x' + y'}{\sqrt{2}}) + (\frac{x' + y'}{\sqrt{2}})^2 + 2(\frac{x' - y'}{\sqrt{2}}) - 4(\frac{x' + y'}{\sqrt{2}}) + 3 = 0.$$

Expanding and simplifying, the equation reduces to $2y'^2 - \sqrt{2}x' - 3\sqrt{2}y' + 3 = 0$, a parabola with its vertex at $(\frac{3\sqrt{2}}{8}, \frac{3\sqrt{2}}{4})$ and its axis parallel to the new x'-axis.

4. Determine the angle through which the axes must be rotated to remove the xy term in the equation $7x^2 - 6\sqrt{3}\,xy + 13y^2 = 16$.

In the given equation, substitute $x = x'\cos\theta - y'\sin\theta$
and $y = x'\sin\theta + y'\cos\theta$. Then,

$$7(x'\cos\theta - y'\sin\theta)^2 - 6\sqrt{3}(x'\cos\theta - y'\sin\theta)(x'\sin\theta + y'\cos\theta)$$
$$+ 13(x'\sin\theta + y'\cos\theta)^2 = 16.$$

Expanding and collecting coefficients of the different terms,

$$(7\cos^2\theta - 6\sqrt{3}\sin\theta\cos\theta + 13\sin^2\theta)x'^2 + [12\sin\theta\cos\theta - 6\sqrt{3}(\cos^2\theta - \sin^2\theta)]x'y'$$
$$+ (7\sin^2\theta + 6\sqrt{3}\sin\theta\cos\theta + 13\cos^2\theta)y'^2 = 16.$$

To eliminate the $x'y'$ term set its coefficient equal to zero and solve for θ.

$$12\sin\theta\cos\theta - 6\sqrt{3}(\cos^2\theta - \sin^2\theta) = 0, \quad \text{or}$$
$$6\sin 2\theta - 6\sqrt{3}(\cos 2\theta) = 0.$$

Then $\tan 2\theta = \sqrt{3}$, $2\theta = 60°$, and $\theta = 30°$.

Substituting this value of θ, the equation reduces to $x'^2 + 4y'^2 = 4$. This is the equation of an ellipse with its center at the origin and its axes along the new axes. The semi-axes are $a = 2$, $b = 1$.

THE MOST GENERAL FORM of the equation of the second degree is

$$Ax^2 + Bxy + Cy^2 + Dx + Ey + F = 0.$$

It is shown in the general discussion of this equation that the angle θ, through which it is necessary to rotate the axes to eliminate the xy-term can be found from the equation

$$\tan 2\theta = \frac{B}{A - C}.$$

5. By translation and rotation of the axes reduce the equation

$$5x^2 + 6xy + 5y^2 - 4x + 4y - 4 = 0$$

to its simplest form. Sketch the curve showing all three sets of coordinate axes.

To remove the first degree terms, use $x = x' + h$, $y = y' + k$.

$$5(x' + h)^2 + 6(x' + h)(y' + k) + 5(y' + k)^2 - 4(x' + h) + 4(y' + k) - 4 = 0.$$

Expanding and collecting terms,

$$5x'^2 + 6x'y' + 5y'^2 + (10h + 6k - 4)x' + (10k + 6h + 4)y' + 5h^2 + 6hk + 5k^2 - 4h + 4k - 4 = 0.$$

Set $10h + 6k - 4 = 0$ and $10k + 6h + 4 = 0$, and solve to obtain $h = 1$, $k = -1$. Then the equation reduces to

$$5x'^2 + 6x'y' + 5y'^2 = 8.$$

To find θ, use $\tan 2\theta = \frac{B}{A - C} = \frac{6}{5 - 5} = \infty$. Therefore $2\theta = 90°$, $\theta = 45°$.

The equations for rotation are $x' = \frac{x'' - y''}{\sqrt{2}}$, $y' = \frac{x'' + y''}{\sqrt{2}}$.

Substituting,

$$5(\frac{x''-y''}{\sqrt{2}})^2 + 6(\frac{x''-y''}{\sqrt{2}})(\frac{x''+y''}{\sqrt{2}}) + 5(\frac{x''+y''}{\sqrt{2}})^2 = 8.$$

Expanding and simplifying, the equation reduces to

$$4x''^2 + y''^2 = 4,$$

an ellipse with its axes on the x''-, y''-axes, center at the new origin, semi-major axis 2, semi-minor axis 1.

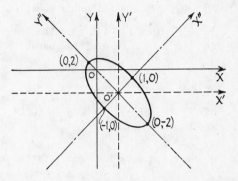

THE GENERAL EQUATION $Ax^2 + Bxy + Cy^2 + Dx + Ey + F = 0$, except in *special cases,* is the equation of a conic section. It can be proved that if

$$B^2 - 4AC < 0, \quad \text{the curve is an ellipse,}$$

$$B^2 - 4AC = 0, \quad \text{the curve is a parabola,}$$

$$B^2 - 4AC > 0, \quad \text{the curve is a hyperbola.}$$

For the special cases which may exist, the locus will be two straight lines, a point, or imaginary.

6. Determine the nature of the locus of the following equation: $4x^2 - 4xy + y^2 - 6x + 3y + 2 = 0.$

Since $B^2 - 4AC = 16 - 16 = 0$, the locus may be a parabola.
By grouping the terms this equation may be factored.

$$(4x^2 - 4xy + y^2) - 3(2x - y) + 2 = 0,$$

$$(2x - y)^2 - 3(2x - y) + 2 = 0,$$

$$(2x - y - 1)(2x - y - 2) = 0.$$

The locus is two parallel lines, $2x - y - 1 = 0$ and $2x - y - 2 = 0$.

7. Determine the nature of the locus of the equation $9x^2 - 12xy + 7y^2 + 4 = 0.$

Here $B^2 - 4AC = (144 - 252) < 0$, which is necessary for the ellipse.
If, however, we write this equation in the form

$$(3x - 2y)^2 + 3y^2 + 4 = 0,$$

we see that no real values of x and y will satisfy the equation. The locus is imaginary.

Another method is to solve the equation for y in terms of x by the quadratic formula.

$$y = \frac{+12x \pm \sqrt{(12x)^2 - 4(7)(9x^2 + 4)}}{2(7)} = \frac{+6x \pm \sqrt{-(27x^2 + 28)}}{7}.$$

The locus is imaginary for all real values of x.

8. Remove the first degree terms in $3x^2 + 4y^2 - 12x + 4y + 13 = 0.$

Completing the square, $3(x^2 - 4x + 4) + 4(y^2 + y + \frac{1}{4}) = 0$, or $3(x - 2)^2 + 4(y + \frac{1}{2})^2 = 0.$

Setting $x - 2 = x'$ and $y + \frac{1}{2} = y'$ gives $3x'^2 + 4y'^2 = 0.$

This is satisfied only by $x' = 0$, $y' = 0$, which is the new origin and its locus.

The locus of the original equation is the point $(2, -\frac{1}{2})$.

9. Simplify the following equation: $4x^2 - 4xy + y^2 - 8\sqrt{5}\,x - 16\sqrt{5}\,y = 0$.

Since $B^2 - 4AC = 0$, the locus may be a parabola.

In the case of the parabola rotate the axes before translating them.

$\tan 2\theta = \dfrac{-4}{4-1} = -\dfrac{4}{3}$.   Therefore $\cos 2\theta = -\dfrac{3}{5}$.

Since $\cos 2\theta = 2\cos^2\theta - 1 = -\dfrac{3}{5}$, $\cos^2\theta = \dfrac{1}{5}$, $\cos\theta = \dfrac{1}{\sqrt{5}}$, and $\sin\theta = \dfrac{2}{\sqrt{5}}$.

The equations for rotation are $x = \dfrac{x' - 2y'}{\sqrt{5}}$, $y = \dfrac{2x' + y'}{\sqrt{5}}$.   Substituting,

$$4\Big(\frac{x' - 2y'}{\sqrt{5}}\Big)^2 - 4\Big(\frac{x' - 2y'}{\sqrt{5}}\Big)\Big(\frac{2x' + y'}{\sqrt{5}}\Big) + \Big(\frac{2x' + y'}{\sqrt{5}}\Big)^2 - 8\sqrt{5}\Big(\frac{x' - 2y'}{\sqrt{5}}\Big) - 16\sqrt{5}\Big(\frac{2x' + y'}{\sqrt{5}}\Big) = 0.$$

Expanding and simplifying, we obtain $y'^2 - 8x' = 0$,   a parabola.

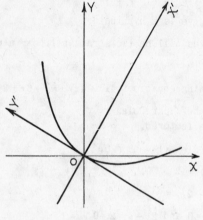

Problem 9.

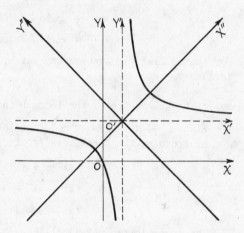

Problem 10.

10. Simplify the equation $xy - 2y - 4x = 0$.   Sketch the curve showing all three sets of axes.

Since $B^2 - 4AC = 1 > 0$, the curve if it exists is a hyperbola.

Substituting $x = x' + h$, $y = y' + k$, the equation becomes

$$(x' + h)(y' + k) - 2(y' + k) - 4(x' + h) = 0, \text{ or}$$
$$x'y' + (k - 4)x' + (h - 2)y' + hk - 2k - 4h = 0.$$

When $k = 4$, $h = 2$, the resulting equation becomes $x'y' = 8$.

To determine the angle for rotation: $\tan 2\theta = \dfrac{1}{0} = \infty$, $2\theta = 90°$, $\theta = 45°$.

Then $x' = \dfrac{x'' - y''}{\sqrt{2}}$, $y' = \dfrac{x'' + y''}{\sqrt{2}}$, and $\Big(\dfrac{x'' - y''}{\sqrt{2}}\Big)\Big(\dfrac{x'' + y''}{\sqrt{2}}\Big) = 8$.

Simplifying, the final equation is $x''^2 - y''^2 = 16$,   an equilateral hyperbola.

11. Find the equation of the conic through the five points: $(1, 1)$, $(2, 3)$, $(3, -1)$, $(-3, 2)$, $(-2, -1)$.

Divide the general equation of the second degree by A and write it
$$x^2 + B'xy + C'y^2 + D'x + E'y + F' = 0.$$

Substitute the coordinates of the points for x and y.

$$B' + C' + D' + E' + F' = -1$$
$$6B' + 9C' + 2D' + 3E' + F' = -4$$
$$-3B' + C' + 3D' - E' + F' = -9$$
$$-6B' + 4C' - 3D' + 2E' + F' = -9$$
$$2B' + C' - 2D' - E' + F' = -4$$

Solving these equations, $B' = \dfrac{8}{9}$, $C' = -\dfrac{13}{9}$, $D' = -\dfrac{1}{9}$, $E' = \dfrac{19}{9}$, $F' = -\dfrac{22}{9}$.

Substituting these values in the original equation and simplifying, the equation is

$$9x^2 + 8xy - 13y^2 - x + 19y - 22 = 0.$$

Since $B^2 - 4AC = (64 + 468) > 0$, the conic is a hyperbola.

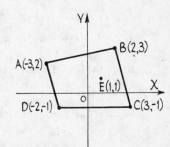

A second method of solving this problem follows.

The equation of the line AB is $x - 5y + 13 = 0$, of line CD is $y + 1 = 0$. The equation of this pair of lines is $(y + 1)(x - 5y + 13) = xy - 5y^2 + x + 8y + 13 = 0$.

In like manner, the equation of the pair of lines AD and BC is $12x^2 + 7xy + y^2 - 5x - 4y - 77 = 0$.

The system of all curves passing through the points of intersection of these lines is

$$xy - 5y^2 + x + 8y + 13 + k(12x^2 + 7xy + y^2 - 5x - 4y - 77) = 0.$$

To find the curve of this system which passes through the fifth point $(1, 1)$, substitute the coordinates of this point for x and y and solve for k. Then $k = 3/11$.

When this value is substituted for k, the equation reduces to

$$9x^2 + 8xy - 13y^2 - x + 19y - 22 = 0.$$

SUPPLEMENTARY PROBLEMS

1. By translating the axes, using $x = x' + h$, $y = y' + k$, reduce each of the following equations to its simplest standard form, and state the nature of the locus.

   (a) $y^2 - 6y - 4x + 5 = 0$.   *Ans.* $y^2 = 4x$. Parabola

   (b) $x^2 + y^2 + 2x - 4y - 20 = 0$.   *Ans.* $x^2 + y^2 = 25$. Circle

   (c) $3x^2 - 4y^2 + 12x + 8y - 4 = 0$.   *Ans.* $3x^2 - 4y^2 = 12$. Hyperbola

   (d) $2x^2 + 3y^2 - 4x + 12y - 20 = 0$.   *Ans.* $2x^2 + 3y^2 = 34$. Ellipse

   (e) $x^2 + 5y^2 + 2x - 20y + 25 = 0$.   *Ans.* $x^2 + 5y^2 + 4 = 0$. Imaginary ellipse

2. Remove the first degree terms in each of the following equations by the method of completing the square.

   (a) $x^2 + 2y^2 - 4x + 6y - 8 = 0$.   *Ans.* $2x^2 + 4y^2 = 33$

   (b) $3x^2 - 4y^2 - 6x - 8y - 10 = 0$.   *Ans.* $3x^2 - 4y^2 = 9$

   (c) $2x^2 + 5y^2 - 12x + 10y - 17 = 0$.   *Ans.* $2x^2 + 5y^2 = 40$

   (d) $3x^2 + 3y^2 - 12x + 12y - 1 = 0$.   *Ans.* $3x^2 + 3y^2 = 25$

3. By translation of axes remove the first degree terms in $2xy - x - y + 4 = 0$.
   Ans. $4xy + 7 = 0$

4. By translation of axes remove the first degree terms in $x^2 + 2xy + 3y^2 + 2x - 4y - 1 = 0$.
   Ans. $2x^2 + 4xy + 6y^2 - 13 = 0$

5. Each of the following is the equation of a conic. Determine the nature of each conic.
   Use $B^2 - 4AC$.

   (a) $3x^2 - 10xy + 3y^2 + x - 32 = 0$.   *Ans.* Hyperbola

   (b) $41x^2 - 84xy + 76y^2 = 168$.   *Ans.* Ellipse

   (c) $16x^2 + 24xy + 9y^2 - 30x + 40y = 0$.   *Ans.* Parabola

   (d) $xy + x - 2y + 3 = 0$.   *Ans.* Hyperbola

   (e) $x^2 - 4xy + 4y^2 = 4$.   *Ans.* Two parallel lines

6. By rotation of axes simplify the equation $9x^2 + 24xy + 16y^2 + 90x - 130y = 0$ and identify
   the locus.   *Ans.* $x^2 - 2x - 6y = 0$. Parabola

7. Rotate the axes through the angle $\theta = \tan^{-1} \frac{4}{3}$ and simplify the equation
   $$9x^2 + 24xy + 16y^2 + 80x - 60y = 0.$$
   Draw the graph showing both sets of axes.   *Ans.* $x^2 - 4y = 0$

8. Simplify each of the following equations by suitable transformations of axes, and draw
   the figure showing the locus and all sets of axes.

   (a) $9x^2 + 4xy + 6y^2 + 12x + 36y + 44 = 0$.   *Ans.* $2x^2 + y^2 = 2$

   (b) $x^2 - 10xy + y^2 + x + y + 1 = 0$.   *Ans.* $32x^2 - 48y^2 = 9$

   (c) $17x^2 - 12xy + 8y^2 - 68x + 24y - 12 = 0$. *Ans.* $x^2 + 4y^2 = 16$

   (d) $2x^2 + 3xy + 4y^2 + 2x - 3y + 5 = 0$.   *Ans.* No locus (imaginary)

9. Determine the equation of the conic through $(5,2)$, $(1,-2)$, $(-1,1)$, $(2,5)$, and $(-1,-2)$.
   Ans. $49x^2 - 55xy + 36y^2 - 110x - 19y - 231 = 0$.   Ellipse

10. Determine the equation of the conic through $(1,1)$, $(-1,2)$, $(0,-2)$, $(-2,-1)$, $(3,-3)$.
    Ans. $16x^2 + 46xy + 49y^2 + 16x + 23y - 150 = 0$.   Ellipse

11. Determine the equation of the conic through $(4,1)$, $(2,2)$, $(3,-2)$, $(4,-1)$, $(1,-3)$.
    Ans. $17x^2 - 16xy + 54y^2 + 11x + 64y - 370 = 0$.   Ellipse

12. Determine the equation of the conic through $(1,6)$, $(-3,-2)$, $(-5,0)$, $(3,4)$, $(0,10)$.
    Ans. $xy - 2x + y - 10 = 0$. Hyperbola

CHAPTER 9

Polar Coordinates

POLAR COORDINATES. Instead of fixing the position of a point in a plane in terms of its distances from two perpendicular lines, it is sometimes preferable to show its location in terms of its distance from a fixed point and its direction from a fixed line through this point. The coordinates of points in this system are called *polar coordinates*.

In this system we have a fixed point *O* called the *pole* and a directed line *OA* called the *polar axis*.

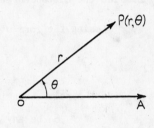

The polar coordinates of the point *P* are written (r, θ), where *r* is the distance *OP*, and θ the vectorial angle *AOP*. The distance *r* measured from *O* to the point *P* on the terminal side of the angle *AOP* is positive. As in trigonometry the vectorial angle θ is positive when measured in a counterclockwise direction, negative if measured clockwise; *r* is positive if measured from the pole to the point along the terminal side of the angle, negative if measured in the opposite direction, *i.e.*, on the terminal side produced through *O*.

If *r* and θ are connected by an equation of any form, we may assign values to θ and determine corresponding values of *r*. The points thus determined lie on a definite curve.

SYMMETRY. When plotting a graph in rectangular coordinates we often make use of symmetry. In polar coordinates there are tests which also may be used.

If the equation is unchanged when θ is replaced by $-\theta$, the curve is symmetric with respect to the polar axis.

The curve is symmetric about the 90°-line if its equation is unchanged when θ is replaced by $\pi - \theta$.

A curve is symmetric with respect to the pole if its equation is unchanged when *r* is replaced by $-r$, or when θ is replaced by $\pi + \theta$.

RELATION BETWEEN RECTANGULAR AND POLAR COORDINATES.

Consider the point $P(r, \theta)$, and suppose that the polar axis *OX* and the pole *O* are respectively the positive *x*-axis and the origin of a system of rectangular coordinates. Let the rectangular coordinates of *P* be (x, y). Then,

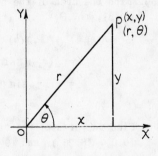

$$x = r \cos \theta,$$

$$y = r \sin \theta,$$

$$r = \sqrt{x^2 + y^2},$$

and

$$\theta = \tan^{-1} \frac{y}{x}.$$

SOLVED PROBLEMS

1. Find the distance between the points $P_1(r_1, \theta_1)$ and $P_2(r_2, \theta_2)$.

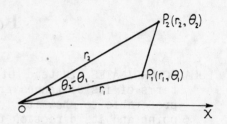

Since we have given two sides and the included angle of a triangle, the third side can be found by the Law of Cosines.

$$P_1P_2 = \sqrt{r_1^2 + r_2^2 - 2r_1r_2 \cos(\theta_2 - \theta_1)}.$$

2. Find the distance between the points $(6, 15°)$ and $(8, 75°)$.

Using the form of Problem 1 above, $\quad d = \sqrt{6^2 + 8^2 - 2(6)(8) \cos(75° - 15°)}$

$$= \sqrt{36 + 64 - 96(\tfrac{1}{2})} = 2\sqrt{13}.$$

3. Find the polar equation of the circle with its center at (r_1, θ_1) and radius a.

Let (r, θ) be any point on the circle.

Using the cosine law, the equation is $\quad a^2 = r^2 + r_1^2 - 2rr_1 \cos(\theta - \theta_1)$

$$\text{or} \quad r^2 - 2r_1r \cos(\theta - \theta_1) + r_1^2 = a^2.$$

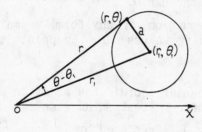

Problem 3.

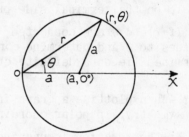

Problem 4.

4. Write the equation of the circle with its center at $(a, 0°)$ and radius a.

Here $\theta_1 = 0°$. By the cosine law, $\quad a^2 = r^2 + a^2 - 2ra \cos\theta$.

Then the required equation is $\quad r^2 = 2ar \cos\theta \quad$ or $\quad r = 2a \cos\theta$.

5. Find the area of the triangle whose vertices are $(0,0)$, (r_1, θ_1), and (r_2, θ_2).

$$\begin{aligned}\text{Area} &= \tfrac{1}{2}(OP_1)(h) \\ &= \tfrac{1}{2}(r_1)r_2 \sin(\theta_2 - \theta_1) \\ &= \tfrac{1}{2}r_1r_2 \sin(\theta_2 - \theta_1).\end{aligned}$$

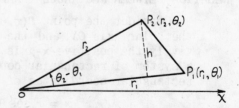

6. Find the area of the triangle whose vertices are $(0,0)$, $(6, 20°)$, and $(9, 50°)$.

$$\text{Area} = \tfrac{1}{2}r_1r_2 \sin(\theta_2 - \theta_1) = \tfrac{1}{2}(6)(9) \sin(50° - 20°) = 13.5 \text{ square units.}$$

7. Find the equation of the line through the point $(2, 30°)$ and perpendicular to OX.

Let (r, θ) be any point on the line.

Then $r \cos \theta = 2 \cos 30° = 2(\frac{\sqrt{3}}{2}) = \sqrt{3}$, or $r \cos \theta = \sqrt{3}$.

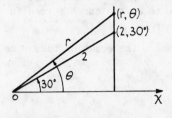 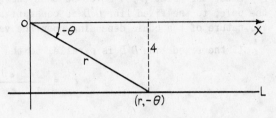

Problem 7.   Problem 8.

8. Find the polar equation of a straight line parallel to the polar axis OX and 4 units below it.

Let $(r, -\theta)$ be any point on the line L.

Then $r \sin(-\theta) = 4$, or $r \sin \theta + 4 = 0$.

Note. $\cos(-\theta) = \cos \theta$; $\sin(-\theta) = -\sin \theta$.

9. A line passes through the point $(4, 30°)$ and makes an angle of $150°$ with the polar axis. Find its equation.

Let (r, θ) be any point on the line.

Then $OA = r \cos(\theta - 60°) = 4 \sin 60°$, or $r \cos(\theta - 60°) = 2\sqrt{3}$.

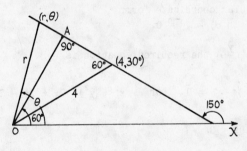 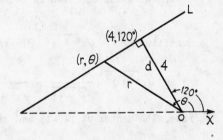

Problem 9.   Problem 10.

10. Find the equation of the line through $(4, 120°)$ which is perpendicular to the line joining $(4, 120°)$ and the pole $(0, 0)$.   Let (r, θ) be any point on the line.

The line L is perpendicular to the line d.   Hence $d = r \cos(\theta - 120°) = 4$, and the equation of L is $r \cos(\theta - 120°) = 4$.

The equation $r \cos(\theta - 120°) = 4$ is the polar form of the normal form of the equation of the line in rectangular coordinates where $p = 4$ and $\omega = 120°$.

11. Find the locus of $P(r, \theta)$ so that $\dfrac{OP}{MP} = e$ (a constant).

$MP = NO + OQ = p + r \cos \theta$.

Since $OP = e(MP)$, $r = e(p + r \cos \theta)$

or $r = \dfrac{ep}{1 - e \cos \theta}$.

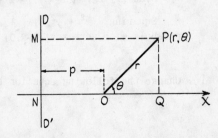

If $D'D$ were to the right of the pole O, the equation would have the form

$$r = \frac{ep}{1 + e \cos \theta}.$$

Since the point (r, θ) moves so that the ratio of its distance from the fixed point O, the pole, to the fixed line $D'D$ is constant and equal to e, the curve formed is a conic, the nature of the conic depending upon the value of e.

If the fixed line $D'D$ is parallel to the polar axis, the equation takes the form

$$r = \frac{ep}{1 \pm e \sin \theta}.$$

12. Determine the nature of the conic defined by the equation $r = \dfrac{12}{4 + 3 \cos \theta}$.

Dividing numerator and denominator by 4, the equation becomes $r = \dfrac{3}{1 + \frac{3}{4} \cos \theta}$.

Hence $e = \frac{3}{4}$ and the locus is an ellipse.

Since $ep = 3$ or $\frac{3}{4} p = 3$, then $p = 4$, and the directrix $D'D$ is perpendicular to the polar axis and 4 units to the right of the pole.

13. Find the polar equation of the ellipse $9x^2 + 4y^2 = 36$.

Using the relations $x = r \cos \theta$, $y = r \sin \theta$ and substituting, the required equation is
$$9r^2 \cos^2 \theta + 4r^2 \sin^2 \theta = 36, \quad \text{or} \quad r^2(4 + 5 \cos^2 \theta) = 36.$$

14. Write the following equation in rectangular coordinate form:
$$r^2 - 2r(\cos \theta - \sin \theta) - 7 = 0.$$

Substituting $r = \sqrt{x^2 + y^2}$, $\theta = \tan^{-1} \dfrac{y}{x}$, the required equation is

$$x^2 + y^2 - 2\sqrt{x^2 + y^2} \left(\frac{x}{\sqrt{x^2 + y^2}} - \frac{y}{\sqrt{x^2 + y^2}} \right) - 7 = 0 \quad \text{or} \quad x^2 + y^2 - 2x + 2y - 7 = 0,$$

a circle with center at $(1, -1)$ and radius $= 3$.

15. Write the following equation in rectangular coordinate form:
$$r = \frac{4}{1 - \cos \theta}, \quad \text{or} \quad r(1 - \cos \theta) = 4.$$

Substitute $r = \sqrt{x^2 + y^2}$ and $\cos \theta = \dfrac{x}{\sqrt{x^2 + y^2}}$ to get $\sqrt{x^2 + y^2} \left(1 - \dfrac{x}{\sqrt{x^2 + y^2}} \right) = 4$.

Simplifying, $\sqrt{x^2 + y^2} - x = 4$ or $\sqrt{x^2 + y^2} = x + 4$.

Squaring, $x^2 + y^2 = x^2 + 8x + 16$ or $y^2 - 8x - 16 = 0$, which is the equation of a parabola with vertex at $(-2, 0)$ and symmetrical about the x-axis.

16. Change the following equation to rectangular coordinates and identify the curve:
$$r = \frac{1}{1 - 2 \sin \theta}.$$

Substituting, $\quad \sqrt{x^2 + y^2} = \dfrac{1}{1 - \dfrac{2y}{\sqrt{x^2 + y^2}}}$.

Simplifying, $\quad \sqrt{x^2 + y^2} = \dfrac{\sqrt{x^2 + y^2}}{\sqrt{x^2 + y^2} - 2y}$ $\quad$ or $\quad \sqrt{x^2 + y^2}\,(\sqrt{x^2 + y^2} - 2y - 1) = 0$.

But $\sqrt{x^2 + y^2} = 0$ only when $x = y = 0$.

Squaring and simplifying $\sqrt{x^2 + y^2} - 2y - 1 = 0$ gives $x^2 - 3y^2 - 4y - 1 = 0$, a hyperbola.

17. Find the coordinates of the points of intersection of the following pair of curves:

$$(1) \quad r = 1 - \cos\theta$$
$$(2) \quad r = \sin\tfrac{1}{2}\theta.$$

By trigonometry, $\quad 1 - \cos\theta = 2\sin^2\tfrac{1}{2}\theta$.

Then $\quad 2\sin^2\tfrac{1}{2}\theta = \sin\tfrac{1}{2}\theta$, $\quad$ or $\quad \sin\tfrac{1}{2}\theta\,(2\sin\tfrac{1}{2}\theta - 1) = 0$. $\quad$ Solving, $\sin\theta = 0,\ \tfrac{1}{2}$.

For $\sin\tfrac{1}{2}\theta = 0$, $\theta = 0°$; for $\sin\tfrac{1}{2}\theta = \tfrac{1}{2}$, $\tfrac{1}{2}\theta = 30°,\ 150°$, and $\theta = 60°,\ 300°$.

Hence the coordinates of the points of intersection are $(0, 0°)$, $(\tfrac{1}{2}, 60°)$, $(\tfrac{1}{2}, 300°)$.

18. Find the center and radius of the circle $r^2 + 4r\cos\theta - 4\sqrt{3}\,r\sin\theta - 20 = 0$.

Use the equation of the circle in Problem 3. When expanded, this becomes

$$r^2 - 2r(r_1\cos\theta_1\cos\theta + r_1\sin\theta_1\sin\theta) + r_1^2 - a^2 = 0$$

or $\quad r^2 - 2r_1\cos\theta_1\,r\cos\theta - 2r_1\sin\theta_1\,r\sin\theta + r_1^2 - a^2 = 0$.

By comparing the given equation with the expanded form, we have

$$(1)\ -2r_1\cos\theta_1 = 4, \quad (2)\ 2r_1\sin\theta_1 = 4\sqrt{3}, \quad \text{and} \quad (3)\ r_1^2 - a^2 = -20.$$

Dividing equation (2) by equation (1) gives $\tan\theta_1 = -\sqrt{3}$, $\theta_1 = 120°$.

Substituting in (1), $-2r_1(-\tfrac{1}{2}) = 4$ or $r_1 = 4$. Then from (3), $16 - a^2 = -20$, $a = 6$.

Hence the circle has center $(4, 120°)$ and radius 6.

19. Find the equation of the locus of a point which moves so that the product of its distances from $(-a, 0°)$ and $(a, 0°)$ is a^2.

In triangle AOP, by the cosine law,

$$AP = \sqrt{a^2 + r^2 - 2ar\cos(180° - \theta)} = \sqrt{a^2 + r^2 + 2ar\cos\theta}.$$

In triangle BOP, $\quad PB = \sqrt{a^2 + r^2 - 2ar\cos\theta}$.

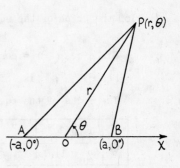

$(AP)(PB) = \sqrt{(a^2 + r^2)^2 - 4a^2r^2\cos^2\theta} = a^2$.

Squaring, $\quad a^4 + 2a^2r^2 + r^4 - 4a^2r^2\cos^2\theta = a^4$.

Simplifying, $\quad r^4 + 2a^2r^2 - 4a^2r^2\cos^2\theta = 0 \quad$ or $\quad r^2(r^2 + 2a^2 - 4a^2\cos^2\theta) = 0$.

Hence the required equation is $r^2 + 2a^2 - 4a^2\cos^2\theta = 0$,

or $\quad r^2 = 2a^2(2\cos^2\theta - 1) = 2a^2\cos 2\theta$. (The lemniscate. See Problem 25 below.)

20. A line of length $2a$ has its extremities on two fixed perpen-
    dicular lines. Find the locus of the foot of the perpendicu-
    lar from the intersection of the two fixed lines to the given
    line.

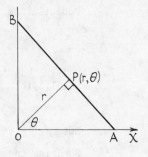

   Let one of the fixed lines be the polar axis, and let the
point of intersection of the two fixed lines be the pole.

   Then $OA = OP \sec \theta = AB \cos(90° - \theta)$

   or $r \sec \theta = 2a \cos(90° - \theta)$

   or $\dfrac{r}{\cos \theta} = 2a \sin \theta.$

   Hence $r = 2a \sin \theta \cos \theta$ or $r = a \sin 2\theta$. (The four-leaved rose.)

21. Discuss and plot the locus of the equation $r = 10 \cos \theta$.

   Since $\cos(-\theta) = \cos \theta$, the locus is symmetric with respect to the polar axis.
   θ may have any value, but r varies from 0 to ± 10; hence the curve is a closed curve.
   To determine points on the graph, assume values for θ and determine corresponding
values for r. From Problem 4 we know that the locus is a circle, with the radius $a = 5$
and center on the polar axis.

θ	$0°$	$30°$	$45°$	$60°$	$90°$	$120°$	$135°$	$150°$	$180°$
r	10	8.7	7.1	5	0	−5	−7.1	−8.7	−10

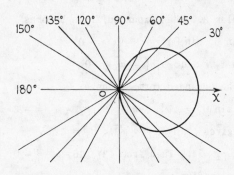

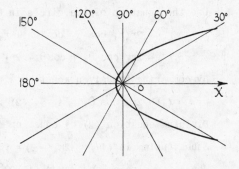

Problem 21.   Problem 22.

22. Plot the graph of the equation $r = \dfrac{2}{1 - \cos \theta}$.

   Since $\cos(-\theta) = \cos \theta$, the curve is symmetric with respect to the polar axis.

   For $\theta = 0°$, r is infinite; and for $\theta = 180°$, $r = 1$. The curve is an open curve.

   The locus of the equation is a parabola. Refer to Problem 11 above.

θ	$0°$	$30°$	$60°$	$90°$	$120°$	$150°$	$180°$	$210°$	$240°$	$270°$	$300°$	$330°$	$360°$
r	∞	14.9	4	2	1.3	1.1	1	1.1	1.3	2	4	14.9	∞

23. Plot the graph of the three-leaved rose $r = 10 \sin 3\theta$.

   Since the sine is positive in quadrants 1 and 2, and negative in quadrants 3 and 4,
this curve will be symmetric about a line passing through the pole and perpendicular to

the polar axis.

r will be zero when 3θ is 0°, 180°, or some multiple of 180°, that is, when θ = 0°, 60°, 120°, ... r will have its greatest numerical value when 3θ = 90°, 270°, or some odd multiple of 90°, that is, when θ = 30°, 90°, 150°, ...

θ	0°	30°	60°	90°	120°	150°	180°	210°	240°	270°	300°	330°
r	0	10	0	–10	0	10	0	–10	0	10	0	–10

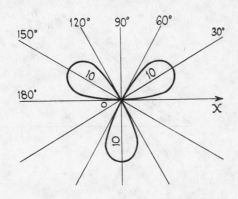

Problem 23.   *Problem 24.*

24. Plot the graph of the cardiod r = 5(1 + cos θ).

This curve is symmetric about the polar axis.
Since cos θ varies from 1 to –1, r cannot be negative.
The value of r varies from 10 to 0 as θ varies from 0° to 180°

θ	0°	30°	45°	60°	90°	120°	135°	150°	180°
r	10	9.3	8.5	7.5	5	2.5	1.5	.67	0

25. Discuss and plot the graph of the lemniscate
$$r^2 = 9 \cos 2\theta.$$

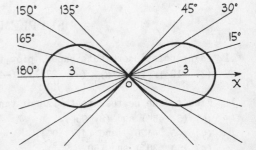

If we replace r by –r and θ by –θ, the equation reduces to its original form since cos (–2θ) = cos 2θ and (–r)² = r². The locus is therefore symmetric about the pole and also about the polar axis.

r has its greatest numerical value for θ = 0° since cos 0° = 1, and r = 3.

For 45° < θ < 135°, and for 225° < θ < 315°, r is imaginary. For θ = ± 45°, cos 2θ = 0; therefore r = 0, and the lines θ = ± π/4 are tangents to the curve at the origin.

$r = \pm 3\sqrt{\cos 2\theta}.$

θ	2θ	cos 2θ	r
0	0	1	±3
15°	30°	.866	±2.8
30°	60°	.5	±2.1
45°	90°	0	0

26. Find the locus of a point such that its radius vector is proportional to its vectorial angle.

The equation is $r = a\theta$. This curve is called the Spiral of Archimedes.

θ	0	$\pi/6$	$\pi/3$	$\pi/2$	π	$3\pi/2$	2π
r	0	$.52a$	$1.0a$	$1.6a$	$3.1a$	$4.7a$	$6.3a$

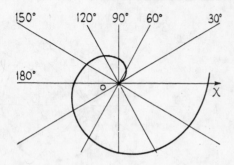

Problem 26.

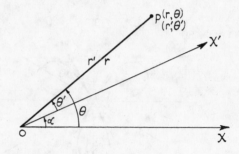

Problem 27.

27. Let $P(r,\theta)$ be any point. Show that when the polar axis is rotated about the pole O through any angle α such that the new coordinates are (r',θ'), then $r' = r$ and $\theta' = \theta - \alpha$.

The formula for rotation in polar coordinates is $\theta = \theta' + \alpha$ and $r = r'$.

28. By rotating the polar axis through $90°$ in a counterclockwise direction, show that the equation of the cardiod in Problem 24 becomes $r = 5(1 - \sin\theta)$.

Substitute for θ the value $90° + \theta'$.

Then $r' = 5[1 + \cos(90° + \theta')] = 5(1 - \sin\theta')$, since $\cos(90° + \theta') = -\sin\theta'$.

SUPPLEMENTARY PROBLEMS

1. Using polar coordinate paper, plot the following points: $(2, 30°)$, $(-3, 30°)$, $(5, 75°)$, $(3, 210°)$, $(2, \pi/2)$, $(-2, 270°)$, $(-4, 300°)$, $(-3, -5\pi/6)$, $(4, 0°)$, $(0, 30°)$, $(0, 60°)$.

2. Find to the nearest tenth the distance between each of the following pairs of points.
   (a) $(5, 45°)$ and $(8, 90°)$.   *Ans.* 5.7
   (b) $(-5, -120°)$ and $(4, 150°)$.   *Ans.* 6.4
   (c) $(50, 30°)$ and $(50, -90°)$.   *Ans.* 86.6
   (d) $(3, 150°)$ and $(-2, 60°)$.   *Ans.* 3.6

3. Find the area of each triangle whose vertices are the pole and each pair of points in Problem 2.   *Ans.* (a) 14.14; (b) 10; (c) 1082.5; (d) 3.

4. Write the polar equation of the straight line through $(4, 120°)$ and perpendicular to OX.
Ans. $r\cos\theta + 2 = 0$.

5. Write the polar equation of the straight line through $(3, -30°)$ and parallel to OX.
Ans. $2r\sin\theta + 3 = 0$.

6. Write the polar equation of the line through $(2, 120°)$ and the pole.   *Ans.* $\theta = 2\pi/3$.

7. Find the polar equation of the line passing through $(4, 2\pi/3)$ and perpendicular to the line joining the origin to this point.   *Ans.* $r \cos(\theta - 2\pi/3) = 4$.

8. Find the polar equation of a line passing through $(3, 0°)$ and making an angle of $3\pi/4$ with the initial line. Find r when $\theta = -\pi/4$ and explain your answer.
    Ans. $\sqrt{2}\, r \cos(\theta - \pi/4) = 3$.

9. Find the polar equation of a line passing through $(4, 20°)$ and making an angle of $140°$ with the polar axis.   *Ans.* $r \cos(\theta - 50°) = 2\sqrt{3}$.

10. Find the polar equation of the circle with its center at the pole and radius 5.
    Ans. $r = 5$.

11. Find the equation of the circle with its center at $(4, 30°)$ and radius 5.
    Ans. $r^2 - 8r \cos(\theta - \pi/6) - 9 = 0$.

12. Find the equation of each of the following circles:
    (a) Center at $(3, 0°)$ and passing through the pole.   *Ans.* $r = 6 \cos \theta$
    (b) Center at $(4, 45°)$ and passing through the pole.   *Ans.* $r = 8 \cos(\theta - 45°)$
    (c) Center at $(5, 90°)$ and passing through the pole.   *Ans.* $r - 10 \sin \theta = 0$
    (d) Passing through the pole, $(3, 90°)$, and $(4, 0°)$.   *Ans.* $r = 4 \cos \theta + 3 \sin \theta$

13. Find the equation of the circle with center at $(8, 120°)$, which passes through $(4, 60°)$.
    Ans. $r^2 - 16r \cos(\theta - 120°) + 16 = 0$.

14. By comparison with the equation in Solved Problem 3, find the center and radius of the circle $r^2 - 4r \cos(\theta - \pi/4) - 12 = 0$.   *Ans.* $(2, \pi/4)$, radius 4.

15. Given the circle $r^2 - 4\sqrt{3}\, r \cos \theta - 4r \sin \theta + 15 = 0$, find the coordinates of the center and the radius.   *Ans.* Center $(4, \pi/6)$, radius 1.

16. Find the equation of the circle with its center at $(8, \pi/4)$ and tangent to the polar axis.   *Ans.* $r^2 - 16r \cos(\theta - \pi/4) + 32 = 0$.

17. Find the equation of the circle with its center at $(4, 30°)$ and tangent to OX.
    Ans. $r^2 - 8r \cos(\theta - \pi/6) + 12 = 0$.

18. Show that the equation of the circle through the pole and the points $(a, 0°)$ and $(b, 90°)$ is $r = a \cos \theta + b \sin \theta$.

19. Find the center and radius of the circle $r = 5 \cos \theta - 5\sqrt{3} \sin \theta$.
    Ans. $(5, -60°)$, 5.

20. In Solved Problem 11, the equation of a conic section with its focus at the pole and its directrix perpendicular to the polar axis and p units to the left of the pole, was shown to be

$$r = \frac{ep}{1 - e \cos \theta}.$$

If the directrix is p units to the right of the pole, the equation becomes

$$r = \frac{ep}{1 + e \cos \theta}.$$

Show that the polar equation of the conic with its focus at the pole and directrix parallel to the polar axis and p units from it is

$$r = \frac{ep}{1 \pm e \sin \theta},$$

where the sign is plus if the directrix is above the polar axis and minus if below.

21. Determine the nature of each of the following conics with one focus at the pole. Determine e and give the location of the directrix in terms of its direction with respect to the polar axis and its distance from the pole.

(a) $r = \dfrac{4}{2 - 3 \cos \theta}$.   Ans. Hyperbola, $e = 3/2$; one directrix perpendicular to the polar axis and 4/3 units from the corresponding focus.

(b) $r = \dfrac{2}{1 - \cos \theta}$.   Ans. Parabola; $e = 1$; directrix perpendicular to the polar axis and 2 units to left of focus.

(c) $r = \dfrac{6}{2 - \sin \theta}$.   Ans. Ellipse; $e = \frac{1}{2}$; directrix parallel to the polar axis and 6 units below the pole.

22. Identify and sketch the locus of each of the following conics:

(a) $r = \dfrac{4}{2 + \cos \theta}$;   (b) $r = \dfrac{5}{1 - \cos \theta}$;   (c) $r = \dfrac{2}{2 + 3 \sin \theta}$.

23. Find the polar equation of the ellipse $9x^2 + 16y^2 = 144$.
    Ans. $r^2(9 \cos^2\theta + 16 \sin^2\theta) = 144$.

24. Change to polar coordinates: $2x^2 - 3y^2 - x + y = 0$.   Ans. $r = \dfrac{\cos \theta - \sin \theta}{2 \cos^2\theta - 3 \sin^2\theta}$.

In Problems 25-30, change the equations to polar coordinates.

25. $(x^2 + y^2)^2 = 2a^2xy$.   Ans. $r^2 = a^2 \sin 2\theta$.

26. $y^2 = \dfrac{x^3}{2a - x}$.   Ans. $r = 2a \sin \theta \tan \theta$.

27. $(x^2 + y^2)^3 = 4x^2y^2$.   Ans. $r^2 = \sin^2 2\theta$.

28. $x - 3y = 0$.   Ans. $\theta = \arctan 1/3$.

29. $x^4 + x^2y^2 - (x + y)^2 = 0$.   Ans. $r = \pm(1 + \tan \theta)$.

30. $(x^2 + y^2)^3 = 16x^2y^2(x^2 - y^2)^2$.   Ans. $r = \pm \csc 4\theta$.

31. By transforming the two point form of the equation of a straight line to polar coordinates, show that the polar equation of the line through (r_1, θ_1) and (r_2, θ_2) is

$$rr_1 \sin(\theta - \theta_1) + r_1r_2 \sin(\theta_1 - \theta_2) + r_2r \sin(\theta_2 - \theta) = 0.$$

32. Change to polar coordinates and simplify, eliminating the radicals: $\dfrac{(x - 4)^2}{25} + \dfrac{y^2}{9} = 1$.
    Ans. $r = \dfrac{9}{5 - 4 \cos \theta}$ or $r = \dfrac{-9}{5 + 4 \cos \theta}$. Why are these equations identical?

In Problems 33-39, change the equations to rectangular coordinates.

33. $r = 3 \cos \theta$.   Ans. $x^2 + y^2 - 3x = 0$.

34. $r = 1 - \cos \theta$.   Ans. $(x^2 + y^2 + x)^2 = x^2 + y^2$.

35. $r = 2 \cos \theta + 3 \sin \theta$. 　　　*Ans.* $x^2 + y^2 - 2x - 3y = 0$.

36. $\theta = 45°$. 　　　　　　　　*Ans.* $x - y = 0$.

37. $r = \dfrac{3}{2 + 3 \sin \theta}$. 　　　　*Ans.* $4x^2 - 5y^2 + 18y - 9 = 0$.

38. $r = a\theta$. 　　　　　　　　*Ans.* $\sqrt{x^2 + y^2} = a \tan^{-1} \dfrac{y}{x}$.

39. $r^2 = 9 \cos 2\theta$. 　　　　　*Ans.* $(x^2 + y^2)^2 = 9(x^2 - y^2)$.

40. Find the points of intersection of the following pair of curves: $r - 4(1 + \cos \theta) = 0$, $r(1 - \cos \theta) = 3$. 　　*Ans.* $(6, 60°)$, $(2, 120°)$, $(2, 240°)$, $(6, 300°)$.

41. Find the points of intersection of the curves: $r = \sqrt{2} \cos \theta$, $r = \sin 2\theta$.
    Ans. $(1, 45°)$, $(0, 90°)$, $(-1, 135°)$.

42. Find the points of intersection of the curves: $r = 1 + \cos \theta$, $r = \dfrac{1}{2(1 - \cos \theta)}$.
    Ans. $(1 + \dfrac{\sqrt{2}}{2}, \pm 45°)$, $(1 - \dfrac{\sqrt{2}}{2}, \pm 135°)$.

43. Find the points of intersection of the curves: $r = \sqrt{6} \cos \theta$, $r^2 = 9 \cos 2\theta$.
    Ans. $(\dfrac{3\sqrt{2}}{2}, 30°)$, $(-\dfrac{3\sqrt{2}}{2}, 150°)$, $(-\dfrac{3\sqrt{2}}{2}, 210°)$, $(\dfrac{3\sqrt{2}}{2}, 330°)$.

44. Discuss and plot the graph of the curve $r = 4 \sin 2\theta$.

45. Discuss and plot the graph of the curve $r = \dfrac{9}{4 - 5 \cos \theta}$.

46. Sketch the curve $r = 2(1 + \sin \theta)$.

47. Sketch the curve $r^2 = 4 \sin 2\theta$.

48. Sketch the curve $r = 1 + 2 \sin \theta$.

49. Sketch the spiral $r\theta = 4$.

50. Derive the polar equation of the ellipse when the pole is at the center. Hint: Use the law of cosines and the fact that the sum of the focal radii is $2a$.
    Ans. $r^2(1 - e^2 \cos^2\theta) = b^2$.

51. A line of length 20 units has its extremities on two fixed perpendicular lines. Find the locus of the foot of the perpendicular from the point of intersection of the fixed lines to the line of constant length. Let one of the fixed lines be the polar axis.
    Ans. $r = 10 \sin 2\theta$.

52. Find the locus of the vertex of a triangle whose base is a fixed line of length $2b$ and the product of whose other two sides is b^2. Take the polar axis on the base of the triangle and the pole as the midpoint of the base.
    Ans. $r^2 = 2b^2 \cos 2\theta$. This is the lemniscate.

CHAPTER 10

Tangents and Normals

TANGENTS AND NORMALS. The tangent to a curve at a point on the curve is defined as follows.

Let P and Q be two points on the curve as shown. Draw the secant line PQ. Now let the point Q move along the curve toward P. The secant line PQ turns about the point P, and as Q approaches coincidence with P the secant line PQ approaches coincidence with the limiting line PT. The limiting line PT is called the *tangent line* to the curve at the point P.

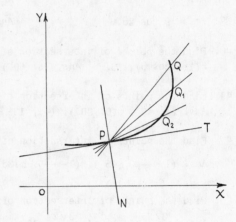

The *normal PN* to the curve is the line perpendicular to the tangent to the curve at the point of contact P.

To find the equation of the tangent to the curve at a point $P_1(x_1, y_1)$ on a curve whose equation is given, we must find the slope of the tangent.

Example: Find the slope of the tangent to the circle $x^2 + y^2 = r^2$ at the point $P_1(x_1, y_1)$ which lies on the circle.

Choose another point $Q(x_1 + h, y_1 + k)$. Draw the secant $P_1 Q$. The slope of the secant is k/h. As the secant turns about P_1 the point Q approaches P_1 and k and h each approach zero. The limit of the ratio k/h as k and h approach zero determines the slope m of the tangent.

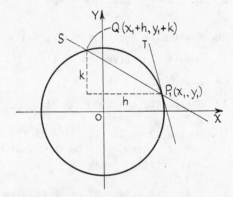

Since (x_1, y_1) and $(x_1 + h, y_1 + k)$ each lie on the circle these coordinates must satisfy the equation of the circle. Substituting, we have

$$(1) \quad x_1^2 + y_1^2 = r^2$$

and $(2) \quad (x_1 + h)^2 + (y_1 + k)^2 = r^2 \quad$ or $\quad x_1^2 + 2hx_1 + h^2 + y_1^2 + 2ky_1 + k^2 = r^2$.

Subtracting (1) from (2), we get $\quad 2hx_1 + h^2 + 2ky_1 + k^2 = 0$
$$\text{or} \quad k(2y_1 + k) = -h(2x_1 + h).$$

Hence, $\dfrac{k}{h} = -\dfrac{2x_1 + h}{2y_1 + k}$. The limit of this expression as h and k each approach zero becomes $-\dfrac{2x_1}{2y_1}$, or $m = -\dfrac{x_1}{y_1}$.

Since the tangent passes through $P_1(x_1, y_1)$, its equation is

$$y - y_1 = -\frac{x_1}{y_1}(x - x_1).$$

Clearing of fractions, $\quad y_1 y - y_1^2 = -x_1 x + x_1^2 \quad$ or
$$x_1 x + y_1 y = x_1^2 + y_1^2 = r^2.$$

The equation of the normal is $y - y_1 = \dfrac{y_1}{x_1}(x - x_1)$

or $x_1 y - y_1 x = x_1 y_1 - x_1 y_1 = 0.$

SOLVED PROBLEMS

1. Find the equations of the tangent and normal to the ellipse $\dfrac{x^2}{a^2} + \dfrac{y^2}{b^2} = 1$ at the point $P_1(x_1, y_1)$.

Let the coordinates of Q be $(x_1 + h,\ y_1 + k)$. Substituting the coordinates of P_1 and Q in the equation, we have

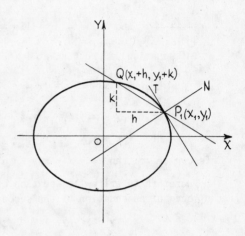

(1) $\dfrac{x_1^2}{a^2} + \dfrac{y_1^2}{b^2} = 1$ and

(2) $\dfrac{(x_1 + h)^2}{a^2} + \dfrac{(y_1 + k)^2}{b^2} = 1.$

Expanding (2) and subtracting (1) from (2),

$2b^2 h x_1 + b^2 h^2 + 2a^2 k y_1 + k^2 a^2 = 0.$

Solving, $\dfrac{k}{h} = -\dfrac{b^2(2x_1 + h)}{a^2(2y_1 + k)}$,   and $\lim \dfrac{k}{h} = -\lim \dfrac{2b^2 x_1 + b^2 h}{2a^2 y_1 + a^2 k} = -\dfrac{b^2 x_1}{a^2 y_1}.$

Using $y - y_1 = m(x - x_1)$, we have $y - y_1 = -\dfrac{b^2 x_1}{a^2 y_1}(x - x_1)$

or $a^2 y_1 y - a^2 y_1^2 = -b^2 x_1 x + b^2 x_1^2.$

Since $b^2 x_1^2 + a^2 y_1^2 = a^2 b^2$, this becomes $b^2 x_1 x + a^2 y_1 y = a^2 b^2$ or $\dfrac{x_1 x}{a^2} + \dfrac{y_1 y}{b^2} = 1$, the equation of the tangent.

The slope of the normal is $\dfrac{a^2 y_1}{b^2 x_1}$, and its equation is $a^2 y_1 x - b^2 x_1 y = (a^2 - b^2) x_1 y_1.$

2. Find the equations of the tangent and normal to the parabola $y^2 = 4ax$ at the point $P_1(x_1, y_1)$.

Substituting the coordinates of $P_1(x_1, y_1)$ and $Q(x_1 + h,\ y_1 + k)$ in the equation of the parabola, we have

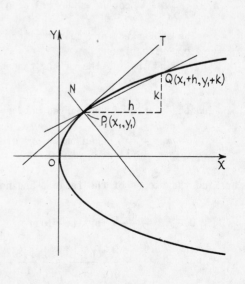

$y_1^2 = 4ax_1$ and $(y_1 + k)^2 = 4a(x_1 + h).$

Expanding and solving for k/h, we find

$\dfrac{k}{h} = \dfrac{4a}{2y_1 + k}$,   and $\lim \dfrac{k}{h} = \lim \dfrac{4a}{2y_1 + k} = \dfrac{2a}{y_1}.$

The equation of the tangent then is

$y - y_1 = \dfrac{2a}{y_1}(x - x_1)$ or $y_1 y - y_1^2 = 2ax - 2ax_1.$

Since $y_1^2 = 4ax_1$, this equation can be written $y_1 y = 2a(x + x_1).$

The slope of the normal is $-\dfrac{y_1}{2a}$ and its equation is $y_1 x + 2ay = x_1 y_1 + 2ay_1$.

3. Find the equation of the tangent to the curve $xy = a^2$ at the point $P_1(x_1, y_1)$ on the curve.

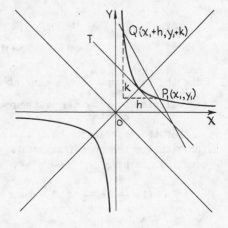

Substituting the coordinates of points $P_1(x_1, y_1)$ and $Q(x_1 + h,\ y_1 + k)$ and solving for k/h, we have

$$\frac{k}{h} = -\frac{y_1 + k}{x_1} \quad \text{and} \quad \lim \frac{k}{h} = -\lim \frac{y_1 + k}{x_1} = -\frac{y_1}{x_1}.$$

The equation of the tangent is therefore

$$y - y_1 = -\frac{y_1}{x_1}(x - x_1)$$

$$\text{or} \quad x_1 y - x_1 y_1 = -y_1 x + x_1 y_1$$

$$\text{or} \quad y_1 x + x_1 y = 2x_1 y_1 = 2a^2.$$

This can be written $\frac{1}{2}(y_1 x + x_1 y) = a^2$.

Therefore to write the equation of the tangent at the point of contact $P_1(x_1, y_1)$ to the locus of an equation of the second degree, replace

$$x^2 \text{ by } x_1 x, \quad y^2 \text{ by } y_1 y, \quad xy \text{ by } \tfrac{1}{2}(y_1 x + x_1 y), \quad x \text{ by } \tfrac{1}{2}(x + x_1) \text{ and } y \text{ by } \tfrac{1}{2}(y + y_1).$$

4. Let $P_1 T$ be the length of the tangent and $P_1 N$ the length of the normal to the curve at P_1. The projections ST and SN are called respectively the subtangent and the subnormal at P_1.

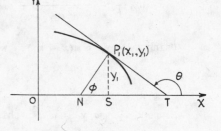

If m is the slope of the tangent at $P_1(x_1, y_1)$,

then $\qquad -\dfrac{y_1}{m} = $ length of subtangent,

and $\qquad y_1 m = $ length of subnormal.

This is seen to be true since $\dfrac{ST}{y_1} = -\cot\theta = -\dfrac{1}{m}$ and $ST = -\dfrac{y_1}{m}$.

Also, $\dfrac{SN}{y_1} = -\cot\phi = -\cot(\theta - 90°) = \tan\theta = m$ and $SN = my_1$.

The subtangent and subnormal are measured in opposite directions; therefore they will have opposite signs.

To find the lengths of the tangent and normal apply the right triangle theorem.

5. Find the slopes of the tangent and normal to the circle $x^2 + y^2 = 5$ at the point $(2, 1)$.

Use $m = -\dfrac{x_1}{y_1}$. Then the slope of the tangent is $-\dfrac{2}{1}$ and the slope of the normal is $\dfrac{1}{2}$.

6. Find the slopes of the tangent and normal to the ellipse $\dfrac{x^2}{9} + \dfrac{y^2}{16} = 1$ at point $\left(2, \dfrac{4\sqrt{5}}{3}\right)$.

Use $m = -\dfrac{a^2 x_1}{b^2 y_1}$ for the slope of the tangent. Substituting the coordinates of the given point, $m = -\dfrac{16(2)}{9(4\sqrt{5}/3)} = -\dfrac{8\sqrt{5}}{15}$, and the slope of the normal is $\dfrac{3\sqrt{5}}{8}$.

7. Show that the slope of the tangent to the curve $4x^2 + 4xy + y^2 - 9 = 0$ at any point on the curve is $m = -2$.

   Use the two points $P_1(x_1, y_1)$ and $Q(x_1 + h, \ y_1 + k)$ and determine the $\lim \dfrac{k}{h}$.

   Substituting,   (1) $4(x_1 + h)^2 + 4(x_1 + h)(y_1 + k) + (y_1 + k)^2 - 9 = 0$ and

   (2) $4x_1^2 + 4x_1 y_1 + y_1^2 - 9 = 0$.

   Expanding (1) and subtracting (2) from the expansion, we have

   $$\lim \frac{k}{h} = -\frac{8x_1 + 4y_1}{4x_1 + 2y_1} = -2.$$

   Another Method. The original equation can be written $(2x + y)^2 - 9 = 0$.

   Factoring, $(2x + y + 3)(2x + y - 3) = 0$,   two parallel lines of slope -2.

8. Find the slope of the tangent to the hyperbola $9x^2 - 4y^2 = 36$ at the point $(3, \dfrac{3\sqrt{5}}{2})$.

   Use the two points $P_1(x_1, y_1)$ and $Q(x_1 + h, \ y_1 + k)$ and determine the $\lim \dfrac{k}{h}$.

   Substituting,   (1) $9(x_1 + h)^2 - 4(y_1 + k)^2 = 36$ and   (2) $9x_1^2 - 4y_1^2 = 36$.

   Expanding and solving for $\dfrac{k}{h}$, we have $\dfrac{4k}{9h} = \dfrac{2x_1 + h}{2y_1 + k}$ and $\lim \dfrac{k}{h} = \dfrac{9x_1}{4y_1} = m$.

   The slope at $(3, \dfrac{3\sqrt{5}}{2})$ is $m = \dfrac{27}{6\sqrt{5}} = \dfrac{9\sqrt{5}}{10}$.

9. Find the slopes of the tangent and normal to the curve $y^2 = 2x^3$ at the point $(2, 4)$.

   Use the points on the curve $P_1(x_1, y_1)$ and $Q(x_1 + h, \ y_1 + k)$.

   Substituting,   (1) $(y_1 + k)^2 = 2(x_1 + h)^3$ or $y_1^2 + 2ky_1 + k^2 = 2x_1^3 + 6x_1^2 h + 6x_1 h^2 + 2h^3$

   and   (2) $y_1^2 = 2x_1^3$.

   Subtracting (2) from expanded (1) gives $2ky_1 + k^2 = 6x_1^2 h + 6x_1 h^2 + 2h^3$.

   Then $\dfrac{k}{h} = \dfrac{6x_1^2 + 6x_1 h + 2h^2}{2y_1 + k}$ and $\lim \dfrac{k}{h} = \dfrac{6x_1^2}{2y_1} = \dfrac{3x_1^2}{y_1}$.

   At point $(2, 4)$, $m = \lim \dfrac{k}{h} = \dfrac{12}{4} = 3$.   The slope of the normal is $-\dfrac{1}{3}$.

10. Find the equations of the tangent and normal to the curve $y^2 = 2x^3$ at the point $(2, 4)$.

    In Problem 9, the slope of this curve at point $(2, 4)$ was found to be 3.
    Hence, the equation of the tangent is $y - 4 = 3(x - 2)$ or $y = 3x - 2$.

    The equation of the normal is $y - 4 = -\dfrac{1}{3}(x - 2)$ or $x + 3y = 14$.

11. Write the equations of the tangent and normal to the curve $x^2 + 3xy - 4y^2 + 2x - y + 1 = 0$ at the point $(2, -1)$.

    Using the rule which follows Problem 3, we have

    $$x_1 x + 3(\frac{x_1 y + y_1 x}{2}) - 4y_1 y + 2(\frac{x + x_1}{2}) - (\frac{y + y_1}{2}) + 1 = 0.$$

Substituting $x_1 = 2$, $y_1 = -1$, we have $3x + 13y + 7 = 0$, the equation of the tangent with slope $-3/13$.

The equation of the normal is $y + 1 = \dfrac{13}{3}(x - 2)$ or $13x - 3y - 29 = 0$.

12. Find the equations of the lines of slope m which are tangent to the ellipse

$$(1)\quad b^2 x^2 + a^2 y^2 = a^2 b^2 .$$

The equations will have the form, $(2)\ y = mx + k$.

Solving (1) and (2) gives $b^2 x^2 + a^2 (mx + k)^2 = a^2 b^2$.

Expanding and collecting terms, $(3)\ (b^2 + a^2 m^2) x^2 + 2a^2 mkx + a^2 k^2 - a^2 b^2 = 0$.

When the lines are tangent to the curve, the roots of (3) must be equal, that is, the discriminant must be equal to zero. Hence,

$4a^4 m^2 k^2 - 4(b^2 + a^2 m^2)(a^2 k^2 - a^2 b^2) = 0$ or $k^2 = a^2 m^2 + b^2$, and $k = \pm \sqrt{a^2 m^2 + b^2}$.

The equations of the lines with slope m and tangent to the ellipse are

$$y = mx \pm \sqrt{a^2 m^2 + b^2} .$$

13. Write the equations of the lines which are tangent to the ellipse $x^2 + 4y^2 = 100$ and parallel to the line $3x + 8y = 7$.

The slope of the given line is $-3/8$. Therefore the required equations of the tangents are of the form $y = -\dfrac{3}{8}x + k$, where k is to be determined.

By solving this equation and the equation of the ellipse simultaneously and imposing the condition for equal roots, we can determine k. Thus,

$$x^2 + 4(-\tfrac{3}{8}x + k)^2 - 100 = 0 \quad\text{or}\quad 25x^2 - 48kx + (64k^2 - 1600) = 0.$$

For equal roots the discriminant must be zero, or $(-48k)^2 - 4(25)(64k^2 - 1600) = 0$.

Solving, $16k^2 = 625$, $k = \pm \dfrac{25}{4}$. Hence the required equations are

$$y = -\tfrac{3}{8}x \pm \tfrac{25}{4} \quad\text{or}\quad 3x + 8y \pm 50 = 0.$$

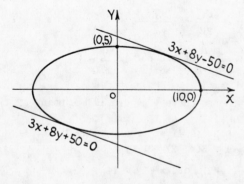

Problem 13.

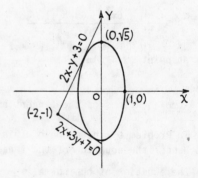

Problem 14.

14. Find the equations of the lines through the point $(-2,-1)$ which are tangent to the ellipse $5x^2 + y^2 = 5$.

Let $P_1(x_1, y_1)$ be a point of contact. The equation of the tangent is of the form $5x_1 x + y_1 y = 5$ and the point $(-2,-1)$ lies on this tangent; hence, $-10x_1 - y_1 = 5$. Also, the point (x_1, y_1) is on the ellipse; hence, $5x_1^2 + y_1^2 = 5$.

Solving these equations simultaneously to find (x_1, y_1), we have two points of contact, $(-\frac{2}{3}, \frac{5}{3})$ and $(-\frac{2}{7}, -\frac{15}{7})$. Substituting these values in $5x_1x + y_1y = 5$, we get $2x - y + 3 = 0$ and $2x + 3y + 7 = 0$.

15. Find the lengths of the subtangent, subnormal, tangent and normal for the point $(-1,3)$ on the ellipse
$$9x^2 + y^2 = 18.$$

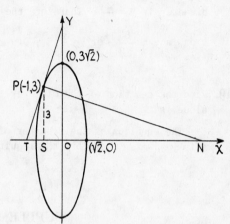

For the tangent use the form
$$9x_1x + y_1y = 18.$$

Substituting the coordinates of the point, $-9x + 3y = 18$ or $3x - y + 6 = 0$. Hence $m = 3$.

Subtangent $ST = -y_1/m = -3/3 = -1$.
Subnormal $SN = my_1 = 3(3) = 9$.

Length of tangent, $PT = \sqrt{3^2 + 1^2} = \sqrt{10}$.

Length of normal, $PN = \sqrt{9^2 + 3^2} = 3\sqrt{10}$.

DEFINITION. The locus of the middle points of a system of parallel chords of any conic is called a *diameter* of the conic.

If the slope of the parallel chords is m, the equation of the diameter determined by the midpoints of these chords is:

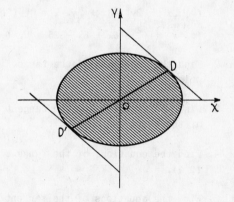

For the ellipse $\dfrac{x^2}{a^2} + \dfrac{y^2}{b^2} = 1$, $\quad y = -\dfrac{b^2x}{a^2m}$.

For the parabola $y^2 = 4ax$, $\quad y = \dfrac{2a}{m}$.

For the hyperbola $\dfrac{x^2}{a^2} - \dfrac{y^2}{b^2} = 1$, $\quad y = \dfrac{b^2x}{a^2m}$.

For the hyperbola $xy = a^2$, $\quad y = -mx$.

For the general case of the conic $ax^2 + 2hxy + by^2 + 2gx + 2fy + c = 0$, the equation of the diameter has the form $(ax + hy + g) + m(hx + by + f) = 0$.

16. Find the equation of that diameter of the ellipse $\dfrac{x^2}{9} + \dfrac{y^2}{4} = 1$ which bisects all chords of slope $\dfrac{1}{3}$.

Using $y = -\dfrac{b^2x}{a^2m}$, the equation of the diameter is $y = -\dfrac{4x}{9(1/3)}$ or $4x + 3y = 0$.

17. Find the equation of the diameter of the conic $3x^2 - xy - y^2 - x - y = 5$ which bisects the chords of slope 4.

Using $(ax + hy + g) + m(hx + by + f) = 0$, where $a = 3$, $h = -\frac{1}{2}$, $b = -1$, $g = -\frac{1}{2}$, $f = -\frac{1}{2}$ and $c = -5$, we get $3x - \frac{1}{2}y - \frac{1}{2} + 4(-\frac{1}{2}x - y - \frac{1}{2}) = 0$ or $2x - 9y - 5 = 0$.

18. Find the equation of the diameter of the parabola $y^2 = 16x$ which bisects chords parallel to the line $2x - 3y = 5$.

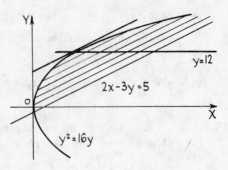

     The slope of the line $2x - 3y - 5 = 0$ is $\frac{2}{3}$.

     For the parabola $y^2 = 4ax$ the equation of the diameter is $y = \frac{2a}{m}$. Therefore the required equation is $y = \frac{8}{2/3}$ or $y - 12 = 0$.

19. Find the equation of the diameter of the hyperbola $xy = 16$ which bisects chords whose slope is 2.

     The equation of the diameter of the hyperbola $xy = a^2$ which bisects chords of slope m is $y = -mx$. Therefore the required equation is $y = -2x$.

SUPPLEMENTARY PROBLEMS

1. Write the equations of the tangent and normal to each of the following circles at the points given:

   (a) $x^2 + y^2 = 25$, $(3,4)$.   *Ans.* $3x + 4y = 25$; $4x - 3y = 0$

   (b) $2x^2 + 2y^2 - 3x + 5y - 2 = 0$, $(2,0)$.   *Ans.* $x + y - 2 = 0$; $x - y - 2 = 0$

   (c) $x^2 + y^2 - 6x + 8y - 25 = 0$, $(-2,1)$.   *Ans.* $x - y + 3 = 0$; $x + y + 1 = 0$

2. Find the equations of the tangent and normal to the ellipse $2x^2 + 3y^2 - 30 = 0$ at the point $(-3,2)$.   *Ans.* $x - y + 5 = 0$; $x + y + 1 = 0$

3. Find the equations of the tangent and normal to the ellipse $3x^2 + 4y^2 - 6x + 8y - 45 = 0$ at the point $(-3,-2)$.   *Ans.* $3x + y + 11 = 0$; $x - 3y - 3 = 0$

4. Find the equations of the tangent and normal to the parabola $x^2 - 4y = 0$ at the point $(2,1)$.   *Ans.* $x - y - 1 = 0$; $x + y - 3 = 0$

5. Find the equations of the tangent and normal to each of the following hyperbolas at the points given:

   (a) $6x^2 - 9y^2 - 8x + 3y + 16 = 0$, $(-1,2)$.   *Ans.* $20x + 33y - 46 = 0$; $33x - 20y + 73 = 0$

   (b) $x^2 - 2xy - y^2 - 2x + 4y + 4 = 0$, $(2,-2)$.   *Ans.* $3x + 2y - 2 = 0$; $2x - 3y - 10 = 0$

   (c) $xy - 4 = 0$, $(2,2)$.   *Ans.* $x + y - 4 = 0$; $x - y = 0$

6. Find the equations of the tangents to the hyperbola $5x^2 - 4y^2 = 4$ at the points where it is cut by the line $5x - 2y - 4 = 0$.   *Ans.* $5x - 2y - 4 = 0$

7. Find the equations of the tangents to the hyperbola $x^2 - 4y^2 - 12 = 0$ which pass through the point $(1,4)$.   *Ans.* $x - y + 3 = 0$; $19x + 11y - 63 = 0$

8. Find the points on the hyperbola $x^2 - 2y^2 - 8 = 0$ at which the tangents are perpendicular to the line $4x + 5y - 2 = 0$.   *Ans.* $(\frac{10\sqrt{34}}{17}, \frac{4\sqrt{34}}{17})$, $(\frac{-10\sqrt{34}}{17}, \frac{-4\sqrt{34}}{17})$

9. Find the slope of the curve $y^2 = x^3 + 2x^2$ at (x_1, y_1). Ans. $\lim \dfrac{k}{h} = \dfrac{3x_1^2 + 4x_1}{2y_1}$.

10. Find the equations of the tangent and normal to the curve in the preceding problem at the point $(2, -4)$. Ans. $5x + 2y - 2 = 0$; $2x - 5y - 24 = 0$

11. (a) Find the lengths of the subtangent and subnormal at the point $(2, -4)$ on the curve $y^2 = x^3 + 2x^2$. Ans. $-8/5$, 10

    (b) Find the lengths of the tangent and normal. Ans. $\dfrac{4\sqrt{29}}{5}$, $2\sqrt{29}$

12. Find the equation of the tangents to the hyperbola $2xy + y^2 - 8 = 0$ which have the slope $m = -2/3$. Ans. $2x + 3y - 8 = 0$; $2x + 3y + 8 = 0$

13. Find the equations of the tangent and normal, and the lengths of the subtangent and subnormal to the curve $y^2 - 6y - 8x - 31 = 0$ at the point $(-3, -1)$.
    Ans. $x + y + 4 = 0$; $x - y + 2 = 0$; $-1, 1$

14. Find the slope of the tangent to the curve $4x^2 - 12xy + 9y^2 - 2x + 3y - 6 = 0$ at any point (x_1, y_1) on the curve. Ans. $m = 2/3$. Interpret your answer.

15. Find the equations of the tangents to the curve $4x^2 - 2y^2 - 3xy + 2x - 3y - 10 = 0$ which are parallel to the line $x - y + 5 = 0$. Ans. $x - y - 1 = 0$; $41x - 41y + 39 = 0$

16. Find the equations of the tangents to the hyperbola $xy = 2$ which are perpendicular to the line $x - 2y = 7$. Ans. $2x + y - 4 = 0$; $2x + y + 4 = 0$

17. At what points on the ellipse $x^2 + xy + y^2 - 3 = 0$ are the tangents parallel to the x-axis? parallel to the y-axis? Ans. $(1, -2)$, $(-1, 2)$; $(2, -1)$, $(-2, 1)$

18. At what points on the curve $x^2 - 2xy + y + 1 = 0$ are the tangents parallel to the line $2x + y = 5$. Ans. $(1, 2)$ and $(0, -1)$

19. Find the equations of the lines through the point $(5, 6)$ which are tangent to the parabola $y^2 = 4x$. Ans. $x - y + 1 = 0$; $x - 5y + 25 = 0$

20. Show that the tangents to the parabola $y^2 = 4ax$ at the ends of the latus rectum are perpendicular, that is, show that the slopes are ± 1.

21. Find the equations of the tangent and normal to the parabola $x^2 = 5y$ at the point whose abscissa is 3. Ans. $6x - 5y - 9 = 0$; $25x + 30y - 129 = 0$

22. Show that the equations of tangents of slope m for the parabola $y^2 = 4ax$ are
$$y = mx + \frac{a}{m}, \quad (m \neq 0).$$

23. Show that the equations of tangents of slope m for the circle $x^2 + y^2 = a^2$ are
$$y = mx \pm a\sqrt{m^2 + 1}.$$

24. Show that the equations of tangents of slope m for the hyperbola $b^2x^2 - a^2y^2 = a^2b^2$ are
$y = mx \pm \sqrt{a^2m^2 - b^2}$, and for $b^2x^2 - a^2y^2 = -a^2b^2$ are $y = mx \pm \sqrt{b^2 - a^2m^2}$.

25. Find the equations of the tangents to the ellipse $5x^2 + 7y^2 = 35$ which are perpendicular to the line $3x + 4y - 12 = 0$. Ans. $3y = 4x \pm \sqrt{157}$

26. Find the equations of the tangents to the hyperbola $16x^2 - 9y^2 = 144$ which are parallel to the line $4x - y - 14 = 0$. Ans. $y = 4x \pm 8\sqrt{2}$

27. The parabola $y^2 = 4ax$ passes through the point $(-8,4)$. Find the equation of the tangent to this parabola which is parallel to the line $3x + 2y - 6 = 0$.   *Ans.* $9x + 6y = 2$

28. Find the equation of the tangent to the curve $x^3 + y^3 = 3axy$ at the point $P_1(x_1, y_1)$.
    Ans. $(y_1^2 - ax_1)y + (x_1^2 - ay_1)x = ax_1y_1$

29. Find the value of b if the line $y = mx + b$ is tangent to the parabola $x^2 = 4ay$.
    Ans. $b = -am^2$

30. Using the result in Problem 29, find the equation of the tangent to the parabola $x^2 = -2y$ which is parallel to the line $x - 2y - 4 = 0$.   *Ans.* $4x - 8y + 1 = 0$

31. Find the equation of the diameter of the hyperbola $x^2 - 4y^2 = 9$ which bisects chords:
    (a) of slope 4.   *Ans.* $x - 16y = 0$
    (b) parallel to $3x - 5y - 2 = 0$.   *Ans.* $5x - 12y = 0$
    (c) parallel to the tangent at $(5,2)$.   *Ans.* $2x - 5y = 0$
    (d) parallel to the asymptote with a positive slope. *Ans.* $x - 2y = 0$

32. Find the equation of the diameter conjugate to the diameter $x - 16y = 0$ in Problem 31(a).
    Ans. $4x - y = 0$

33. Find the equation of the diameter of the ellipse $9x^2 + 25y^2 = 225$ bisecting chords having slope 3.   *Ans.* $3x + 25y = 0$

34. Find the equation of the diameter of the parabola $y^2 = 8x$ bisecting chords having slope 2/3.   *Ans.* $y = 6$

35. Find the equation of the diameter of the ellipse $x^2 + 4y^2 = 4$ conjugate to the diameter $y = 3x$.   *Ans.* $x + 12y = 0$

36. Find the equation of the diameter of the conic $xy + 2y^2 - 4x - 2y + 6 = 0$ bisecting chords which have slope 2/3.   *Ans.* $2x + 11y = 16$

37. Find the diameter of the conic $x^2 - 3xy - 2y^2 - x - 2y - 1 = 0$ bisecting chords of slope 3.   *Ans.* $7x + 15y + 7 = 0$

38. Find the equation of the diameter of the ellipse $4x^2 + 5y^2 = 20$ which bisects chords:
    (a) of slope $-2/3$.   *Ans.* $6x - 5y = 0$
    (b) parallel to the line $3x - 5y = 6$.   *Ans.* $4x + 3y = 0$

39. Find the equation of the diameter of the hyperbola $xy = 16$ which bisects chords parallel to the line $x + y = 1$.   *Ans.* $y = x$

CHAPTER 11

Higher Plane Curves

HIGHER PLANE CURVES. An *algebraic curve* is one which can be represented by a polynomial in x and y set equal to zero. Curves whose equations cannot be so represented, as for example $y = \sin x$, $y = e^x$, $y = \log x$, are called *transcendental curves*.

Algebraic curves of higher degree than the second and transcendental curves are called *higher plane curves*.

See Chapter 2 for discussion of intercepts, symmetry, and extent of a curve.

SOLVED PROBLEMS

1. Plot the curve $y^2 = (x-1)(x-3)(x-4)$.

This curve is symmetric with respect to the x-axis, since the equation remains unchanged when y is replaced by $-y$.

Its x-intercepts are 1, 3, 4. When $x = 0$, $y^2 = -12$; hence the curve does not intersect the y-axis.

For $x < 1$ each factor in the right member is negative and y is imaginary.

For $1 \leqq x \leqq 3$, y is real. For $3 < x < 4$, y^2 is negative and hence y is imaginary.

For $x \geqq 4$, y^2 is positive and y is real and increases numerically without limit as x increases without limit.

Form a table of values to determine points on the curve. $\quad y = \pm\sqrt{(x-1)(x-3)(x-4)}$.

x	1	1.5	2	2.5	3	4	4.5	5	5.5	6
y	0	±1.37	±1.41	±1.06	0	0	±1.62	±2.83	±4.11	±5.48

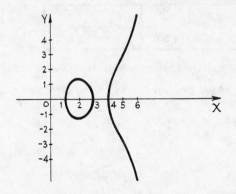

Problem 1.

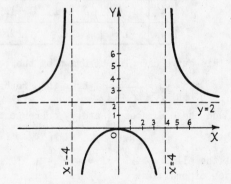

Problem 2.

2. Plot the curve $x^2y - 2x^2 - 16y = 0$.

Intercepts. If $y = 0$, $x = 0$; and if $x = 0$, $y = 0$.

Symmetry. The curve is symmetric about the y-axis, since the equation remains unchanged when $-x$ is substituted for x. It is not symmetric about the x-axis or the origin.

93

Solving for y and x, we find (1) $y = \dfrac{2x^2}{x^2 - 16} = \dfrac{2x^2}{(x - 4)(x + 4)}$

and (2) $x = \pm 4 \sqrt{\dfrac{y}{y - 2}}$.

From (1) we see that y becomes infinite as x approaches the values 4 and –4 through values both greater and smaller than 4 and –4. No values of x are excluded.

From (2), values of $0 < y < 2$ must be excluded. As y approaches 2 through values greater than 2, x becomes infinite.

The lines $x = \pm 4$ and $y = 2$ are asymptotes.

x	0	± 1	± 2	± 3	± 4	± 5	± 6	± 7	± 8	$\pm \infty$
y	0	–.13	–.67	–2.6	$\pm \infty$	5.6	3.6	3.0	2.7	2

3. Sketch the curve $x^3 - x^2 y + y = 0$.

Solving for y, $y = \dfrac{x^3}{x^2 - 1}$.

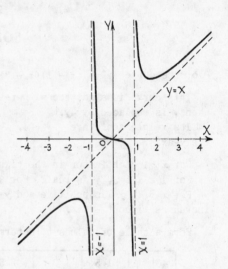

For $x = \pm 1$, y becomes infinite; hence $x = 1$ and $x = -1$ are vertical asymptotes.

Express $y = \dfrac{x^3}{x^2 - 1}$ as $y = x + \dfrac{x}{x^2 - 1}$. As x increases without limit, y increases without limit and the fraction $\dfrac{x}{x^2 - 1}$ approaches zero. Hence the curve approaches the line $y = x$ as an asymptote. For $x > 1$ one branch of the curve lies above the line $y = x$; for $x < -1$ another branch lies below $y = x$.

The curve passes through the origin and is symmetric about the origin. A few values of x and y are shown in the table below.

x	$\pm 1/2$	0	± 1	± 1.5	± 2	± 2.5	± 3	± 4
y	$\mp 1/6$	0	∞	± 2.7	± 2.67	± 3.0	± 3.4	± 4.3

This curve can also be drawn by the method of *addition of ordinates*. To do this, let $y_1 = x$ and $y_2 = \dfrac{x}{x^2 - 1}$. Draw the graphs of these two equations on the same set of axes and add the ordinates y_1 and y_2 corresponding to the same abscissas.

4. Plot the ellipse $2x^2 + 2xy + y^2 - 1 = 0$ by the addition of ordinates.

Solving for y, $y = \dfrac{-2x \pm \sqrt{4x^2 - 8x^2 + 4}}{2}$

$= -x \pm \sqrt{1 - x^2}$.

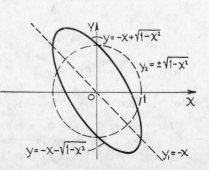

Draw the line $y_1 = -x$ and the circle $y_2 = \pm \sqrt{1 - x^2}$ or $x^2 + y_2^2 = 1$. The resulting ellipse is symmetric with respect to the origin.

5. *The Trigonometric Function.*   Plot the graph of $y = \sin x$.

Express the angle x in terms of radians. (π radians = 180°.)

x	0	$\pm\dfrac{\pi}{6}$	$\pm\dfrac{\pi}{3}$	$\pm\dfrac{\pi}{2}$	$\pm\dfrac{2\pi}{3}$	$\pm\dfrac{5\pi}{6}$	$\pm\pi$	$\pm\dfrac{7\pi}{6}$	$\pm\dfrac{4\pi}{3}$	$\pm\dfrac{3\pi}{2}$	$\pm\dfrac{5\pi}{3}$	$\pm\dfrac{11\pi}{6}$	$\pm2\pi$
$\sin x$	0	$\pm.5$	$\pm.87$	±1	$\pm.87$	$\pm.5$	0	$\mp.5$	$\mp.87$	∓1	$\mp.87$	$\mp.5$	0

Since the values of $\sin x$ repeat themselves, $\sin x$ is a periodic function with period 2π; therefore the graph of $y = \sin x$ is exactly similar in any interval of 2π radians. Since $\sin(-x) = -\sin x$, the graph is symmetric with respect to the origin. No value of x is excluded, but the curve lies between $y = 1$ and $y = -1$.

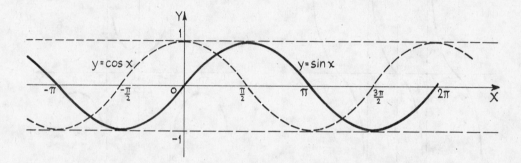

The graph of $y = \cos x$ may be obtained as in the case of $y = \sin x$ by plotting points. See the dash line graph in the figure above.

x	0	$\pm\dfrac{\pi}{6}$	$\pm\dfrac{\pi}{3}$	$\pm\dfrac{\pi}{2}$	$\pm\dfrac{2\pi}{3}$	$\pm\dfrac{5\pi}{6}$	$\pm\pi$	$\pm\dfrac{7\pi}{6}$	$\pm\dfrac{4\pi}{3}$	$\pm\dfrac{3\pi}{2}$	$\pm\dfrac{5\pi}{3}$	$\pm2\pi$
$\cos x$	1	.87	.5	0	$-.5$	$-.87$	-1	$-.87$	$-.5$	0	.5	1

Since $\cos x = \sin(x + \pi/2)$, any point on the cosine curve has the same ordinate as any point on the sine curve $\pi/2$ units farther to the right. Since $\cos(-x) = \cos x$, the graph is symmetric about the vertical axis.

6. Plot the graph of $y = \sin 3x$.

Since $\sin x$ goes through its entire set of values while x varies from 0 to 2π, the function $\sin nx$ (where n is any constant) goes through its entire set of values while nx varies from 0 to 2π or while x varies from 0 to $2\pi/n$.

Here $n = 3$; hence the period of $\sin 3x$ is $2\pi/3$.

The curve is symmetric about the origin.

The range of y is $-1 \leqq y \leqq 1$, and no value of x is excluded.

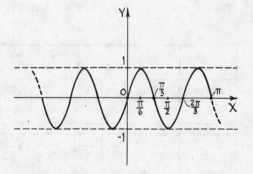

x	0	$\dfrac{\pi}{6}$	$\dfrac{\pi}{3}$	$\dfrac{\pi}{2}$	$\dfrac{2\pi}{3}$	$\dfrac{5\pi}{6}$	π
$\sin 3x$	0	1	0	-1	0	1	0

7. Plot the graph of $y = \tan x$.

Since $\tan(-x) = -\tan x$, the curve is symmetric about the origin.
The period of the function is π.
The value of the function is infinite when x is an odd multiple of $\pi/2$, and the curve passes through all values of y between $x = -\pi/2$ and $\pi/2$. No value of x or y is excluded.

x	$-\dfrac{\pi}{2}$	$-\dfrac{\pi}{4}$	$-\dfrac{\pi}{6}$	0	$\dfrac{\pi}{6}$	$\dfrac{\pi}{4}$	$\dfrac{\pi}{3}$	$\dfrac{\pi}{2}$
$\tan x$	∞	-1	$-.58$	0	$.58$	1	1.73	∞

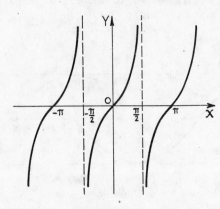

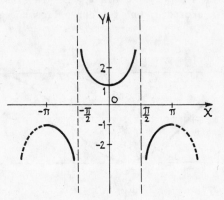

Problem 7.   *Problem 8.*

8. Plot the graph of $y = \sec x$.

Since $\sec(-x) = \sec x$, the curve is symmetric about the y-axis.
The period of the function is 2π.
Since $\sec x = 1/\cos x$, the values of $\sec x$ can easily be found from a table for $\cos x$.
Since the range of $\cos x$ is -1 to $+1$, the range of $\sec x$ is all values from $-\infty$ to -1 and from 1 to $+\infty$.

x	0	$\pm\dfrac{\pi}{3}$	$\pm\dfrac{\pi}{2}$	$\pm\dfrac{2\pi}{3}$	$\pm\pi$
$\sec x$	1	2	∞	-2	-1

9. Plot the graph of $y = \sin x + \sin 3x$. Use the method of addition of ordinates.

x	0	$\dfrac{\pi}{6}$	$\dfrac{\pi}{3}$	$\dfrac{\pi}{2}$	$\dfrac{2\pi}{3}$	$\dfrac{5\pi}{6}$	π
$\sin x$	0	$.5$	$.87$	1	$.87$	$.5$	0
$\sin 3x$	0	1	0	-1	0	1	0

x	$\dfrac{7\pi}{6}$	$\dfrac{4\pi}{3}$	$\dfrac{3\pi}{2}$	$\dfrac{5\pi}{3}$	$\dfrac{11\pi}{6}$	2π
$\sin x$	$-.5$	$-.87$	-1	$-.87$	$-.5$	0
$\sin 3x$	-1	0	1	0	-1	0

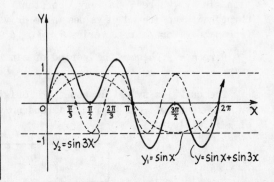

10. *The Exponential Function.*   Plot the graph of $y = a^x$, where a is any positive constant greater than unity.

For definiteness, let $a = 5$.   Then the equation becomes $y = 5^x$.

For $x = 0$, $y = 5^0 = 1$.   As x increases in value, y increases. For negative values of x, 5^x is positive but decreasing in value. Hence the curve lies entirely above the x-axis.

The curve is not symmetrical with respect to either axis or the origin. For negative values of x, as x increases numerically the curve approaches the negative x-axis as an asymptote.

x	0	1	2	−1	−2	−3	−4
y	1	5	25	.2	.04	.008	.0016

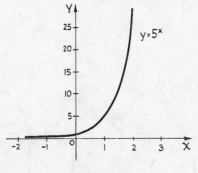

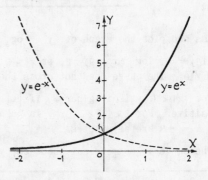

Problem 10.   *Problem 11.*

11. Plot the graph of $y = e^x$.

The quantity $e = 2.718$ is the base of the natural or Naperian logarithms.

x	−3	−2	−1	−.5	0	.5	1	2
e^x	.050	.135	.368	.606	1	1.65	2.72	7.39

The graph of $y = e^{-x}$ shown in the figure above may be described as the reflection in the y-axis of the curve $y = e^x$.

12. Discuss the equation and sketch the graph of $y = e^{-x^2}$, the probability curve.

The intercept on the y-axis is unity. There is no x-intercept.

The curve is symmetrical with respect to the y-axis. The x-axis is an asymptote; as $x \to \pm\infty$, $y \to 0$.

The curve lies entirely above the x-axis, since $e^{-x^2} > 0$ for all values of x.

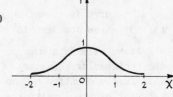

x	0	±.5	±1	±1.5	±2
y	1	.78	.37	.11	.02

13. *The Logarithmic Function.*

The locus of the equation $y = \log_a x$, called the logarithmic curve, differs from the locus of $y = a^x$ only in its relation to the axes. In fact both equations can be written in exponential form or logarithmic form. Let $a = 10$ and plot the graph of

$$y = \log_{10} x, \quad (\text{or } x = 10^y).$$

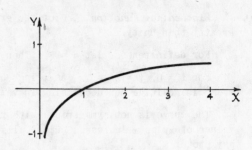

Since x cannot have negative values, the curve will be entirely to the right of the y-axis. For positive values of $x < 1$, y will be negative. For $x = 1$, $y = 0$. As x increases in value y will increase. There is no symmetry. The negative y-axis is an asymptote.

x	.1	.5	1	2	3	4	5	10
y	-1	$-.30$	0	.30	.48	.60	.70	1

14. Discuss and plot the graph of $y = \log_e (x^2 - 9)$.

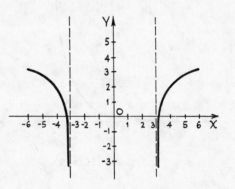

If $y = 0$, $\log_e (x^2 - 9) = 0$, or $x^2 - 9 = 1$, and $x = \pm\sqrt{10}$. The curve does not cross the y-axis.

For $|x| < 3$, y is imaginary. If $|x| > \sqrt{10}$, y is positive. For $3 < |x| < \sqrt{10}$, y is negative. The lines $x = \pm 3$ are asymptotes.

The curve is symmetrical about the y-axis.

x	± 3.1	± 3.2	± 3.5	± 4	± 5	± 6
y	$-.49$	0.22	1.18	1.95	2.77	3.29

15. *Parametric Equations.* It is sometimes advantageous to express x and y each in terms of a third variable. The third variable is called a parameter. The two equations expressing x and y in terms of the parameter are called parametric equations. By giving successive values to the parameter, successive pairs of values of x and y are determined. These points when plotted are joined by a smooth curve which represents the locus of the parametric equations.

Plot the curve $x = 2t$, $y = \dfrac{2}{t}$.

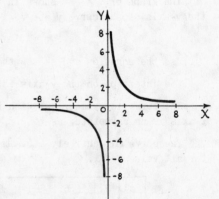

t	$\pm 1/4$	$\pm 1/2$	± 1	± 2	± 3	± 4
x	$\pm 1/2$	± 1	± 2	± 4	± 6	± 8
y	± 8	± 4	± 2	± 1	$\pm 2/3$	$\pm 1/2$

The curve is symmetric about the origin. The x- and y-axes are asymptotes.

If we eliminate the parameter t, we obtain the rectangular equation of the curve, $xy = 4$. This is the equation of an equilateral or rectangular hyperbola with the coordinate axes as asymptotes.

To eliminate the parameter t, substitute $t = \dfrac{x}{2}$ in $y = \dfrac{2}{t}$ to get $y = \dfrac{2}{x/2}$ or $xy = 4$.

16. Plot the curve whose parametric equations are $x = \frac{1}{2}t^2$, $y = \frac{1}{4}t^3$.

t	-3	-2	-1	0	1	2	3
x	4.5	2	0.5	0	0.5	2	4.5
y	-6.75	-2	-0.25	0	0.25	2	6.75

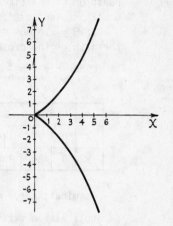

Eliminating t, the rectangular equation of the curve is $2y^2 = x^3$, the semicubical parabola. The curve is symmetric about the x-axis.

To eliminate the parameter t:
From $x = \frac{1}{2}t^2$ or $2x = t^2$, we get $(2x)^3 = (t^2)^3$.

From $y = \frac{1}{4}t^3$ or $4y = t^3$, we get $(4y)^2 = (t^3)^2$.

Then $(2x)^3 = t^6 = (4y)^2$, or $x^3 = 2y^2$.

17. Plot the curve whose parametric equations are $x = t + 1$, $y = t(t + 4)$.

t	-5	-4	-3	-2	-1	0	1	2
x	-4	-3	-2	-1	0	1	2	3
y	5	0	-3	-4	-3	0	5	12

When the parameter t is eliminated the rectangular equation obtained is $y = x^2 + 2x - 3$, a parabola.

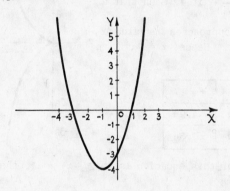

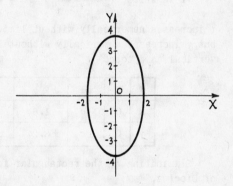

Problem 17.   Problem 18.

18. Plot the curve whose parametric equations are $x = 2 \cos \theta$, $y = 4 \sin \theta$.

θ	$0°$	$30°$	$60°$	$90°$	$120°$	$150°$	$180°$	$210°$	$240°$	$270°$	$300°$	$330°$	$360°$
x	2	1.7	1	0	-1	-1.7	-2	-1.7	-1	0	1	1.7	2
y	0	2	3.5	4	3.5	2	0	-2	-3.5	-4	-3.5	-2	0

If the parameter θ is eliminated, the rectangular equation is $\frac{x^2}{4} + \frac{y^2}{16} = 1$, an ellipse.

To eliminate the parameter θ:
$\cos \theta = \frac{x}{2}$ and $\sin \theta = \frac{y}{4}$. Hence $\cos^2\theta + \sin^2\theta = 1 = \frac{x^2}{4} + \frac{y^2}{16}$.

19. If a projectile is fired at an angle θ with the horizontal with an initial speed V_0, its position at any time t is given by $x = (V_0 \cos \theta) t$, $y = (V_0 \sin \theta) t - \frac{1}{2} g t^2$, where g is usually taken as 32 ft/sec^2, x and y are expressed in feet and t in seconds.

If $\theta = $ arc cos 3/5 and $V_0 = 120$ ft/sec, plot the path of the projectile.

Since $\sin \theta = \frac{4}{5}$, we have $x = 72t$,
$$y = 96t - 16t^2.$$

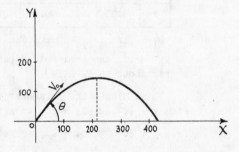

t	0	1	2	3	4	5	6
x	0	72	144	216	288	360	432
y	0	80	128	144	128	80	0

Eliminating t, we have $y = \frac{4x}{3} - \frac{x^2}{324}$, a

parabola with a vertical axis of symmetry. The maximum height is 144 feet, and the range $x = 432$ feet.

20. Discuss and plot the curve whose parametric equations are $x = \dfrac{2at^2}{t^2 + 1}$, $y = \dfrac{2at^3}{t^2 + 1}$.

When $t = 0$, $x = 0$ and $y = 0$. For all values of t, positive or negative, x is positive or zero; but y is positive for $t > 0$ and negative for $t < 0$. The curve is symmetric about the x-axis.

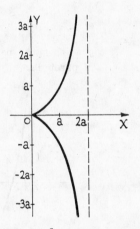

If we write $x = \dfrac{2at^2}{t^2 + 1} = 2a - \dfrac{2a}{t^2 + 1}$, it is seen that as

t increases numerically without limit, x approaches the value $2a$, but y increases numerically without limit. Hence $x = 2a$ is a vertical asymptote.

t	0	± 1	± 2	± 3	± 4
x	0	a	$1.6a$	$1.8a$	$1.9a$
y	0	$\pm a$	$\pm 3.2a$	$\pm 5.4a$	$\pm 7.5a$

Eliminating t, the rectangular form of this equation is $y^2(2a - x) = x^3$, the Cissoid of Diocles.

21. Discuss and plot the graph of the curve
$$x = a \cos^3 \theta, \quad y = a \sin^3 \theta.$$

Since $\cos(-\theta) = \cos \theta$, while $\sin(-\theta) = -\sin \theta$, this curve is symmetric about the x-axis. Since $\sin(180° - \theta) = \sin \theta$, but $\cos(180° - \theta) = -\cos \theta$, this curve is symmetric about the y-axis. Since the numerical value of the sine and cosine is never greater than unity,
$$-a \leqq x \leqq a, \quad \text{and} \quad -a \leqq y \leqq a.$$

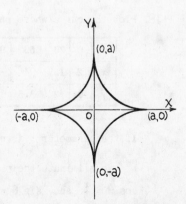

θ	0°	30°	60°	90°	120°	150°	180°
x	a	$.65a$	$.13a$	0	$-.13a$	$-.65a$	$-a$
y	0	$.13a$	$.65a$	a	$.65a$	$.13a$	0

When θ is eliminated the rectangular equation of this curve is $x^{2/3} + y^{2/3} = a^{2/3}$, the hypocycloid of four cusps.

To eliminate the parameter θ:

$$(x/a)^{2/3} + (y/a)^{2/3} = (\cos^3\theta)^{2/3} + (\sin^3\theta)^{2/3} = \cos^2\theta + \sin^2\theta = 1, \text{ or } x^{2/3} + y^{2/3} = a^{2/3}.$$

22. Discuss and plot the graph of the curve
$$x = a(\theta - \sin\theta),$$
$$y = a(1 - \cos\theta).$$

When $\theta = 0$, $x = 0$, $y = 0$.
When $\theta = 180°$, $x = \pi a$, $y = 2a$.
When $\theta = 360°$, $x = 2\pi a$, $y = 0$.

θ	0°	30°	60°	90°	120°	150°	180°	210°	240°	270°	300°	330°	360°
x	0	.02a	.18a	.57a	1.2a	2.1a	πa	4.2a	5.1a	5.7a	6.1a	6.3a	2πa
y	0	.13a	.5a	a	1.5a	1.9a	2a	1.9a	1.5a	a	.5a	.13a	0

When θ is eliminated, the equation of this curve is $x = a\cos^{-1}\dfrac{a-y}{a} - \sqrt{2ay-y^2}$, the cycloid.

To eliminate the parameter θ:

From $y = a(1 - \cos\theta)$ obtain $\cos\theta = \dfrac{a-y}{a}$. Then $\theta = \cos^{-1}\dfrac{a-y}{a}$ and $\sin\theta = \dfrac{\sqrt{2ay-y^2}}{a}$.

Substituting in $x = a\theta - a\sin\theta$, we have $x = a\cos^{-1}\dfrac{a-y}{a} - \sqrt{2ay-y^2}$.

23. Change the following equation into parametric equations: $x^2 + 3xy + 3y^2 - ax = 0$.

Let $y = tx$. Then $x^2 + 3x^2 t + 3x^2 t^2 - ax = 0$.

Dividing through by x, we obtain $x + 3xt + 3xt^2 - a = 0$.

Solving for x, $\quad x = \dfrac{a}{3t^2 + 3t + 1}, \quad y = tx = \dfrac{at}{3t^2 + 3t + 1}$.

SUPPLEMENTARY PROBLEMS

In each of Problems 1-14, discuss the given equation and draw the curve.

1. $(y^2 - 4)x - 9y = 0$

2. $y = (x + 1)(x + 2)(x - 2)$

3. $y^2 = (x + 1)(x + 2)(x - 2)$

4. $y^2(4 - x) = x^3$. Cissoid.

5. $x^3 - x^2 y + 4y = 0$

6. $x^2 y - 3x^2 - 9y = 0$

7. $x^2 y + 4y - 8 = 0$. Witch of Agnesi.

8. $x^2 + 2xy - 4 + y^2 = 0$.

9. $y = \dfrac{x^2 - 4}{x^2 - 3x - 4}$

10. $y^2 = \dfrac{x - 4}{x^2 + 2x - 8}$

11. $4x^2 - 12x - 4xy + y^2 + 6y - 7 = 0$

12. $x^3 + 4x^2 + xy^2 - 4y^2 = 0$. Strophoid.

13. $xy^2 - xy - 2x - 4 = 0$

14. $x^{2/3} + y^{2/3} = a^{2/3}$. Hypocycloid of four cusps.

In each of Problems 15-22, draw the graph of the given equation.

15. $y = 2 \sin 3x$ 18. $y = \cos(x - \pi/4)$ 21. $y = 3 \cos \frac{\pi}{2}(x - 1)$

16. $y = 2 \sin x/3$ 19. $y = 2 \sec x/2$

17. $y = \tan 2x$ 20. $y = \cot(x + \pi/3)$ 22. $y = \frac{1}{3} \csc 3x$

In each of Problems 23-28, draw the graph of the given equation.

23. $y = \text{arc sin } x$ 25. $y = 3 \text{ arc cos } x/3$ 27. $y = \text{arc csc } 2x$

24. $y = 2 \text{ arc tan } 2x$ 26. $y = \text{arc sec } x$ 28. $y = \text{arc cot } x/2$

In each of Problems 29-35, draw the graph of the given equation.

29. $y = 2e^{x/2}$ 31. $y = 10^{x/3}$ 33. $y = \log_{10} \sqrt{x^2 - 16}$ 35. $y = \frac{e^x + e^{-x}}{2}$.

30. $y = 4^{-x}$ 32. $y = \log_e(3 + x)$ 34. $y = \log_e \sqrt{27 - x^3}$ Catenary.

In each of Problems 36-49, draw the graph of the given equation by the method of compounding ordinates. In Problems 46, 47, 48, show the boundary curves.

36. $4x^2 - 4xy + y^2 - x = 0$ 43. $y = x/2 + \cos 2x$

37. $x^2 - 2xy + y^2 + x - 1 = 0$ 44. $y = e^{-x} + 2e^{x/2}$

38. $3x^2 - 2xy + y^2 - 5x + 4y + 3 = 0$ 45. $y = \sin 2x + 2 \cos x$

39. $x^2 + 2xy + y^2 - 4x - 2y = 0$ 46. $y = x \sin x$

40. $2x^2 + y^2 - 2xy - 4 = 0$ 47. $y = e^{-x/2} \cos \frac{\pi x}{2}$

41. $y = 2 \cos x + \sin 2x$ 48. $y = xe^{-x^2}$

42. $y = e^{x/2} + x^2$ 49. $y = x - \sin \frac{\pi x}{3}$

In each of Problems 50-55, find parametric equations. Use the value given for x or y.

50. $x - xy = 2$, $y = 1 - t$.   *Ans.* $x = \frac{2}{t}$, $y = 1 - t$

51. $x^2 - 4y^2 = K^2$, $x = K \sec \theta$.   *Ans.* $x = K \sec \theta$, $y = \frac{K \tan \theta}{2}$

52. $x^3 + y^3 = 6xy$, $y = tx$.   *Ans.* $x = \frac{6t}{1 + t^3}$, $y = \frac{6t^2}{1 + t^3}$

53. $x^2 - 2xy + 2y^2 = 2a^2$, $x = 2a \cos t$.   *Ans.* $x = 2a \cos t$, $y = a(\cos t \pm \sin t)$

54. $x^2y + b^2y - a^2x = 0$, $x = b \cot \frac{t}{2}$.   *Ans.* $x = b \cot \frac{t}{2}$, $y = \frac{a^2}{2b} \sin t$

55. $x^{2/3} + y^{2/3} = a^{2/3}$, $y = a \sin^3 \theta$.   *Ans.* $x = a \cos^3 \theta$, $y = a \sin^3 \theta$

In each of Problems 56-59, eliminate the parameter and obtain the rectangular equation.

56. $x = a \sec \theta$, $y = b \tan \theta$.   *Ans.* $\frac{x^2}{a^2} - \frac{y^2}{b^2} = 1$

57. $x = 2 \cos \theta - 1$, $y = 3 \sin \theta - 2$.   *Ans.* $\frac{(x + 1)^2}{4} + \frac{(y + 2)^2}{9} = 1$

58. $x = \frac{1}{2} \cos t$, $y = \cos 2t$.   *Ans.* $y = 8x^2 - 1$

59. $x = \frac{3am}{1 + m^3}$, $y = \frac{3am^2}{1 + m^3}$.   *Ans.* $x^3 + y^3 = 3axy$

60. A projectile is fired from a point A with an initial velocity of 3000 ft/sec at an angle of 35° with the horizontal. Find to three significant figures the horizontal distance from A to the point B where the projectile hits the ground, and the time taken.
Ans. 264,000 ft, 108 sec

61. At what angle must a gun be elevated to strike a target on the ground $7500\sqrt{3}$ yards distant if the muzzle velocity of the gun is 1200 ft/sec? What is the time of flight?
Ans. 30°, 37.5 sec

62. A projectile is shot at an angle of elevation of 60° and with an initial velocity of 2500 ft/sec. Find the range and the maximum height the projectile attains.
Ans. 169,000 ft, 73,200 ft

Plot the curve of each of the parametric equations in Problems 63-70.

63. $x = 4 \cos t$, $y = 4 \sin t$.

64. $x = t + \dfrac{1}{t}$, $y = t - \dfrac{1}{t}$.

65. $x = t^2 + 2$, $y = t^3 - 1$.

66. $x = 4 \tan \theta$, $y = 4 \sec \theta$.

67. $x = \dfrac{1}{1 + t}$, $y = \dfrac{1}{1 + t^2}$.

68. $x = 1 + t^2$, $y = 4t - t^3$.

69. $x = \sin t + \cos t$, $y = \cos 2t$.

70. $x = \theta - \sin \theta$, $y = 1 - \cos \theta$.

71. Plot the curve whose parametric equations are $x = 8 \cos^3\theta$, $y = 8 \sin^3\theta$.

72. Plot the curve whose parametric equations are $x = \dfrac{6t}{1 + t^3}$, $y = \dfrac{6t^2}{1 + t^3}$.

73. Plot the graph of the curve whose parametric equations are $x = 4 \tan \theta$, $y = 4 \cos^2\theta$.

74. Plot the curve whose parametric equations are $x = 4 \sin \theta$, $y = 4 \tan \theta (1 + \sin \theta)$.

CHAPTER 12

Introduction to Solid Analytic Geometry

CARTESIAN COORDINATES. In plane analytic geometry the position of any point in the plane was determined in terms of its perpendicular distances from two intersecting lines, usually perpendicular. In solid analytic geometry one method of locating a point in space is in terms of its perpendicular distances from three mutually perpendicular planes. These planes are called coordinate planes, and the three perpendicular distances are called the *coordinates* of the point.

The lines of intersection of these coordinate planes are the three axes *OX, OY* and *OZ*, called the *coordinate axes*, with positive directions shown by arrows. The coordinate planes divide all space into eight parts called *octants*, numbered as follows: Number I is the octant whose bounding edges are the positive directions of the three coordinate axes; then II, III and IV lie above the *xy*-plane in counter-clockwise order about *OZ*. Numbers V, VI, VII and VIII lie below the *xy*-plane, number V lying under number I.

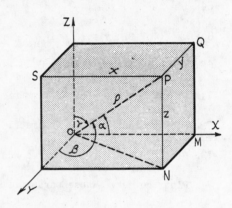

In the figure the distances *SP, QP* and *NP* are respectively the *x, y* and *z* coordinates of the point *P*, and the point is denoted by (x,y,z) or $P(x,y,z)$.

The distance *OP* of the point *P* from the origin *O* is

$$OP = \sqrt{\overline{ON}^2 + \overline{NP}^2} = \sqrt{\overline{OM}^2 + \overline{MN}^2 + \overline{NP}^2} = \sqrt{x^2 + y^2 + z^2}.$$

Hence if $OP = \rho$, then $\rho^2 = x^2 + y^2 + z^2$.

DIRECTION ANGLES AND DIRECTION COSINES.

Let the angles between *OP* and *OX, OY, OZ* be respectively α, β, γ. Then,

$$x = \rho \cos \alpha, \quad y = \rho \cos \beta, \quad z = \rho \cos \gamma.$$

Squaring these relations and adding,

$$x^2 + y^2 + z^2 = \rho^2 = \rho^2 \cos^2\alpha + \rho^2 \cos^2\beta + \rho^2 \cos^2\gamma,$$

or

$$1 = \cos^2\alpha + \cos^2\beta + \cos^2\gamma.$$

We have also the relations $\cos \alpha = \dfrac{x}{\rho}$, $\cos \beta = \dfrac{y}{\rho}$, $\cos \gamma = \dfrac{z}{\rho}$,

or $\cos \alpha = \dfrac{x}{\sqrt{x^2 + y^2 + z^2}}$, $\cos \beta = \dfrac{y}{\sqrt{x^2 + y^2 + z^2}}$, $\cos \gamma = \dfrac{z}{\sqrt{x^2 + y^2 + z^2}}$.

The angles α, β, γ of the line *OP* are called the *direction angles* of *OP*, and the cosines of these angles are called the *direction cosines* of *OP*.

If a line does not pass through the origin O, then its direction angles α, β, γ are the angles between the axes and a line drawn through O parallel to the given line and having the same direction.

DIRECTION NUMBERS. Any three numbers a, b and c proportional to the direction cosines of a line are called *direction numbers* of the line. To find the direction cosines of a line whose direction numbers a, b and c are known, divide the numbers by $\pm\sqrt{a^2 + b^2 + c^2}$. Use the sign in front of the radical which will cause the resulting direction cosines to have the proper sign.

DISTANCE BETWEEN TWO POINTS. The distance between any two points $P_1(x_1,y_1,z_1)$ and $P_2(x_2,y_2,z_2)$ is

$$d = \sqrt{(x_2 - x_1)^2 + (y_2 - y_1)^2 + (z_2 - z_1)^2}.$$

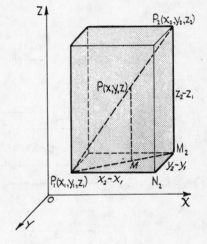

DIRECTION OF A LINE. The direction cosines of P_1P_2 are

$$\cos\alpha = \frac{x_2 - x_1}{\sqrt{(x_2 - x_1)^2 + (y_2 - y_1)^2 + (z_2 - z_1)^2}}$$

$$\cos\beta = \frac{y_2 - y_1}{\sqrt{(x_2 - x_1)^2 + (y_2 - y_1)^2 + (z_2 - z_1)^2}}$$

$$\cos\gamma = \frac{z_2 - z_1}{\sqrt{(x_2 - x_1)^2 + (y_2 - y_1)^2 + (z_2 - z_1)^2}}.$$

POINT OF DIVISION. If the point $P(x,y,z)$ divides the line from $P_1(x_1,y_1,z_1)$ to $P_2(x_2,y_2,z_2)$ in the ratio $\dfrac{P_1P}{PP_2} = \dfrac{r}{1}$, then

$$x = \frac{x_1 + rx_2}{1 + r}, \qquad y = \frac{y_1 + ry_2}{1 + r}, \qquad z = \frac{z_1 + rz_2}{1 + r}.$$

ANGLE BETWEEN TWO LINES. The angle between two lines that do not meet is defined as the angle between two intersecting lines, each of which is parallel to one of the given lines.

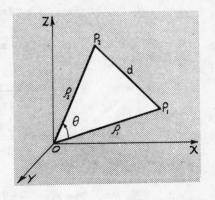

Let OP_1 and OP_2 be two lines through the origin parallel to the two given lines, and let θ be the angle between the lines. By the cosine law, $\cos\theta = \dfrac{\rho_1^2 + \rho_2^2 - d^2}{2\rho_1\rho_2}$.

Now $\rho_1^2 = x_1^2 + y_1^2 + z_1^2$, $\quad\rho_2^2 = x_2^2 + y_2^2 + z_2^2$, and $d^2 = (x_2 - x_1)^2 + (y_2 - y_1)^2 + (z_2 - z_1)^2$. Substituting and simplifying,

$$\cos\theta = \frac{x_1x_2 + y_1y_2 + z_1z_2}{\rho_1\rho_2}.$$

But $\dfrac{x_1}{\rho_1} = \cos \alpha_1$, $\dfrac{x_2}{\rho_2} = \cos \alpha_2$, etc.   Hence

$$\cos \theta = \cos \alpha_1 \cos \alpha_2 + \cos \beta_1 \cos \beta_2 + \cos \gamma_1 \cos \gamma_2 .$$

If the two lines are parallel, $\cos \theta = 1$ and hence

$$\alpha_1 = \alpha_2, \quad \beta_1 = \beta_2, \quad \gamma_1 = \gamma_2 .$$

If the two lines are perpendicular, $\cos \theta = 0$ and hence

$$\cos \alpha_1 \cos \alpha_2 + \cos \beta_1 \cos \beta_2 + \cos \gamma_1 \cos \gamma_2 = 0 .$$

SOLVED PROBLEMS

1. Plot the following points and find the distance of each from the origin and their perpendicular distances from the axes: $A(6,2,3)$, $B(8,-2,4)$.

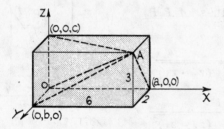

$OA = \sqrt{6^2 + 2^2 + 3^2} = 7$

$Aa = \sqrt{3^2 + 2^2} = \sqrt{13}$

$Ab = \sqrt{6^2 + 3^2} = 3\sqrt{5}$

$Ac = \sqrt{6^2 + 2^2} = 2\sqrt{10}$

$OB = \sqrt{8^2 + (-2)^2 + 4^2} = 2\sqrt{21}$

$Ba = \sqrt{4^2 + (-2)^2} = 2\sqrt{5}$

$Bb = \sqrt{8^2 + 4^2} = 4\sqrt{5}$

$Bc = \sqrt{8^2 + (-2)^2} = 2\sqrt{17}$

2. Find the distance between the points $P_1(5,-2,3)$ and $P_2(-4,3,7)$.

$$d = \sqrt{(x_2 - x_1)^2 + (y_2 - y_1)^2 + (z_2 - z_1)^2} = \sqrt{(-4-5)^2 + (3+2)^2 + (7-3)^2} = \sqrt{122}$$

3. Find the direction cosines and the direction angles of the line drawn from the origin to the point $(-6,2,3)$.

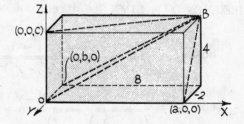

$$\cos \alpha = \frac{x_2 - x_1}{\sqrt{(x_2 - x_1)^2 + (y_2 - y_1)^2 + (z_2 - z_1)^2}}$$

$$= \frac{-6 - 0}{\sqrt{(-6-0)^2 + (2-0)^2 + (3-0)^2}} = \frac{-6}{7},$$

and $\alpha = 149°$.

$$\cos \beta = \frac{y_2 - y_1}{7} = \frac{2-0}{7} = \frac{2}{7}, \text{ and } \beta = 73°24'.$$

$$\cos \gamma = \frac{z_2 - z_1}{7} = \frac{3-0}{7} = \frac{3}{7}, \text{ and } \gamma = 64°37'.$$

4. Prove that the geometrical center or centroid or center of area, *i.e.*, the intersection of the medians, of any triangle $A(x_1, y_1, z_1)$, $B(x_2, y_2, z_2)$, $C(x_3, y_3, z_3)$ is

$$\left(\frac{x_1 + x_2 + x_3}{3}, \frac{y_1 + y_2 + y_3}{3}, \frac{z_1 + z_2 + z_3}{3}\right).$$

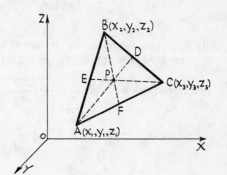

The medians of the triangle ABC intersect in a point $P(x, y, z)$ such that $\dfrac{AP}{PD} = \dfrac{BP}{PF} = \dfrac{CP}{PE} = \dfrac{2}{1} = r$.

The coordinates of the point D are

$$\left(\frac{x_2 + x_3}{2}, \frac{y_2 + y_3}{2}, \frac{z_2 + z_3}{2}\right).$$

Then the coordinates of point P, which divides AD in the ratio $r = \dfrac{AP}{PD} = \dfrac{2}{1}$, are

$$x = \frac{x_1 + r\left(\frac{x_2 + x_3}{2}\right)}{1 + r} = \frac{x_1 + x_2 + x_3}{3}. \text{ Similarly, } y = \frac{y_1 + y_2 + y_3}{3}, \quad z = \frac{z_1 + z_2 + z_3}{3}.$$

5. Find the direction cosines and direction angles of a line directed upward if its direction numbers are 2, –3, 6.

$$\cos \alpha = \frac{2}{\sqrt{4 + 9 + 36}} = \frac{2}{7}, \quad \alpha = 73°24'. \quad \cos \beta = \frac{-3}{7}, \quad \beta = 115°23'. \quad \cos \gamma = \frac{6}{7}, \quad \gamma = 31°.$$

6. Show that the lines $A(5, 2, -3)$ to $B(6, 1, 4)$, and $C(-3, -2, -1)$ to $D(-1, -4, 13)$, are parallel.

The direction numbers of AB are 6–5, 1–2, 4+3, or 1, –1, 7.
The direction numbers of CD are –1+3, –4+2, 13+1, or 2, –2, 14.

If two lines whose direction numbers are a, b, c and a', b', c' are parallel, $\dfrac{a}{a'} = \dfrac{b}{b'} = \dfrac{c}{c'}$.

Hence since $\dfrac{2}{1} = \dfrac{-2}{-1} = \dfrac{14}{7}$, these two lines are parallel.

7. Show that the lines AB and BC are perpendicular to each other. $A(-11, 8, 4)$, $B(-1, -7, -1)$, $C(9, -2, 4)$.

The direction numbers of AB are –1+11, –7–8, –1–4, or 10, –15, –5, or 2, –3, –1.
The direction numbers of BC are 9+1, –2+7, 4+1, or 10, 5, 5, or 2, 1, 1.

If two lines whose direction numbers are a, b, c and a', b', c' are perpendicular, then $aa' + bb' + cc' = 0$. Substituting, $(2)(2) + (-3)(1) + (-1)(1) = 0$. Hence the lines AB and BC are perpendicular.

8. Find the angle θ between the lines $A(-3, 2, 4)$, $B(2, 5, -2)$ and $C(1, -2, 2)$, $D(4, 2, 3)$.

The direction numbers of AB are $2+3$, $5-2$, $-2-4$, or 5, 3, –6.
The direction numbers of CD are $4-1$, $2+2$, $3-2$, or 3, 4, 1.

The direction cosines of AB are $\cos \alpha = \dfrac{5}{\sqrt{25 + 9 + 36}} = \dfrac{5}{\sqrt{70}}$, $\cos \beta = \dfrac{3}{\sqrt{70}}$, $\cos \gamma = \dfrac{-6}{\sqrt{70}}$.

The direction cosines of CD are $\cos \alpha_1 = \dfrac{3}{\sqrt{9 + 16 + 1}} = \dfrac{3}{\sqrt{26}}$, $\cos \beta_1 = \dfrac{4}{\sqrt{26}}$, $\cos \gamma_1 = \dfrac{1}{\sqrt{26}}$.

Then $\cos \theta = \cos \alpha \cos \alpha_1 + \cos \beta \cos \beta_1 + \cos \gamma \cos \gamma_1$

$$= \frac{5}{\sqrt{70}} \cdot \frac{3}{\sqrt{26}} + \frac{3}{\sqrt{70}} \cdot \frac{4}{\sqrt{26}} - \frac{6}{\sqrt{70}} \cdot \frac{1}{\sqrt{26}} = 0.49225, \text{ and } \theta = 60°30.7'.$$

9. Find the interior angles of the triangle whose vertices are $A(3,-1,4)$, $B(1,2,-4)$, $C(-3,2,1)$.

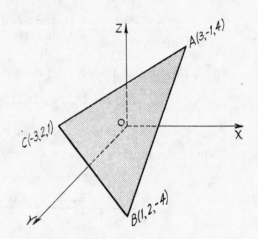

Direction cosines of $AB = (\frac{-2}{\sqrt{77}}, \frac{3}{\sqrt{77}}, \frac{-8}{\sqrt{77}})$.

Direction cosines of $BC = (\frac{-4}{\sqrt{41}}, 0, \frac{5}{\sqrt{41}})$.

Direction cosines of $AC = (\frac{-2}{\sqrt{6}}, \frac{1}{\sqrt{6}}, \frac{-1}{\sqrt{6}})$.

Note. The direction cosines of a line AB are the negative of the direction cosines of BA.

$$\cos A = \frac{-2}{\sqrt{77}} \cdot \frac{-2}{\sqrt{6}} + \frac{3}{\sqrt{77}} \cdot \frac{1}{\sqrt{6}} + \frac{-8}{\sqrt{77}} \cdot \frac{-1}{\sqrt{6}} = \frac{15}{\sqrt{462}}. \qquad A = 45°44.7'.$$

$$\cos B = \frac{2}{\sqrt{77}} \cdot \frac{-4}{\sqrt{41}} + \frac{-3}{\sqrt{77}} \cdot 0 + \frac{8}{\sqrt{77}} \cdot \frac{5}{\sqrt{41}} = \frac{32}{\sqrt{3157}}. \qquad B = 55°16.9'.$$

$$\cos C = \frac{4}{\sqrt{41}} \cdot \frac{2}{\sqrt{6}} + 0 + \frac{-5}{\sqrt{41}} \cdot \frac{1}{\sqrt{6}} = \frac{3}{\sqrt{246}}. \qquad C = 78°58.4'. \qquad A + B + C = 180°.$$

10. Find the area of the triangle in Problem 9.

From trigonometry, if two sides and the included angle of a triangle are known, say the sides b, c and angle A, then area $= \frac{1}{2}bc \sin A$.

Length of AB or $c = \sqrt{77}$, of AC or $b = 3\sqrt{6}$.

Hence, area $= \frac{1}{2}(3\sqrt{6})(\sqrt{77}) \sin 45°44.7' = 23.1$ square units.

11. Find the equation of the locus of a point at a distance r units from point (x_0, y_0, z_0).

$$\sqrt{(x-x_0)^2 + (y-y_0)^2 + (z-z_0)^2} = r \quad \text{or} \quad (x-x_0)^2 + (y-y_0)^2 + (z-z_0)^2 = r^2, \quad \text{the}$$
equation of a sphere with its center at (x_0, y_0, z_0) and radius r.

The general form of the equation of a sphere is $x^2 + y^2 + z^2 + dx + ey + fz + g = 0$.

12. Find the equation of the sphere with its center at $(2,-2,3)$ and tangent to the XY plane.

Since the sphere is tangent to the XY plane its radius is 3. Hence,

$$\sqrt{(x-2)^2 + (y+2)^2 + (z-3)^2} = 3. \quad \text{Squaring and simplifying, } x^2 + y^2 + z^2 - 4x + 4y - 6z + 8 = 0.$$

13. Find the coordinates of the center and radius of the sphere $x^2 + y^2 + z^2 - 6x + 4y - 8z = 7$.

Completing the square, $x^2 - 6x + 9 + y^2 + 4y + 4 + z^2 - 8z + 16 = 36$

or $(x-3)^2 + (y+2)^2 + (z-4)^2 = 36.$

Comparing with $(x - x_0)^2 + (y - y_0)^2 + (z - z_0)^2 = r^2$, we find that the center is point $(3, -2, 4)$ and the radius is 6.

14. Find the equation of the locus of a point twice as far from $(2, -3, 4)$ as from $(-1, 2, -2)$.

Let $P(x, y, z)$ be any point on the locus.   Then

$$\sqrt{(x - 2)^2 + (y + 3)^2 + (z - 4)^2} = 2\sqrt{(x + 1)^2 + (y - 2)^2 + (z + 2)^2}.$$

Squaring and simplifying, $3x^2 + 3y^2 + 3z^2 + 12x - 22y + 24z + 7 = 0$, a sphere with center $(-2, \frac{11}{3}, -4)$ and radius $r = \frac{2}{3}\sqrt{70}$.

15. Find the equation of the perpendicular bisector of the line joining the points $(2, -1, 3)$ and $(-4, 2, 2)$.

Let $P(x, y, z)$ be any point on the locus.   Then

$$\sqrt{(x + 4)^2 + (y - 2)^2 + (z - 2)^2} = \sqrt{(x - 2)^2 + (y + 1)^2 + (z - 3)^2}.$$

Squaring and simplifying, $6x - 3y + z + 5 = 0$.   This is the equation of a plane every point of which is equidistant from the two given points. The plane cuts the axes in the points $(-5/6, 0, 0)$, $(0, 5/3, 0)$ and $(0, 0, -5)$, and cuts the line at $(-1, 1/2, 5/2)$.

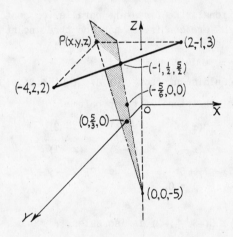

Problem 15.

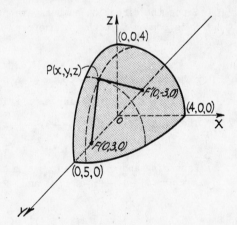

Problem 16.

16. Find the equation of the locus of a point the sum of whose distances from $(0, 3, 0)$ and $(0, -3, 0)$ is 10.

Let $P(x, y, z)$ be any point on the locus.   Then $FP + PF' = 10$,   or

$$\sqrt{(x - 0)^2 + (y - 3)^2 + (z - 0)^2} + \sqrt{(x - 0)^2 + (y + 3)^2 + (z - 0)^2} = 10.$$

Transpose one radical and square both members of the equation and collect terms. This gives $3y + 25 = 5\sqrt{x^2 + y^2 + 6y + 9 + z^2}$.

Squaring and simplifying, $25x^2 + 16y^2 + 25z^2 = 400$, an ellipsoid with center at origin.

17. Find the equation of the locus of a point the difference of whose distances from $(4, 0, 0)$ and $(-4, 0, 0)$ is 6.

Let (x,y,z) be any point on the locus. Then

$$\sqrt{(x-4)^2 + (y-0)^2 + (z-0)^2} \; - \; \sqrt{(x+4)^2 + (y-0)^2 + (z-0)^2} \;=\; 6,$$

or $\sqrt{(x^2 - 8x + 16 + y^2 + z^2} \;=\; 6 + \sqrt{x^2 + 8x + 16 + y^2 + z^2}.$

Squaring and simplifying, $4x + 9 \;=\; -3\sqrt{x^2 + 8x + 16 + y^2 + z^2}.$

Squaring again and simplifying, $7x^2 - 9y^2 - 9z^2 = 63,$ a hyperboloid of revolution about the x-axis.

18. Find the equation of the locus of a point whose distance from the z-axis is 3 times its distance from the point $(-1,2,-3)$.

Distance from the z-axis = distance from the point $(-1,2,-3)$.

Hence, $\sqrt{x^2 + y^2} \;=\; 3\sqrt{(x+1)^2 + (y-2)^2 + (z+3)^2}.$

Squaring and simplifying, $8x^2 + 8y^2 + 9z^2 + 18x - 36y + 54z + 126 = 0,$ an ellipsoid.

19. Show that the points $A(-2,0,3)$, $B(3,10,-7)$, $C(1,6,-3)$ lie in a straight line.

Direction numbers of AB = 5, 10, -10 or 1, 2, -2; of BC = -2, -4, 4 or -1, -2, 2.

Since the direction numbers are proportional, the lines are parallel or continuous. But B is common to both. Therefore AB and BC are continuous and the three points lie in a straight line.

20. Find the equation of the locus of a point equidistant from $(1,3,8)$, $(-6,-4,2)$, $(3,2,1)$.

Let (x,y,z) be any point satisfying the conditions of the problem.

Then   (1) $(x-1)^2 + (y-3)^2 + (z-8)^2 \;=\; (x+6)^2 + (y+4)^2 + (z-2)^2,$

and   (2) $(x-1)^2 + (y-3)^2 + (z-8)^2 \;=\; (x-3)^2 + (y-2)^2 + (z-1)^2.$

Expanding and simplifying, we get (1) $7x + 7y + 6z - 9 = 0$ and (2) $2x - y - 7z + 30 = 0$. Answer: $7x + 7y + 6z - 9 = 0$ and $2x - y - 7z + 30 = 0$.

21. Show that the triangle $A(3,5,-4)$, $B(-1,1,2)$, $C(-5,-5,-2)$ is isosceles.

Length of AB = $\sqrt{(3 + 1)^2 + (5 - 1)^2 + (-4 - 2)^2}$ = $2\sqrt{17}$.

Length of BC = $\sqrt{(-5 + 1)^2 + (-5 - 1)^2 + (-2 - 2)^2}$ = $2\sqrt{17}$.

Length of AC = $\sqrt{(-5 - 3)^2 + (-5 - 5)^2 + (-2 + 4)^2}$ = $2\sqrt{42}$.

Since $AB = BC = 2\sqrt{17}$, the triangle is isosceles.

22. Show by two different methods that the points $A(5,1,5)$, $B(4,3,2)$ and $C(-3,-2,1)$ are the vertices of a right triangle.

1. Use the Pythagorean Theorem. $AB = \sqrt{(5 - 4)^2 + (1 - 3)^2 + (5 - 2)^2}$ = $\sqrt{14}$.

$$BC = \sqrt{(4 + 3)^2 + (3 + 2)^2 + (2 - 1)^2} \;=\; \sqrt{75}.$$

$$CA = \sqrt{(-3 - 5)^2 + (-2 - 1)^2 + (1 - 5)^2} \;=\; \sqrt{89}.$$

$$(AB)^2 + (BC)^2 = (CA)^2, \text{ or } 14 + 75 = 89.$$

2. Show that AB and BC are perpendicular.

Direction cosines of AB, $\dfrac{1}{\sqrt{14}}$, $\dfrac{-2}{\sqrt{14}}$, $\dfrac{3}{\sqrt{14}}$. Direction cosines of BC, $\dfrac{7}{5\sqrt{3}}$, $\dfrac{5}{5\sqrt{3}}$, $\dfrac{1}{5\sqrt{3}}$.

$$\cos B = \frac{1}{\sqrt{14}} \cdot \frac{7}{5\sqrt{3}} - \frac{2}{\sqrt{14}} \cdot \frac{5}{5\sqrt{3}} + \frac{3}{\sqrt{14}} \cdot \frac{1}{5\sqrt{3}} = \frac{7 - 10 + 3}{5\sqrt{42}} = 0.$$

Otherwise: The sum of the products of the direction numbers of the two lines equals zero. $7(1) + 5(-2) + 1(3) = 0$.

SUPPLEMENTARY PROBLEMS

1. Plot the points $(2,2,3)$, $(4,-1,2)$, $(-3,2,4)$, $(3,4,-5)$, $(-4,-3,-2)$, $(0,4,-4)$, $(4,0,-2)$, $(0,0,-3)$, $(-4,0,-2)$, $(3,4,0)$.

2. Find the distance from the origin to each of the points in Problem 1 above.
   Ans. $\sqrt{17}$, $\sqrt{21}$, $\sqrt{29}$, $5\sqrt{2}$, $\sqrt{29}$, $4\sqrt{2}$, $2\sqrt{5}$, 3, $2\sqrt{5}$, 5.

3. Find the distance between each of the following pairs of points.
   (a) $(2,5,3)$ and $(-3,2,1)$.   *Ans.* $\sqrt{38}$
   (b) $(0,3,0)$ and $(6,0,2)$.   *Ans.* 7
   (c) $(-4,-2,3)$ and $(3,3,5)$.   *Ans.* $\sqrt{78}$

4. Find the perimeter of each of the following triangles.
   (a) $(4,6,1)$, $(6,4,0)$, $(-2,3,3)$.   *Ans.* $10 + \sqrt{74}$
   (b) $(-3,1,-2)$, $(5,5,-3)$, $(-4,-1,-1)$.   *Ans.* $20 + \sqrt{6}$
   (c) $(8,4,1)$, $(6,3,3)$, $(-3,9,5)$.   *Ans.* $14 + 9\sqrt{2}$

5. Plot each of the following points and for each point find the radius vector from the origin and the direction cosines.
   (a) $(-6,2,3)$.   *Ans.* 7, $\cos \alpha = -6/7$, $\cos \beta = 2/7$, $\cos \gamma = 3/7$.
   (b) $(6,-2,9)$.   *Ans.* 11, $\cos \alpha = 6/11$, $\cos \beta = -2/11$, $\cos \gamma = 9/11$.
   (c) $(-8,4,8)$.   *Ans.* 12, $\cos \alpha = -2/3$, $\cos \beta = 1/3$, $\cos \gamma = 2/3$.
   (d) $(3,4,0)$.   *Ans.* 5, $\cos \alpha = 3/5$, $\cos \beta = 4/5$, $\cos \gamma = 0$.
   (e) $(4,4,4)$.   *Ans.* $4\sqrt{3}$, $\cos \alpha = 1/\sqrt{3}$, $\cos \beta = 1/\sqrt{3}$, $\cos \gamma = 1/\sqrt{3}$.

6. Find the direction angles for the points in Problem 5 (a), (d), and (e).
   Ans. (a) $\alpha = 148°59.8'$, $\beta = 73°23.9'$, $\gamma = 64°37.4'$
          (d) $\alpha = 53°7.8'$, $\beta = 36°52.2'$, $\gamma = 90°$
          (e) $\alpha = \beta = \gamma = 54°44.1'$

7. Find the lengths of the medians of the following triangles. Answers are expressed for medians taken in order through A, B, C.
   (a) $A(2,-3,1)$, $B(-6,5,3)$, $C(8,7,-7)$.   *Ans.* $\sqrt{91}$, $\sqrt{166}$, $\sqrt{217}$
   (b) $A(7,5,-4)$, $B(3,-9,-2)$, $C(-5,3,6)$.   *Ans.* $2\sqrt{41}$, $\sqrt{182}$, $\sqrt{206}$
   (c) $A(-7,4,6)$, $B(3,6,-2)$, $C(1,-8,8)$.   *Ans.* $\sqrt{115}$, $\sqrt{181}$, $\sqrt{214}$

8. In each of the following examples find the direction cosines of the line drawn from the first point to the second.
   (a) $(-4,1,7)$, $(2,-3,2)$.   (c) $(-6,5,-4)$, $(-5,-2,-4)$.   (e) $(3,-5,4)$, $(-6,1,2)$.
   (b) $(7,1,-4)$, $(5,-2,-3)$.   (d) $(5,-2,3)$, $(-2,3,7)$.

   Ans. (a) $\dfrac{6\sqrt{77}}{77}$, $-\dfrac{4\sqrt{77}}{77}$, $-\dfrac{5\sqrt{77}}{77}$ (d) $-\dfrac{7\sqrt{10}}{30}$, $\dfrac{\sqrt{10}}{6}$, $\dfrac{2\sqrt{10}}{15}$

         (b) $-\dfrac{\sqrt{14}}{7}$, $-\dfrac{3\sqrt{14}}{14}$, $\dfrac{\sqrt{14}}{14}$ (e) $-\dfrac{9}{11}$, $\dfrac{6}{11}$, $-\dfrac{2}{11}$

         (c) $\dfrac{\sqrt{2}}{10}$, $-\dfrac{7\sqrt{2}}{10}$, 0

9. Find a set of direction numbers for the line through the two points in each of the following.
   (a) $(4,7,3)$, $(-5,-2,6)$.   *Ans.* 3, 3, -1
   (b) $(-2,3,-4)$, $(1,3,2)$.   *Ans.* -3, 0, -6
   (c) $(11,2,-3)$, $(4,-5,4)$.   *Ans.* 1, 1, -1

10. Find the angle not greater than $90°$ between the lines joining the following pairs of points.
    (a) $(8,2,0)$, $(4,6,-7)$; $(-3,1,2)$, $(-9,-2,4)$.   *Ans.* $88°10.8'$
    (b) $(4,-2,3)$, $(6,1,7)$; $(4,-2,3)$, $(5,4,-2)$.   *Ans.* $90°$
    (c) From $(6,-2,0)$ to $(5,4,2\sqrt{3})$ and from $(5,3,1)$ to $(7,-1,5)$.   *Ans.* $73°11.6'$

11. Find the interior angles of the triangle whose vertices are $(-1,-3,-4)$, $(4,-2,-7)$, and $(2,3,-8)$.   *Ans.* $86°27.7'$, $44°25.4'$, $49°6.9'$

12. Find the area of the triangle in Problem 11.   *Ans.* 16.17 square units

13. Find the points of intersection of the medians in each of the following triangles.
    (a) $(-1,-3,-4)$, $(4,-2,-7)$, $(2,3,-8)$.   *Ans.* $(5/3, -2/3, -19/3)$
    (b) $(2,1,4)$, $(3,-1,2)$, $(5,0,6)$.   *Ans.* $(10/3, 0, 4)$
    (c) $(4,3,-2)$, $(7,-1,4)$, $(-2,1,-4)$.   *Ans.* $(3, 1, -2/3)$

14. Show that the triangle with vertices $(6,10,10)$, $(1,0,-5)$, $(6,-10,0)$ is a right triangle and find its area.   *Ans.* Area = $25\sqrt{21}$ square units

15. Show that the triangle with vertices $(4,2,6)$, $(10,-2,4)$, $(-2,0,2)$ is isosceles and find its area.   *Ans.* Area = $6\sqrt{19}$ square units

16. Prove by two methods that the points $(-11,8,4)$, $(-1,-7,-1)$, $(9,-2,4)$ are the vertices of a right triangle.

17. Prove that the points $(2,-1,0)$, $(0,-1,-1)$, $(1,1,-3)$, $(3,1,-2)$ are the vertices of a rectangle.

18. Show that $(4,2,4)$, $(10,2,-2)$ and $(2,0,-4)$ are the vertices of an equilateral triangle.

19. Show by two different methods that the points $(1,-1,3)$, $(2,-4,5)$ and $(5,-13,11)$ are in a straight line.

20. Derive the equation of the locus of a point equidistant from points $(1,-2,3)$ and $(-3,4,2)$.
    Ans. $8x - 12y + 2z + 15 = 0$

21. Derive the equation of the locus of a point twice as far from $(-2,3,4)$ as from $(3,-1,-2)$.
    Ans. $3x^2 + 3y^2 + 3z^2 - 28x + 14y + 24z + 27 = 0$, a sphere

22. Find the equation of the sphere with radius 5 and center $(-2,3,5)$.
    Ans. $x^2 + y^2 + z^2 + 4x - 6y - 10z + 13 = 0$

23. The direction numbers of two lines are $2,-1,4$ and $-3,2,2$. Show that the lines are perpendicular.

24. Determine k so that the lines joining the points $P_1(k,1,-1)$ and $P_2(2k,0,2)$ shall be perpendicular to the line from P_2 to $P_3(2+2k, k, 1)$.   *Ans.* $k = 3$

25. The direction numbers (or parameters) of a line perpendicular to two lines whose direction numbers are a_1, b_1, c_1, and a_2, b_2, c_2, are given by the following three determinants.

$$\begin{matrix} b_1 \ c_1 & c_1 \ a_1 & a_1 \ b_1 \\ b_2 \ c_2, & c_2 \ a_2, & a_2 \ b_2. \end{matrix}$$

Find the direction numbers of a line perpendicular to two lines whose direction numbers are:
   (a) $1,3,-2$ and $-2,2,4$.   *Ans.* 16, 0, 8 or 2,0,1
   (b) $-3,4,1$ and $2,-6,5$.   *Ans.* 26, 17, 10
   (c) $0,-2,1$ and $4,0,-3$.   *Ans.* 3, 2, 4
   (d) $5,3,-3$ and $-1,1,-2$.   *Ans.* -3, 13, 8

26. Find the direction numbers of a line perpendicular to the two lines determined by the pairs of points $(2,3,-4),(-3,3,-2)$ and $(-1,4,2),(3,5,1)$.   *Ans.* $-2,3,-5$

27. Find the direction cosines of a line perpendicular to each of two lines with direction numbers $3,4,1$ and $6,2,-1$.   *Ans.* $2/7,\ -3/7,\ 6/7$

28. The angle between the line L_1 with direction numbers $x,3,5$ and L_2 with direction numbers $2,-1,2$ is $45°$. Find x.   *Ans.* $4,\ 52$

29. For what value of x will the line through $(4,1,2)$ and $(5,x,0)$ be parallel to the line through $(2,1,1)$ and $(3,3,-1)$.   *Ans.* $x = 3$

30. For what value of x will the lines in Problem 29 be perpendicular?   *Ans.* $x = -3/2$

31. Show that the points $(3,3,3)$, $(1,2,-1)$, $(4,1,1)$, $(6,2,5)$ are the vertices of a parallelogram.

32. Show that the points $(4,2,-6)$, $(5,-3,1)$, $(12,4,5)$, $(11,9,-2)$ are vertices of a rectangle.

33. Show that the line through $(5,1,-2)$ and $(-4,-5,13)$ is a perpendicular bisector of the line segment joining $(-5,2,0)$ and $(9,-4,6)$.

34. Find the angle between the lines joining $(3,1,-2)$, $(4,0,-4)$ and $(4,-3,3)$, $(6,-2,2)$. *Ans.* $\pi/3$ radians

35. Direction numbers of two lines are $3,-2,k$ and $-2,k,4$. Find k if the lines are at right angles.   *Ans.* $k = 3$

36. Find the equation of the locus of a point whose distance from the y-axis is equal to its distance from $(2,1,-1)$.   *Ans.* $y^2 - 2y - 4x + 2z + 6 = 0$

37. Find the equation of the locus of a point whose distance from the xy-plane is equal to its distance from $(-1,2,-3)$.   *Ans.* $x^2 + y^2 + 2x - 4y + 6z + 14 = 0$

38. A point moves so that the difference of the squares of its distances from the x- and y-axes is constant. Find the equation of its locus.   *Ans.* $y^2 - x^2 = a$

39. Find the equation of the locus of a point whose distance from the z-axis is equal to its distance from the xy-plane.   *Ans.* $x^2 + y^2 - z^2 = 0$, a cone

40. Find the equation of a sphere with its center at $(3,-1,2)$ and tangent to the yz-plane. *Ans.* $x^2 + y^2 + z^2 - 6x + 2y - 4z + 5 = 0$

41. Find the equation of a sphere with radius a and tangent to all three coordinate planes, the center being in the first octant.   *Ans.* $x^2 + y^2 + z^2 - 2ax - 2ay - 2az + 2a^2 = 0$

42. Find the equation of a sphere with center $(2,-2,3)$ and passing through $(7,-3,5)$. *Ans.* $x^2 + y^2 + z^2 - 4x + 4y - 6z - 13 = 0$

43. Find the equation of the locus of a point equidistant from $(-2,1,-2)$ and $(2,-2,3)$. *Ans.* $4x - 3y + 5z - 4 = 0$

44. Find the equation of a plane which bisects perpendicularly the line segment joining the points $(-2,3,2)$ and $(6,5,-6)$.   *Ans.* $4x + y - 4z - 20 = 0$

45. Given the points $A(3,2,0)$ and $B(2,1,-5)$, find the locus of a point $P(x,y,z)$ which moves so that PA is perpendicular to PB.   *Ans.* $x^2 + y^2 + z^2 - 5x - 3y + 5z + 8 = 0$

46. What equation must the point (x,y,z) satisfy if the distance from (x,y,z) to the point $(2,-1,3)$ is 4.   *Ans.* $x^2 + y^2 + z^2 - 4x + 2y - 6z - 2 = 0$

47. Find the equation of the locus of a point (x,y,z) which moves so that its distance from $(1,3,2)$ is three times its distance from the xz-plane. *Ans.* $x^2 - 8y^2 + z^2 - 2x - 6y - 4z + 14 = 0$

48. Find the center and the radius of the sphere $x^2 + y^2 + z^2 - 2x + 6y + 2z - 14 = 0$. *Ans.* Center $(1,-3,-1)$, radius 5

49. Find the coordinates of the center and the radius of the sphere:

    (a) $16x^2 + 16y^2 + 16z^2 - 24x + 48y - 5 = 0$.   *Ans.* Center $(\frac{3}{4}, -\frac{3}{2}, 0)$; $r = \frac{5\sqrt{2}}{4}$

    (b) $x^2 + y^2 + z^2 - 2x - 6y + 4z + 14 = 0$.   *Ans.* Center $(1, 3, -2)$; $r = 0$

    (c) $x^2 + y^2 + z^2 + 4x - 2y - 6z = 0$.   *Ans.* Center $(-2, 1, 3)$; $r = \sqrt{14}$

50. Find the equation of a sphere with its center at $(4, -3, 2)$ and tangent to the plane $x + 2 = 0$.
    Ans. $x^2 + y^2 + z^2 - 8x + 6y - 4z - 7 = 0$

51. Write the equation of the locus of a point which moves so that it is
    (a) 4 units in front of the xz-plane.   *Ans.* $y = 4$
    (b) 6 units back of the yz-plane.   *Ans.* $x = -6$
    (c) 3 units back of the plane $y - 1 = 0$.   *Ans.* $y + 2 = 0$
    (d) 3 units from the z-axis.   *Ans.* $x^2 + y^2 = 9$

52. Find the equation of the locus of a point the sum of whose distances from $(3, 0, 0)$ and $(-3, 0, 0)$ is 8.   *Ans.* $7x^2 + 16y^2 + 16z^2 = 112$, an ellipsoid

53. Find the equation of the locus of a point whose distance from $(-1, 2, -2)$ is equal to its distance from the z-axis.   *Ans.* $z^2 + 4z + 2x - 4y + 9 = 0$, a paraboloid

54. Find the equation of the locus of a point whose distance from $(3, -2, 1)$ is 3 times its distance from the xy-plane.
    Ans. $x^2 + y^2 - 8z^2 - 6x + 4y - 2z + 14 = 0$, a hyperboloid

55. Find the equation of the locus of a point the difference of whose distances from $(0, 0, -4)$ and $(0, 0, 4)$ is 6.   *Ans.* $9x^2 + 9y^2 - 7z^2 + 63 = 0$, a hyperboloid

56. Find the equation of the locus of a point whose distance from the yz-plane is twice its distance from $(4, -2, 1)$.   *Ans.* $3x^2 + 4y^2 + 4z^2 - 32x + 16y - 8z + 84 = 0$, ellipsoid

57. Find the equation of the locus of a point whose distance from $(0, 0, -2)$ is one-third its distance from the plane $z + 18 = 0$.
    Ans. $9x^2 + 9y^2 + 8z^2 - 288 = 0$, an ellipsoid

58. Find the equation of the locus of a point equidistant from the plane $z = 5$ and the point $(0, 0, 3)$.   *Ans.* $x^2 + y^2 + 4z - 16 = 0$, a paraboloid

CHAPTER 13

The Plane

EVERY PLANE is represented by an equation of the first degree in one or more of the variables x, y, z. The converse statement also is true. Every equation of the first degree in one or more of the variables x, y, z represents a plane.

The general equation of a plane is $Ax + By + Cz + D = 0$, provided that A, B, and C are not all zero.

The equation of a system of planes passing through point (x_0, y_0, z_0) is
$$A(x - x_0) + B(y - y_0) + C(z - z_0) = 0.$$

LINE PERPENDICULAR TO A PLANE. If a, b, c are the direction numbers of a line, the line will be perpendicular to a plane $Ax + By + Cz + D = 0$ if and only if these direction numbers are proportional to the coefficients of x, y, z in the equation of the plane. If a, b, c, A, B, C are all different from zero, then $\dfrac{a}{A} = \dfrac{b}{B} = \dfrac{c}{C}$ must be true if the line and plane are perpendicular to each other.

PARALLEL AND PERPENDICULAR PLANES.

Two planes $A_1 x + B_1 y + C_1 z + D_1 = 0$ and $A_2 x + B_2 y + C_2 z + D_2 = 0$ are *parallel* if and only if the coefficients of x, y, z in their equations are proportional, *i.e.*, if $\dfrac{A_1}{A_2} = \dfrac{B_1}{B_2} = \dfrac{C_1}{C_2}$.

Two planes $A_1 x + B_1 y + C_1 z + D_1 = 0$ and $A_2 x + B_2 y + C_2 z + D_2 = 0$ are *perpendicular* to each other if and only if $A_1 A_2 + B_1 B_2 + C_1 C_2 = 0$.

NORMAL FORM. The normal form of the equation of a plane is
$$x \cos \alpha + y \cos \beta + z \cos \gamma - p = 0,$$
where p is the perpendicular distance from the origin to the plane and α, β, γ are the direction angles of that perpendicular.

The normal form of the equation of the plane $Ax + By + Cz + D = 0$ is
$$\frac{Ax + By + Cz + D}{\pm \sqrt{A^2 + B^2 + C^2}} = 0,$$
the sign of the radical being taken opposite to that of D so that the normal distance p shall be positive.

INTERCEPT FORM. The intercept form of the equation of a plane is $\dfrac{x}{a} + \dfrac{y}{b} + \dfrac{z}{c} = 1$, where a, b, c are the x-, y-, and z-intercepts respectively.

DISTANCE OF A POINT FROM A PLANE. The perpendicular distance between a point (x_1, y_1, z_1) and a plane $Ax + By + Cz + D = 0$ is $d = \left| \dfrac{Ax_1 + By_1 + Cz_1 + D}{\sqrt{A^2 + B^2 + C^2}} \right|$.

115

ANGLE BETWEEN TWO PLANES. The angle θ between two planes $A_1 x + B_1 y + C_1 z + D_1 = 0$ and $A_2 x + B_2 y + C_2 z + D_2 = 0$ is determined by

$$\cos \theta = \left| \frac{A_1 A_2 + B_1 B_2 + C_1 C_2}{\sqrt{A_1^2 + B_1^2 + C_1^2} \ \sqrt{A_2^2 + B_2^2 + C_2^2}} \right|, \quad \text{for the acute angle.}$$

SPECIAL PLANES. The planes
$$Ax + By + D = 0,$$
$$By + Cz + D = 0,$$
$$Ax + Cz + D = 0, \quad \text{represent planes perpendicular,}$$
respectively, to the xy-, yz-, and xz-planes.

The planes $Ax + D = 0$, $By + D = 0$, $Cz + D = 0$, represent planes perpendicular, respectively, to the x-, y-, and z-axes.

SOLVED PROBLEMS

1. Find the equation of the plane through the point $(4, -2, 1)$ and perpendicular to the line whose direction numbers are $7, 2, -3$.

Use the equation $A(x - x_0) + B(y - y_0) + C(z - z_0) = 0$ and the condition that the coefficients are proportional to the direction numbers.

Then, $7(x - 4) + 2(y + 2) - 3(z - 1) = 0$ or $7x + 2y - 3z - 21 = 0$.

2. Find the equation of the plane perpendicular to the line segment from $(-3, 2, 1)$ to $(9, 4, 3)$ at the midpoint of the segment.

The direction numbers of the line segment are $12, 2, 2$ or $6, 1, 1$. The midpoint of the segment is $(3, 3, 2)$. Hence the equation of the plane is

$$6(x - 3) + (y - 3) + (z - 2) = 0 \quad \text{or} \quad 6x + y + z - 23 = 0.$$

3. Find the equation of the plane through the point $(1, -2, 3)$ and parallel to the plane
$$x - 3y + 2z = 0.$$

The required plane must have an equation of the form $x - 3y + 2z = k$. To determine k substitute the coordinates $(1, -2, 3)$, since this point lies in the required plane.

Then $1 - 3(-2) + 2(3) = k$, or $k = 13$. The required equation is $x - 3y + 2z = 13$.

4. Find the equation of the plane through $(1, 0, -2)$ and perpendicular to each of the planes
$$2x + y - z = 2 \quad \text{and}$$
$$x - y - z = 3.$$

The family of planes through point $(1, 0, -2)$ is $A(x - 1) + B(y - 0) + C(z + 2) = 0$. For this plane to be perpendicular to the two given planes,
$$2A + B - C = 0 \quad \text{and}$$
$$A - B - C = 0. \qquad \text{Solving, } A = -2B \text{ and } C = -3B.$$

The required equation is $-2B(x - 1) + B(y - 0) - 3B(z + 2) = 0$, or $2x - y + 3z + 4 = 0$.

5. Find the equation of the plane through $(1, 1, -1)$, $(-2, -2, 2)$, $(1, -1, 2)$.

Use $Ax + By + Cz + D = 0$. Substituting the coordinates of these points, we have
$$A + B - C + D = 0,$$
$$-2A - 2B + 2C + D = 0,$$
$$A - B + 2C + D = 0.$$

Solving for A, B, C and D, we find $D = 0$, $A = -C/2$, $B = 3C/2$, $C = C$.
Substituting these values and dividing through by C, the equation becomes
$$x - 3y - 2z = 0.$$

Another Method. The equation of a plane through three points (x_1, y_1, z_1), (x_2, y_2, z_2), and (x_3, y_3, z_3) is found by expanding the determinant

$$\begin{vmatrix} x & y & z & 1 \\ x_1 & y_1 & z_1 & 1 \\ x_2 & y_2 & z_2 & 1 \\ x_3 & y_3 & z_3 & 1 \end{vmatrix} = 0.$$

6. Discuss the locus of the equation $2x + 3y + 6z = 12$.

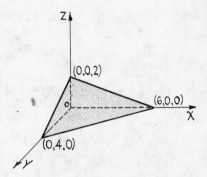

Since the equation is of the first degree it represents a plane.

The direction numbers of the normal to the plane are 2,3,6. The direction cosines of the normal are $\cos \alpha = \frac{2}{7}$, $\cos \beta = \frac{3}{7}$, $\cos \gamma = \frac{6}{7}$.

The intercepts on the axes are $(6,0,0)$, $(0,4,0)$, and $(0,0,2)$.

The lines in which the plane intersects the coordinate planes are called the *traces* of the plane. To find the equations of the traces: In the xy-plane, $z = 0$; hence the equation of this trace is $2x + 3y = 12$. Similarly, to find the trace on the xz-plane, set $y = 0$; hence this trace is $2x + 6z = 12$ or $x + 3z = 6$, and the equation of the yz-trace is $3y + 6z = 12$ or $y + 2z = 4$. The intercepts and traces are shown in the figure above.

To find the length of the normal, that is, the distance from the origin to the plane:

$$d = \frac{Ax_1 + By_1 + Cz_1 + D}{\pm \sqrt{A^2 + B^2 + C^2}}.   \qquad |d| = \left| \frac{2(0) + 3(0) + 6(0) - 12}{7} \right| = \frac{12}{7}.$$

7. Find the perpendicular distance from the point $(-2, 2, 3)$ to the plane $8x - 4y - z - 8 = 0$.

The equation in normal form is $\dfrac{8x - 4y - z - 8}{\sqrt{64 + 16 + 1}} = \dfrac{8x - 4y - z - 8}{9} = 0$.

Substituting the coordinates of the point, $d = \dfrac{8(-2) - 4(2) - 1(3) - 8}{9} = -\dfrac{35}{9}$.

The negative sign shows that the point and the origin are on the same side of the plane.

8. Find the smallest angle between the planes (1) $3x + 2y - 5z - 4 = 0$
and (2) $2x - 3y + 5z - 8 = 0$.

The direction cosines of the normals to the two planes are:

$$\cos \alpha_1 = \frac{3}{\sqrt{38}}, \quad \cos \beta_1 = \frac{2}{\sqrt{38}}, \quad \cos \gamma_1 = -\frac{5}{\sqrt{38}},$$

$$\cos \alpha_2 = \frac{2}{\sqrt{38}}, \quad \cos \beta_2 = \frac{-3}{\sqrt{38}}, \quad \cos \gamma_2 = \frac{5}{\sqrt{38}}.$$

Let θ be the angle between the two normals.

Then $\cos\theta = \left|\dfrac{3}{\sqrt{38}}\cdot\dfrac{2}{\sqrt{38}} - \dfrac{2}{\sqrt{38}}\cdot\dfrac{3}{\sqrt{38}} - \dfrac{5}{\sqrt{38}}\cdot\dfrac{5}{\sqrt{38}}\right| = \dfrac{25}{38}$, and $\theta = 48°51.6'$.

9. Find the point of intersection of the planes: $x + 2y - z = 6,$
$$2x - y + 3z = -13,$$
$$3x - 2y + 3z = -16.$$

Here we have three linear equations. The solution of these simultaneous equations determines the coordinates of the point of intersection of the three planes.
The required point is found to be $(-1, 2, -3)$.

10. Find the equation of the plane passing through the line of intersection of the planes $3x + y - 5z + 7 = 0$ and $x - 2y + 4z - 3 = 0$ and through the point $(-3, 2, -4)$.

The equation of any plane passing through the line of intersection of the two given planes is of the form $3x + y - 5z + 7 + k(x - 2y + 4z - 3) = 0$.

To determine the plane of this pencil of planes which passes through $(-3, 2, -4)$, substitute $-3, 2, -4$ for x, y, z respectively and obtain
$$-9 + 2 + 20 + 7 + k(-3 - 4 - 16 - 3) = 0, \quad \text{or} \quad k = 10/13.$$

Substituting and simplifying, we have $49x - 7y - 25z + 61 = 0$.

11. Find the equations of the planes which bisect the dihedral angles between the planes
$$6x - 6y + 7z + 21 = 0 \quad \text{and}$$
$$2x + 3y - 6z - 12 = 0.$$

Let (x_1, y_1, z_1) be any point on the bisecting plane. Then the distances of (x_1, y_1, z_1) from the two planes must be equal in magnitude. Hence,
$$\frac{6x_1 - 6y_1 + 7z_1 + 21}{-11} = \pm\frac{2x_1 + 3y_1 - 6z_1 - 12}{7}$$

Clearing of fractions and simplifying, we obtain $64x - 9y - 17z + 15 = 0$
and $20x - 75y + 115z + 279 = 0$.

12. Find the equation of the plane through the points $(1, -2, 2)$, $(-3, 1, -2)$ and perpendicular to the plane $2x + y - z + 6 = 0$.

Let $Ax + By + Cz + D = 0$ be the required plane.
Since the two points must lie in the required plane, by substitution
$$A - 2B + 2C + D = 0 \quad \text{and}$$
$$-3A + B - 2C + D = 0.$$

Since the required plane must be perpendicular to $2x + y - z + 6 = 0$, we have
$$2A + B - C = 0.$$
Solving for A, B, D in terms of C, $A = -\dfrac{C}{10}$, $B = \dfrac{6C}{5}$, $D = \dfrac{5C}{10}$.

Substituting these values and dividing through by C, we obtain the required equation
$$x - 12y - 10z - 5 = 0.$$

13. Find the locus of a point equidistant from $(2, -1, 3)$ and $2x - 2y + z - 6 = 0$.

Let (x, y, z) be the point. Then
$$\sqrt{(x-2)^2 + (y+1)^2 + (z-3)^2} = \frac{2x - 2y + z - 6}{3}$$

Squaring and simplifying, $5x^2 + 5y^2 + 8z^2 + 8xy - 4xz + 4yz - 12x - 6y - 42z + 90 = 0$.

14. Find the equations of the planes parallel to $2x - 3y - 6z - 14 = 0$ and distant 5 from the origin.

$2x - 3y - 6z - k = 0$ represents the family of planes parallel to the given plane.

The distance of any point (x_1, y_1, z_1) from $2x - 3y - 6z - k = 0$ is
$$d = \frac{2x_1 - 3y_1 - 6z_1 - k}{7}.$$

Since $d = \pm 5$ from $(0,0,0)$, we have $\pm 5 = \frac{2(0) - 3(0) - 6(0) - k}{7}$ or $k = \pm 35$.

Hence the required equation is $2x - 3y - 6z \pm 35 = 0$.

In the figure below: Plane I is given and Planes II and III are the required planes.

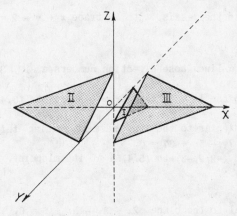

Problem 14.

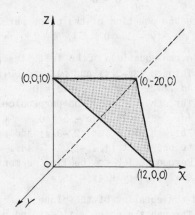

Problem 15.

15. Write the equation of the plane $5x - 3y + 6z = 60$ in the intercept form.

Dividing through by 60, the equation has the intercept form $\frac{x}{12} - \frac{y}{20} + \frac{z}{10} = 1$.
The intercepts are 12, -20, 10.

16. Show that the planes $7x + 4y - 4z + 30 = 0$,
$$36x - 51y + 12z + 17 = 0,$$
$$14x + 8y - 8z - 12 = 0, \text{ and}$$
$$12x - 17y + 4z - 3 = 0$$
form four faces of a rectangular parallelepiped.

The first and third planes are parallel since $\frac{7}{14} = \frac{4}{8} = \frac{-4}{-8}$.

Also, the second and fourth planes are parallel since $\frac{36}{12} = \frac{-51}{-17} = \frac{12}{4}$.

Further, the first and second planes are perpendicular since
$$7(36) + 4(-51) - 4(12) = 252 - 204 - 48 = 0.$$

17. Determine the locus of the equation $x^2 + y^2 - 2xy - 4z^2 = 0$.

Write this equation in the form $x^2 - 2xy + y^2 - 4z^2 = (x - y - 2z)(x - y + 2z) = 0$.

The locus is two planes passing through the origin,
$$x - y - 2z = 0 \quad \text{and}$$
$$x - y + 2z = 0.$$

SUPPLEMENTARY PROBLEMS

1. Write the equation of the plane:
   (a) Parallel to the xy-plane and 3 units below it.   *Ans.* $z = -3$
   (b) Parallel to the yz-plane and having x-intercept 4.   *Ans.* $x = 4$
   (c) Perpendicular to the z-axis at point $(0,0,6)$.   *Ans.* $z = 6$
   (d) Parallel to the xz-plane and 6 units behind it.   *Ans.* $y = -6$ or $y + 6 = 0$

2. Write the equation of the plane which is horizontal and passes through point $(3,-2,-4)$.
   Ans. $z = -4$ or $z + 4 = 0$

3. Write the equation of the plane parallel to the z-axis, x-intercept 2, y-intercept -3.
   Ans. $3x - 2y - 6 = 0$

4. Write the equation of the plane parallel to the z-axis, with xy-trace $x + y - 2 = 0$.
   Ans. $x + y - 2 = 0$

5. Write the equations of the following planes:
   (a) Through $(3,-2,4)$ and perpendicular to a line whose direction numbers are 2, 2, -3.
   Ans. $2x + 2y - 3z + 10 = 0$
   (b) Through $(-1,2,-3)$ and perpendicular to the line segment from $(-3,2,4)$ to $(5,4,1)$.
   Ans. $8x + 2y - 3z - 5 = 0$
   (c) Through the point $(2,-3,4)$ and perpendicular to the line segment joining this point to the point $(4,4,-1)$.   *Ans.* $2x + 7y - 5z + 37 = 0$
   (d) Perpendicular to the line segment from $(-2,2,-3)$ to $(6,4,5)$ at its midpoint.
   Ans. $4x + y + 4z - 15 = 0$

6. Find the equation of the plane:
   (a) Through the point $(-1,2,4)$ and parallel to the plane $2x - 3y - 5z + 6 = 0$.
   Ans. $2x - 3y - 5z + 28 = 0$
   (b) Through the point $(2,-3,6)$ and parallel to the plane $2x - 5y + 7 = 0$.
   Ans. $2x - 5y - 19 = 0$
   (c) Through the origin and parallel to the plane $3x + 7y - 6z + 3 = 0$.
   Ans. $3x + 7y - 6z = 0$
   (d) Parallel to the plane $6x + 3y - 2z - 14 = 0$ and half as far from the origin.
   Ans. $6x + 3y - 2z \pm 7 = 0$
   (e) Parallel to the plane $3x - 6y - 2z - 4 = 0$ at a distance 3 from the origin.
   Ans. $3x - 6y - 2z \pm 21 = 0$

7. Find the equation of the plane:
   (a) Parallel to the plane $6x - 6y + 7z - 44 = 0$ and 2 units farther from the origin.
   Ans. $6x - 6y + 7z \pm 66 = 0$
   (b) Parallel to the plane $4x - 4y + 7z - 3 = 0$ and distant 4 units from point $(4,1,-2)$.
   Ans. $4x - 4y + 7z + 38 = 0$, $4x - 4y + 7z - 34 = 0$
   (c) Parallel to the plane $2x - 3y - 5z + 1 = 0$ and distant 3 units from point $(-1,3,1)$.
   Ans. $2x - 3y - 5z + 16 \pm 3\sqrt{38} = 0$

8. Find the equation of the plane that passes through point $(3,-2,4)$ and is perpendicular to each of the planes $7x - 3y + z - 5 = 0$ and $4x - y - z + 9 = 0$.
   Ans. $4x + 11y + 5z - 10 = 0$

9. Find the equation of the plane through the point $(4,-3,2)$ and perpendicular to the line of intersection of the planes $x - y + 2z - 3 = 0$ and $2x - y - 3z = 0$.
   Ans. $5x + 7y + z - 1 = 0$

10. Find the equation of the plane through $(1,-4,2)$ and perpendicular to each of the planes $2x + 5y - z - 12 = 0$ and $4x - 7y + 3z + 8 = 0$.   *Ans.* $4x - 5y - 17z + 10 = 0$

11. Find the equation of the plane through $(7,0,3)$ and perpendicular to each of the planes $2x - 4y + 3z = 0$ and $7x + 2y + z - 14 = 0$.   *Ans.* $10x - 19y - 32z + 26 = 0$

12. Find the equation of the plane through $(4,1,0)$ and perpendicular to each of the planes $2x - y - 4z - 6 = 0$ and $x + y + 2z - 3 = 0$.   *Ans.* $2x - 8y + 3z = 0$

13. Find the equation of the plane through $(1,1,2)$ and perpendicular to each of the planes $2x - 2y - 4z - 6 = 0$ and $3x + y + 6z - 4 = 0$.   *Ans.* $x + 3y - z - 2 = 0$

14. Find the equation of the plane perpendicular to each of the planes $3x - y + z = 0$ and $x + 5y + 3z = 0$ at a distance $\sqrt{6}$ from the origin.   *Ans.* $x + y - 2z \pm 6 = 0$

15. Find the equation of the plane perpendicular to each of the planes $x - 4y + z = 0$ and $3x + 4y + z - 2 = 0$ at a distance 1 from the origin.   *Ans.* $4x - y - 8z \pm 9 = 0$

16. Find the equation of the plane through $(2,2,2)$ and $(0,-2,0)$ and perpendicular to the plane $x - 2y + 3z - 7 = 0$.   *Ans.* $4x - y - 2z - 2 = 0$

17. Find the equation of the plane through $(2,1,1)$ and $(3,2,2)$ and perpendicular to the plane $x + 2y - 5z - 3 = 0$.   *Ans.* $7x - 6y - z - 7 = 0$

18. Find the equation of the plane through $(2,-1,6)$ and $(1,-2,4)$ and perpendicular to the plane $x - 2y - 2z + 9 = 0$.   *Ans.* $2x + 4y - 3z + 18 = 0$

19. Find the equation of the plane through $(1,2,-2)$ and $(2,0,-2)$ and perpendicular to the plane $3x + y + 2z = 0$.   *Ans.* $4x + 2y - 7z - 22 = 0$

20. Find the equation of the plane through $(1,3,-2)$ and $(3,4,3)$ and perpendicular to the plane $7x - 3y + 5z - 4 = 0$.   *Ans.* $20x + 25y - 13z - 121 = 0$

21. Find the equation of the plane through the points
    (a) $(3,4,1)$, $(-1,-2,5)$, $(1,7,1)$.   *Ans.* $3x + 2y + 6z - 23 = 0$
    (b) $(3,1,4)$, $(2,1,6)$, $(3,2,4)$.   *Ans.* $2x + z - 10 = 0$
    (c) $(2,1,3)$, $(-1,-2,4)$, $(4,2,1)$.   *Ans.* $5x - 4y + 3z - 15 = 0$
    (d) $(3,2,1)$, $(1,3,2)$, $(1,-2,3)$.   *Ans.* $3x + y + 5z - 16 = 0$
    (e) $(4,2,1)$, $(-1,-2,2)$, $(0,4,-5)$.   *Ans.* $11x - 17y - 13z + 3 = 0$

22. Discuss the locus of each of the following planes and sketch, showing intercepts and traces.
    (a) $2x + 4y + 3z - 12 = 0$.   (c) $x + y = 6$.   (e) $2x - z = 0$.
    (b) $3x - 5y + 2z - 30 = 0$.   (d) $2y - 3z = 6$.   (f) $x - 6 = 0$.

23. Use the normal form and write the equations of each of the following planes, given:
    (a) $\alpha = 120°$, $\beta = 45°$, $\gamma = 120°$, $p = 5$.   *Ans.* $x - \sqrt{2}\,y + z + 10 = 0$
    (b) $\alpha = 90°$, $\beta = 135°$, $\gamma = 45°$, $p = 4$.   *Ans.* $y - z + 4\sqrt{2} = 0$
    (c) The foot of the normal from the origin to the plane is point $(2,3,1)$.
              Ans. $2x + 3y + z - 14 = 0$
    (d) $\alpha = 120°$, $\beta = 60°$, $\gamma = 135°$, $p = 2$.   *Ans.* $x - y + \sqrt{2}\,z + 4 = 0$
    (e) $p = 2$, $\dfrac{\cos \alpha}{-1} = \dfrac{\cos \beta}{4} = \dfrac{\cos \gamma}{8}$.   *Ans.* $x - 4y - 8z \pm 18 = 0$

24. Reduce each of the following equations to the normal form and thus determine the direction cosines and length of the normal.
    (a) $2x - 2y + z - 12 = 0$.   *Ans.* $\cos \alpha = 2/3$, $\cos \beta = -2/3$, $\cos \gamma = 1/3$, $p = 4$.
    (b) $9x + 6y - 2z + 7 = 0$.   *Ans.* $\cos \alpha = -9/11$, $\cos \beta = -6/11$, $\cos \gamma = 2/11$, $p = 7/11$.
    (c) $x - 4y + 8z - 27 = 0$.   *Ans.* $\cos \alpha = 1/9$, $\cos \beta = -4/9$, $\cos \gamma = 8/9$, $p = 3$.

25. Determine the perpendicular distance from the point to the plane in each of the following.
    (a) Point $(-2,2,3)$, plane $2x + y - 2z - 12 = 0$.   *Ans.* $-20/3$. Interpret sign.
    (b) Point $(7,3,4)$, plane $6x - 3y + 2z - 13 = 0$.   *Ans.* 4
    (c) Point $(0,2,3)$, plane $6x - 7y - 6z + 22 = 0$.   *Ans.* $10/11$
    (d) Point $(1,-2,3)$, plane $2x - 3y + 2z - 14 = 0$.   *Ans.* 0

26. Find the acute angle between each of the following planes.
    (a) $2x - y + z = 7$, $x + y + 2z - 11 = 0$.   *Ans.* $60°$
    (b) $x + 2y - z = 12$, $x - 2y - 2z - 7 = 0$.   *Ans.* $82°10.7'$

(c) $2x - 5y + 14z = 60$, $2x + y - 2z - 18 = 0$.   *Ans.* $49°52.6'$
(d) $2x + y - 2z = 18$, $4x - 3y - 100 = 0$.   *Ans.* $70°31.7'$

27. Find the point of intersection of the planes $2x - y - 2z = 5$, $4x + y + 3z = 1$, $8x - y + z = 5$.
                                                        Ans. $(3/2, 4, -3)$

28. Find the point of intersection of the planes:
   (a) $2x + y - z - 1 = 0$, $3x - y - z + 2 = 0$, $4x - 2y + z - 3 = 0$. *Ans.* $(1, 2, 3)$
   (b) $2x + 3y + 3 = 0$, $3x + 2y - 5z + 2 = 0$, $3y - 4z + 8 = 0$. *Ans.* $(3/2, -2, 1/2)$
   (c) $x + 2y + 4z = 2$, $2x + 3y - 2z + 3 = 0$, $2x - y + 4z + 8 = 0$. *Ans.* $(-4, 2, 1/2)$

29. Find the equation of the plane passing through the line of intersection of the planes
   $2x - 7y + 4z - 3 = 0$, $3x - 5y + 4z + 11 = 0$,   and the point $(-2, 1, 3)$.
                                                    Ans. $15x - 47y + 28z - 7 = 0$

30. Find the equation of the plane passing through the line of intersection of the planes
   $3x - 4y + 2z - 6 = 0$, $2x + 4y - 2z + 7 = 0$,   and the point $(1, 2, 3)$.
                                                    Ans. $43x - 24y + 12z - 31 = 0$

31. Find the equation of the plane passing through the line of intersection of the planes
   $2x - y + 2z - 6 = 0$, $3x - 6y + 2z - 12 = 0$,   and cutting the x-axis at $(6, 0, 0)$.
                                                    Ans. $x - 5y - 6 = 0$

32. Find the equations of the bisectors of the angles between the planes $2x - y - 2z - 6 = 0$ and
   $3x + 2y - 6z = 12$.   *Ans.* $5x - 13y + 4z - 6 = 0$, $23x - y - 32z - 78 = 0$

33. Find the equations of the bisectors of the angles between the planes $6x - 9y + 2z + 18 = 0$
   and $x - 8y + 4z = 20$.   *Ans.* $65x - 169y + 62z - 58 = 0$, $43x + 7y - 26z + 382 = 0$

34. Find the equations of the bisectors of the angles between the planes $3x + 4y - 6 = 0$ and
   $6x - 6y + 7z + 16 = 0$.   *Ans.* $9x + 2y + 5z + 2 = 0$, $3x + 74y - 35z - 146 = 0$

35. Write the equation of each of the following planes in the intercept form.
   (a) $2x - 3y + 4z = 12$.   (b) $3x + 2y - 5z = 15$.   (c) $x + 3y + 4z = 12$.
   Ans. (a) $\dfrac{x}{6} - \dfrac{y}{4} + \dfrac{z}{3} = 1$.   (b) $\dfrac{x}{5} + \dfrac{y}{7.5} - \dfrac{z}{3} = 1$.   (c) $\dfrac{x}{12} + \dfrac{y}{4} + \dfrac{z}{3} = 1$.

36. Write the equations of the following planes whose intercepts are:

   (a) $(-2, 0, 0)$, $(0, 3, 0)$, $(0, 0, 5)$.   *Ans.* $\dfrac{x}{-2} + \dfrac{y}{3} + \dfrac{z}{5} = 1$.

   (b) $(3, 0, 0)$, $(0, -2, 0)$.   *Ans.* $\dfrac{x}{3} - \dfrac{y}{2} = 1$. (Parallel to the z-axis.)

   (c) $(4, 0, 0)$.   *Ans.* $x = 4$. (Parallel to the yz-plane.)

37. Show that the following planes are the faces of a parallelepiped: $3x - y + 4z - 7 = 0$,
   $x + 2y - z + 5 = 0$, $6x - 2y + 8z + 10 = 0$, $3x + 6y - 3z - 7 = 0$.

38. Write the equation of the locus of a point whose distance from the plane $3x - 2y - 6z = 12$
   is always twice its distance from $x - 2y + 2z + 4 = 0$.
   Ans. $23x - 34y + 10z + 20 = 0$, $5x - 22y + 46z + 92 = 0$

39. Find the perpendicular distance between the parallel planes $2x - 3y - 6z - 14 = 0$ and
   $2x - 3y - 6z + 7 = 0$. Draw a figure.   *Ans.* 3

40. Find the perpendicular distance between the planes $3x + 6y + 2z = 22$ and $3x + 6y + 2z = 27$.
   Draw a figure.   *Ans.* $5/7$

41. Describe the locus of the equation $x^2 + 4y^2 - z^2 + 4xy = 0$.
   Ans. Intersecting planes: $x + 2y + z = 0$, $x + 2y - z = 0$.

42. Describe the locus of the equation $x^2 + y^2 + z^2 + 2xy - 2xz - 2yz - 4 = 0$
   Ans. Parallel planes: $x + y - z + 2 = 0$, $x + y - z - 2 = 0$.

43. Write the equation of the locus of a point whose distance from the plane $6x - 2y + 3z + 4 = 0$
   is equal to its distance from the point $(-1, 1, 2)$.
   Ans. $13x^2 + 45y^2 + 40z^2 + 24xy - 36xz + 12yz + 50x - 82y - 220z + 278 = 0$

CHAPTER 14

The Straight Line in Space

STRAIGHT LINE IN SPACE. The locus of two simultaneous equations of the first degree

$$A_1 x + B_1 y + C_1 z + D_1 = 0$$
$$A_2 x + B_2 y + C_2 z + D_2 = 0$$

is a straight line, the line of intersection of the two planes, except when the planes are parallel.

PARAMETRIC FORM. Let the direction angles of the line L be α, β, γ and let $P_1(x_1, y_1, z_1)$ be a point of the line. Then L is the locus of $P(x, y, z)$ moving so that $x - x_1 = t \cos \alpha$, $y - y_1 = t \cos \beta$, $z - z_1 = t \cos \gamma$, or $x = x_1 + t \cos \alpha$, $y = y_1 + t \cos \beta$, $z = z_1 + t \cos \gamma$, where the parameter t represents the variable length $P_1 P$.

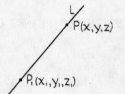

If a, b, c are direction numbers of L, these equations can be written:

$$x = x_1 + at, \quad y = y_1 + bt, \quad z = z_1 + ct.$$

SYMMETRIC FORM. The equations of the line passing through $P_1(x_1, y_1, z_1)$ with direction angles α, β, γ have the form

$$\frac{x - x_1}{\cos \alpha} = \frac{y - y_1}{\cos \beta} = \frac{z - z_1}{\cos \gamma}.$$

If a, b, c are direction numbers of the line, the symmetric equation takes the form

$$\frac{x - x_1}{a} = \frac{y - y_1}{b} = \frac{z - z_1}{c}.$$

If L is perpendicular to one of the coordinate axes, the equation takes one of the following forms:

$$x = x_1, \quad \frac{y - y_1}{b} = \frac{z - z_1}{c} \quad \text{(perpendicular to } x\text{-axis)}.$$

$$y = y_1, \quad \frac{x - x_1}{a} = \frac{z - z_1}{c} \quad \text{(perpendicular to } y\text{-axis)}.$$

$$z = z_1, \quad \frac{x - x_1}{a} = \frac{y - y_1}{b} \quad \text{(perpendicular to } z\text{-axis)}.$$

If L is perpendicular to two axes, two equations are sufficient to determine the line:

$$x = x_1, \quad y = y_1 \quad \text{(perpendicular to } x\text{- and } y\text{-axes)}.$$
$$x = x_1, \quad z = z_1 \quad \text{(perpendicular to } x\text{- and } z\text{-axes)}.$$
$$y = y_1, \quad z = z_1 \quad \text{(perpendicular to } y\text{- and } z\text{-axes)}.$$

TWO POINT FORM. The equations of a straight line through $P_1(x_1, y_1, z_1)$ and $P_2(x_2, y_2, z_2)$ are

$$\frac{x - x_1}{x_2 - x_1} = \frac{y - y_1}{y_2 - y_1} = \frac{z - z_1}{z_2 - z_1}.$$

PROJECTING PLANES.   The equations

$$\frac{x - x_1}{a} = \frac{y - y_1}{b}, \quad \frac{x - x_1}{a} = \frac{z - z_1}{c}, \quad \frac{y - y_1}{b} = \frac{z - z_1}{c},$$

each represents a plane containing the line. Since each of these planes is perpendicular to one of the coordinate planes, it may be thought of as projecting the line upon that plane, and hence it is called a projecting plane of the line.

RELATIVE DIRECTIONS OF A LINE AND A PLANE. A line whose direction numbers are a, b and c, and the plane $Ax + By + Cz + D = 0$ are

(1) parallel when and only when $Aa + Bb + Cc = 0$, and

(2) perpendicular when and only when $\dfrac{A}{a} = \dfrac{B}{b} = \dfrac{C}{c}$.

SYSTEMS OF PLANES CONTAINING A LINE. Given the equations

$$A_1 x + B_1 y + C_1 z + D_1 = 0$$
$$A_2 x + B_2 y + C_2 z + D_2 = 0,$$

the equation

$$A_1 x + B_1 y + C_1 z + D_1 + K(A_2 x + B_2 y + C_2 z + D_2) = 0,$$

where K is a parameter, represents a plane containing the line of intersection of the given planes. Hence this equation is the equation of every plane through the line of intersection of the two given planes.

SOLVED PROBLEMS

1. Given the equations $2x - y + z = 6$, $x + 4y - 2z = 8$,   find
   (a) the point of the line for $z = 1$,
   (b) the points in which the line pierces the coordinate planes,
   (c) the direction numbers, and
   (d) the direction cosines of the line.

   (a) Substituting $z = 1$ in both equations gives $2x - y = 5$, $x + 4y = 10$.

   Solving these equations, $x = \dfrac{10}{3}$, $y = \dfrac{5}{3}$. Hence the required point is $(\dfrac{10}{3}, \dfrac{5}{3}, 1)$.

   (b) Since $z = 0$ in the xy-plane, we proceed as in (a) and find this point to be $(\dfrac{32}{9}, \dfrac{10}{9}, 0)$. Similarly, the other points are found to be $(4, 0, -2)$ and $(0, 10, 16)$.

   (c) The points $(\dfrac{10}{3}, \dfrac{5}{3}, 1)$, $(4, 0, -2)$ lie on the line.

   Hence a set of direction numbers are $4 - \dfrac{10}{3}$, $0 - \dfrac{5}{3}$, $-2 - 1$, or $\dfrac{2}{3}$, $-\dfrac{5}{3}$, -3, or $2, -5, -9$.

   (d) The direction cosines are $\cos \alpha = \dfrac{2}{\sqrt{4 + 25 + 81}} = \dfrac{2}{\sqrt{110}}$, $\cos \beta = \dfrac{-5}{\sqrt{110}}$, $\cos \gamma = \dfrac{-9}{\sqrt{110}}$.

   Another method. The above set of direction numbers may be obtained by observing that the line is perpendicular to the normals to the two planes represented by the given equations. By making use of the determinants obtained from the array

   $$\begin{array}{ccccc} 1 & 4 & -2 & 1 & 4 \\ 2 & -1 & 1 & 2 & -1 \end{array} \quad \text{formed from the coefficients of } x, y, z,$$

   we obtain $\begin{vmatrix} 4 & -2 \\ -1 & 1 \end{vmatrix} = 4 - 2 = 2$, $\begin{vmatrix} -2 & 1 \\ 1 & 2 \end{vmatrix} = -4 - 1 = -5$, $\begin{vmatrix} 1 & 4 \\ 2 & -1 \end{vmatrix} = -9$, or $2, -5, -9$.

2. Find the acute angle between the lines (1) $2x - y + 3z - 4 = 0$, $3x + 2y - z + 7 = 0$
and (2) $x + y - 2z + 3 = 0$, $4x - y + 3z + 7 = 0$.

The direction numbers of the first line are $-5, 11, 7$, and for the second line are -1, $11, 5$, as explained in Problem 1(d) above.

Let θ be the angle between the two lines. Then

$$\cos \theta = \frac{-5}{\sqrt{195}} \cdot \frac{-1}{\sqrt{147}} + \frac{11}{\sqrt{195}} \cdot \frac{11}{\sqrt{147}} + \frac{7}{\sqrt{195}} \cdot \frac{5}{\sqrt{147}} = \frac{23}{3\sqrt{65}}, \text{ and } \theta = 18°1.4'.$$

3. Show that the lines (1) $x - y + z - 5 = 0$, $x - 3y + 6 = 0$
and (2) $2y + z - 5 = 0$, $4x - 2y + 5z - 4 = 0$
are parallel.

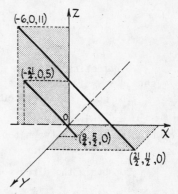

The direction numbers of the first line are:

$$\begin{matrix} 1 & -1 & 1 & -1 \\ 1 & -3 & 0 & -3 \end{matrix} \quad \text{or} \quad 3, 1, -2.$$

The direction numbers of the second line are:

$$\begin{matrix} 0 & 2 & 1 & 2 \\ 4 & -2 & 5 & -2 \end{matrix} \quad \text{or} \quad 12, 4, -8, \quad \text{or} \quad 3, 1, -2.$$

The direction numbers of the two lines are the same. Hence the lines must be parallel.

4. Show that the lines $\dfrac{x+1}{2} = \dfrac{y-5}{3} = \dfrac{z-7}{-1}$ and $\dfrac{x+4}{5} = \dfrac{y-1}{-3} = \dfrac{z-3}{1}$ are perpendicular.

The direction cosines of the first line are $\cos \alpha = \dfrac{2}{\sqrt{14}}$, $\cos \beta = \dfrac{3}{\sqrt{14}}$, $\cos \gamma = \dfrac{-1}{\sqrt{14}}$.

The direction cosines of the second line are $\cos \alpha = \dfrac{5}{\sqrt{35}}$, $\cos \beta = \dfrac{-3}{\sqrt{35}}$, $\cos \gamma = \dfrac{1}{\sqrt{35}}$.

$$\cos \theta = \frac{2}{\sqrt{14}} \cdot \frac{5}{\sqrt{35}} + \frac{3}{\sqrt{14}} \cdot \frac{-3}{\sqrt{35}} + \frac{-1}{\sqrt{14}} \cdot \frac{1}{\sqrt{35}} = \frac{10 - 9 - 1}{\sqrt{14}\sqrt{35}} = 0, \quad \text{and } \theta = 90°.$$

Or, by using the direction numbers ($2, 3, -1$, and $5, -3, 1$) of the lines, we get $2(5) + 3(-3) + (-1)(1) = 0$. Hence the lines are perpendicular.

5. Draw the graph of the line $3x - 2y + 3z - 4 = 0$, $x - 2y - z + 4 = 0$.

Find two of the pierce points of the line, then connect these points.

To find where the line pierces the xy-plane, let $z = 0$. Then

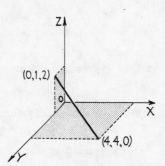

$$3x - 2y = 4$$
$$x - 2y = -4.$$

Solving, $x = 4$, $y = 4$. Hence $(4, 4, 0)$ is the xy-piercing point.

Similarly, the yz-piercing point is $(0, 1, 2)$.

6. Find the point where the line $x + 2y - z - 6 = 0$, $2x - y + 3z + 13 = 0$ pierces the plane $3x - 2y + 3z + 16 = 0$.

Since the piercing point must satisfy all three equations, the problem is the solution of three simultaneous linear equations. If z is eliminated between the equations, we obtain two equations $3x + 2y - 1 = 0,\quad x - y + 3 = 0.$

Solving these two equations, we obtain $x = -1,\ y = 2.$ Substituting these values in $x + 2y - z - 6 = 0$, we find $z = -3$. Hence the line pierces the plane in point $(-1, 2, -3)$.

7. Show that the lines represented by each of the following pairs of planes intersect:

$$x - y - z - 7 = 0,\ 3x - 4y - 11 = 0,\quad \text{and}\quad x + 2y - z - 1 = 0,\ x + y + 1 = 0.$$

Let (x_1, y_1, z_1) be the coordinates of the point of intersection of the two lines. This point must satisfy the equation of each of the planes. Hence:

$$
\begin{aligned}
(1)\ & x_1 - y_1 - z_1 = 7 \\
(2)\ & 3x_1 - 4y_1 = 11 \\
(3)\ & x_1 + 2y_1 - z_1 = 1 \\
(4)\ & x_1 + y_1 = -1.
\end{aligned}
$$

Subtracting (3) from (1) gives $y_1 = -2$. Substituting this value for y_1 in (4), we obtain $x_1 = 1$. Substituting these values in (1), $z_1 = -4$.

The point of intersection is $(1, -2, -4)$.

8. Find the angle between the line $x + 2y - z + 3 = 0,\quad 2x - y + 3z + 5 = 0,$ and the plane $3x - 4y + 2z - 5 = 0$.

To obtain the direction numbers of the line:

$$
\begin{array}{cccc}
1 & 2 & -1 & 2 \\
2 & -1 & 3 & -1
\end{array}
\quad \text{or}\quad 6-1,\ -2-3,\ -1-4,\quad \text{or}\quad 5, -5, -5,\quad \text{or}\quad 1, -1, -1.
$$

The angle between the line and the plane will be the complement of the angle θ between the line and the normal to the plane. The direction numbers of the normal are $3, -4, 2$.

$$\cos\theta = \frac{3(1) - 4(-1) + 2(-1)}{\sqrt{3}\,\sqrt{29}} = \frac{5}{\sqrt{87}}.\quad \text{Hence } \theta = 57°35'.$$

The angle between the line and the plane is $32°25'$.

9. Find the symmetric equations of the line of intersection of the planes

$$
\begin{aligned}
2x - 3y + 3z - 4 &= 0 \\
x + 2y - z + 3 &= 0.
\end{aligned}
$$

Eliminating z and y in turn between the given equations, we obtain

$$5x + 3y + 5 = 0 \quad \text{and}\quad 7x + 3z + 1 = 0.$$

Equating the values of x from the two equations, we obtain

$$x = \frac{3y + 5}{-5} = \frac{3z + 1}{-7} \quad \text{or}\quad \frac{x}{1} = \frac{y + \frac{5}{3}}{-\frac{5}{3}} = \frac{z + \frac{1}{3}}{-\frac{7}{3}} \quad \text{or}\quad \frac{x}{3} = \frac{y + \frac{5}{3}}{-5} = \frac{z + \frac{1}{3}}{-7}.$$

This represents a line through $(0, -\frac{5}{3}, -\frac{1}{3})$ having direction numbers $3, -5, -7$.

10. Write in parametric form the equations of the line of intersection of the planes

$$3x + 3y - 4z + 7 = 0 \quad \text{and}\quad x + 6y + 2z - 6 = 0.$$

Eliminating y and z in turn between the given equations, we obtain

$$x - 2z + 4 = 0 \quad \text{and}\quad x + 3y - 1 = 0.$$

Equating the values of x in the two equations, we have $\dfrac{x}{6} = \dfrac{y - 1/3}{-2} = \dfrac{z - 2}{3}$.

If we now set these ratios equal to a parameter t, we obtain the parametric form of the equations of the given line: $x = 6t,\quad y = \frac{1}{3} - 2t,\quad z = 2 + 3t.$

11. Find the equations of the projection planes of the line of intersection of the planes
$$2x + 3y - 5z + 6 = 0$$
$$3x - 2y + z - 8 = 0.$$

To find the projection planes, eliminate z, y, and x in turn between the two equations to obtain $17x - 7y - 34 = 0$, $13x - 7z - 12 = 0$, and $13y - 17z + 34 = 0$ as the projection planes of the line on the xy-, xz- and yz-planes respectively.

12. Write the equations of the line through $(1, -2, 2)$ with direction angles $60°$, $120°$, $45°$.

Use $\dfrac{x - x_1}{\cos \alpha} = \dfrac{y - y_1}{\cos \beta} = \dfrac{z - z_1}{\cos \gamma}$ to obtain

$\dfrac{x - 1}{\cos 60°} = \dfrac{y + 2}{\cos 120°} = \dfrac{z - 2}{\cos 45°}$,   or $\dfrac{x - 1}{\frac{1}{2}} = \dfrac{y + 2}{-\frac{1}{2}} = \dfrac{z - 2}{\frac{1}{2}\sqrt{2}}$,   or $\dfrac{x - 1}{1} = \dfrac{y + 2}{-1} = \dfrac{z - 2}{\sqrt{2}}$.

13. Write the equations of the line passing through $(-2, 1, 3)$ and $(4, 2, -2)$.

Use $\dfrac{x - x_1}{x_2 - x_1} = \dfrac{y - y_1}{y_2 - y_1} = \dfrac{z - z_1}{z_2 - z_1}$ to get $\dfrac{x + 2}{4 + 2} = \dfrac{y - 1}{2 - 1} = \dfrac{z - 3}{-2 - 3}$ or $\dfrac{x + 2}{6} = \dfrac{y - 1}{1} = \dfrac{z - 3}{-5}$.

14. Find the equations of the line passing through $(1, -3, 4)$ and perpendicular to the plane $x - 3y + 2z = 4$.

The direction numbers of the line are $1, -3, 2$.

The required equations are $\dfrac{x - 1}{1} = \dfrac{y + 3}{-3} = \dfrac{z - 4}{2}$, or $3x + y = 0$, $2y + 3z - 6 = 0$.

15. Find the equation of the plane containing the lines
$$\frac{x - 1}{4} = \frac{y + 1}{2} = \frac{z - 2}{3} \quad \text{and} \quad \frac{x - 1}{5} = \frac{y + 1}{4} = \frac{z - 2}{3}.$$

Note that the lines intersect in the point $(1, -1, 2)$.

Use the equation $Ax + By + Cz + D = 0$. Since each line lies in the plane, it is perpendicular to the normal to the plane. Hence,
$$4A + 2B + 3C = 0,$$
$$5A + 4B + 3C = 0.$$

Also, the point $(1, -1, 2)$ lies in the plane. Therefore,
$$A - B + 2C + D = 0.$$

Since we have four unknowns and only three equations, find three of the unknowns in terms of the fourth.

Determine A, C, D in terms of B: $A = -2B$, $C = 2B$, $D = -B$. Substituting these values in the general equation and dividing through by B, we obtain $2x - y - 2z + 1 = 0$.

SUPPLEMENTARY PROBLEMS

1. Find the coordinates of the point on the line
   (a) $2x - y + z - 5 = 0$, $x + 2y - 2z - 5 = 0$, for $z = 1$.   *Ans.* $(3, 2, 1)$
   (b) $4x - 3y + 2z - 7 = 0$, $x + 4y - z - 5 = 0$, for $y = 2$.   *Ans.* $(7/6, 2, 25/6)$
   (c) $\dfrac{x - 2}{3} = \dfrac{y + 4}{-2} = \dfrac{z - 1}{2}$, for $x = 3$.   *Ans.* $(3, -14/3, 5/3)$
   (d) $2x = 3y - 1$, $3z = 4 - 2y$, for $x = 4$.   *Ans.* $(4, 3, -2/3)$
   (e) $x = 4 - 3t$, $y = -1 + 4t$, $z = 2t - 3$, for $t = 3$.   *Ans.* $(-5, 11, 3)$

2. Find the points in which each of the following lines pierce the coordinate planes. Plot the lines by joining two of the pierce points.

(a) $x - 2y + z = 0$, $\quad 3x + y + 2z = 7$.   *Ans.* $(2,1,0)$, $(7,0,-7)$, $(0, 7/5, 14/5)$

(b) $2x - y + 3z + 1 = 0$, $\quad 5x + 4y - z - 6 = 0$.   *Ans.* $(\frac{2}{13}, \frac{17}{13}, 0)$, $(1,0,-1)$, $(0, \frac{17}{11}, \frac{2}{11})$

(c) $\dfrac{x-1}{2} = \dfrac{y+3}{1} = \dfrac{z-6}{-1}$.   *Ans.* $(13,3,0)$, $(7,0,3)$, $(0,-7/2, 13/2)$

(d) $2x + 3y - 2 = 0$, $\quad y - 3z + 4 = 0$.   *Ans.* $(7,-4,0)$, $(1,0,4/3)$, $(0, 2/3, 14/9)$

(e) $x + 2y - 6 = 0$, $\quad z = 4$.   *Ans.* $(6,0,4)$, $(0,3,4)$

3. Find the direction numbers and the direction cosines for the lines:

(a) $3x + y - z - 8 = 0$, $\quad 4x - 7y - 3z + 1 = 0$.   *Ans.* $2, -1, 5$; $\dfrac{2}{\sqrt{30}}, \dfrac{-1}{\sqrt{30}}, \dfrac{5}{\sqrt{30}}$

(b) $2x - 3y + 9 = 0$, $\quad 2x - y + 8z + 11 = 0$.   *Ans.* $6, 4, -1$; $\dfrac{6}{\sqrt{53}}, \dfrac{4}{\sqrt{53}}, \dfrac{-1}{\sqrt{53}}$

(c) $3x - 4y + 2z - 7 = 0$, $\quad 2x + y + 3z - 11 = 0$.   *Ans.* $14, 5, -11$; $\dfrac{14}{3\sqrt{38}}, \dfrac{5}{3\sqrt{38}}, \dfrac{-11}{3\sqrt{38}}$

(d) $x - y + 2z - 1 = 0$, $\quad 2x - 3y - 5z - 7 = 0$.   *Ans.* $11, 9, -1$; $\dfrac{11}{\sqrt{203}}, \dfrac{9}{\sqrt{203}}, \dfrac{-1}{\sqrt{203}}$

(e) $3x - 2y + z + 4 = 0$, $\quad 2x + 2y - z - 3 = 0$.   *Ans.* $0, 1, 2$; $0, \dfrac{1}{\sqrt{5}}, \dfrac{2}{\sqrt{5}}$

4. Find the acute angle between the lines $x - 2y + z - 2 = 0$, $2y - z - 1 = 0$
         and $x - 2y + z - 2 = 0$, $x - 2y + 2z - 4 = 0$.   *Ans.* $78°27.8'$

5. Find the acute angle between the lines $\dfrac{x-1}{6} = \dfrac{y+2}{-3} = \dfrac{z-4}{6}$ and $\dfrac{x+2}{3} = \dfrac{y-3}{6} = \dfrac{z+4}{-2}$.
                                          Ans. $79°1'$

6. Find the acute angle between the lines
      $2x + 2y + z - 4 = 0$, $x - 3y + 2z = 0$ and $\dfrac{x-2}{7} = \dfrac{y+2}{6} = \dfrac{z-4}{-6}$.   *Ans.* $49°26.5'$

7. Find the acute angle between the line $\dfrac{x+1}{3} = \dfrac{y-1}{6} = \dfrac{z-3}{-6}$ and the plane $2x - 2y + z - 3 = 0$.
                                          Ans. $26°23.3'$

8. Find the acute angle between the line joining the points $(3,4,2)$, $(2,3,-1)$ and the line joining the points $(1,-2,3)$, $(-2,-3,1)$.   *Ans.* $36°19'$

9. Show that the line $\dfrac{x-1}{1} = \dfrac{y+2}{2} = \dfrac{z-3}{4}$ is parallel to the plane $6x + 7y - 5z - 8 = 0$.

10. Write the equations of the line through the point $(2,1,-2)$ and perpendicular to the plane $3x - 5y + 2z + 4 = 0$.   *Ans.* $\dfrac{x-2}{3} = \dfrac{y-1}{-5} = \dfrac{z+2}{2}$

11. Write the equations of the line through
     (a) $(2,-1,3)$ parallel to the x-axis.   *Ans.* $y + 1 = 0$, $\quad z - 3 = 0$
     (b) $(2,-1,3)$ parallel to the y-axis.   *Ans.* $x - 2 = 0$, $\quad z - 3 = 0$
     (c) $(2,-1,3)$ parallel to the z-axis.   *Ans.* $x - 2 = 0$, $\quad y + 1 = 0$
     (d) $(2,-1,3)$, $\cos \alpha = \dfrac{1}{2}$, $\cos \beta = \dfrac{1}{3}$.   *Ans.* $\dfrac{x-2}{3} = \dfrac{y+1}{2} = \dfrac{z-3}{\pm\sqrt{23}}$

12. Write the equations of the line through the point $(-6,4,1)$ and perpendicular to the plane $3x - 2y + 5z + 8 = 0$.   *Ans.* $2x + 3y = 0$, $\quad 5y + 2z - 22 = 0$

13. Write the equations of the line through $(2,0,-3)$ perpendicular to the plane $2x - 3y + 6 = 0$.
                             Ans. $3x + 2y - 6 = 0$, $\quad z + 3 = 0$

14. Write the equations of the line through the point $(1,-2,-3)$ perpendicular to the plane $x - 3y + 2z + 4 = 0$. *Ans.* $\dfrac{x-1}{1} = \dfrac{y+2}{-3} = \dfrac{z+3}{2}$

15. Write the equations of the line through the points $(2,-3,4)$ and $(5,2,-1)$.
Ans. $\dfrac{x-2}{3} = \dfrac{y+3}{5} = \dfrac{z-4}{-5}$

16. Write the equations of the line through the points
    (a) $(1,2,3)$ and $(-2,3,3)$. *Ans.* $x + 3y - 7 = 0$, $z = 3$
    (b) $(-2,2,-3)$ and $(2,-2,3)$. *Ans.* $x + y = 0$, $3y + 2z = 0$
    (c) $(2,3,4)$ and $(2,-3,-4)$. *Ans.* $x - 2 = 0$, $4y - 3z = 0$
    (d) $(1,0,3)$ and $(2,0,3)$. *Ans.* $y = 0$, $z = 3$
    (e) $(2,-1,3)$ and $(6,7,4)$ in parametric form. *Ans.* $x = 2 + \dfrac{4}{9}t$, $y = -1 + \dfrac{8}{9}t$, $z = 3 + \dfrac{1}{9}t$

17. Write the equations of the line through $(1,-2,3)$ and parallel to each of the planes $2x - 4y + z - 3 = 0$ and $x + 2y - 6z + 4 = 0$. *Ans.* $\dfrac{x-1}{22} = \dfrac{y+2}{13} = \dfrac{z-3}{8}$

18. Write the equations of the line through $(1,4,-2)$ and parallel to each of the planes $6x + 2y + 2z + 3 = 0$ and $3x - 5y - 2z - 1 = 0$. *Ans.* $\dfrac{x-1}{1} = \dfrac{y-4}{3} = \dfrac{z+2}{-6}$

19. Write the equations of the line through $(-2,4,3)$ parallel to the line through $(1,3,4)$ and $(-2,2,3)$. *Ans.* $x - 3y + 14 = 0$, $y - z - 1 = 0$

20. Write the equations of the line through $(3,-1,4)$ perpendicular to the lines whose direction numbers are 3, 2, -4 and 2, -3, 2. *Ans.* $\dfrac{x-3}{8} = \dfrac{y+1}{14} = \dfrac{z-4}{13}$

21. Write the equations of the line through $(2,2,-3)$ perpendicular to the lines whose direction numbers are 2, -1, 3 and -1, 2, 0. *Ans.* $x - 2y + 2 = 0$, $y + z + 1 = 0$

22. Write the equations of the line through $(2,-2,4)$ with direction angles $120°$, $60°$, $45°$. *Ans.* $\dfrac{x-2}{-1} = \dfrac{y+2}{1} = \dfrac{z-4}{\sqrt{2}}$

23. Write the equations of the line through $(-2,1,3)$ with direction angles $135°$, $60°$, $120°$. *Ans.* $\dfrac{x+2}{-\sqrt{2}} = \dfrac{y-1}{1} = \dfrac{z-3}{-1}$

24. Write the equations of the line through
    (a) $(0,2,-1)$ with direction numbers 1, -3, 4. *Ans.* $\dfrac{x}{1} = \dfrac{y-2}{-3} = \dfrac{z+1}{4}$
    (b) $(-1,1,-3)$ with direction numbers $\sqrt{2}$, 3, -4. *Ans.* $\dfrac{x+1}{\sqrt{2}} = \dfrac{y-1}{3} = \dfrac{z+3}{-4}$
    (c) $(0,0,0)$ with direction numbers 1, 1, 1. *Ans.* $x = y = z$
    (d) $(-2,3,2)$ with direction numbers 0, 2, 1. *Ans.* $x + 2 = 0$, $y - 2z + 1 = 0$
    (e) $(1,-1,6)$ with direction numbers 2, -1, 1. *Ans.* $x = 2z - 11$, $y = -z + 5$

25. Show that the line $x = \dfrac{2}{7}z + \dfrac{15}{7}$, $y = -\dfrac{5}{7}z - \dfrac{34}{7}$ is perpendicular to the line $x - y - z - 7 = 0$, $3x - 4y - 11 = 0$.

26. Show that the lines $x + 2y - z - 1 = 0$, $x + y + 1 = 0$ and $\dfrac{7x - 15}{2} = \dfrac{7y + 34}{-5} = \dfrac{z}{1}$ are perpendicular.

27. Show that the lines $3x - 2y + 13 = 0$, $y + 3z - 26 = 0$ and $\dfrac{x+4}{5} = \dfrac{y-1}{-3} = \dfrac{z-3}{1}$ are perpendicular.

28. Show that the lines $\dfrac{x-3}{1} = \dfrac{y+8}{-2} = \dfrac{z+6}{-11}$ and $3x + 5y + 7 = 0,\ y + 3z - 10 = 0$ are perpendicular.

29. Show that the lines $x - 2y + 2 = 0,\ 2y + z + 4 = 0$ and $7x + 4y - 15 = 0,\ y + 14z + 40 = 0$ are perpendicular.

30. Show that the line $\dfrac{x-2}{10} = \dfrac{2y-2}{11} = \dfrac{z-5}{7}$ lies in the plane $3x - 8y + 2z - 8 = 0$. To show that a line lies in a plane it is necessary to show that two points on the line lie in the plane, or that one point of the line lies in the plane and the line is perpendicular to the normal to the plane.

31. Show that the line $y - 2x + 5 = 0,\ z - 3x - 4 = 0$ lies in the plane $9x + 3y - 5z + 35 = 0$.

32. Show that the line $x - z - 4 = 0,\ y - 2z - 3 = 0$ lies in the plane $2x + 3y - 8z - 17 = 0$.

33. Show that the line $\dfrac{x-1}{1} = \dfrac{y+2}{2} = \dfrac{z-3}{4}$ lies in the plane $2x + 3y - 2z + 10 = 0$.

34. Find the coordinates of the point in which the line $2x - y - 2z - 5 = 0,\ 4x + y + 3z - 1 = 0$ pierces the plane $8x - y + z - 5 = 0$.   *Ans.* $(3/2, 4, -3)$

35. Find the point in which the line $x = z + 2,\ y = -3z + 1$ pierces the plane $x - 2y - 7 = 0$.   *Ans.* $(3, -2, 1)$

36. Find the point in which the line $\dfrac{x}{1} = \dfrac{2y-3}{1} = \dfrac{2z-1}{5}$ pierces the plane $4x - 2y + z - 3 = 0$.   *Ans.* $(1, 2, 3)$

37. Find the point in which the line $x + 2y + 4z - 2 = 0,\ 2x + 3y - 2z + 3 = 0$ pierces the plane $2x - y + 4z + 8 = 0$.   *Ans.* $(-4, 2, 1/2)$

38. Find the equations of the line lying in the plane $x + 3y - z + 4 = 0$ and perpendicular to the line $x - 2z - 3 = 0,\ y - 2z = 0$ at the point where the line meets the plane.   *Ans.* $3x + 5y + 7 = 0,\ 4x + 5z + 1 = 0$

39. Show that the points $(2, -3, 1),\ (5, 4, -4)$ and $(8, 11, -9)$ lie in a straight line.

40. Find the point of intersection of the lines $2x + y - 5 = 0,\ 3x + z - 14 = 0$ and $x - 4y - 7 = 0,\ 5x + 4z - 35 = 0$.   *Ans.* $(3, -1, 5)$

41. Find the point of intersection of the lines $x - y - z + 8 = 0,\ 5x + y + z + 10 = 0$ and $x + y + z - 2 = 0,\ 2x + y - 3z + 9 = 0$.   *Ans.* $(-3, 3, 2)$

42. Find the point of intersection of the lines $x + 5y - 7z + 1 = 0,\ 10x - 23y + 40z - 27 = 0$ and $x - y + z + 1 = 0,\ 2x + y - 2z + 2 = 0$.   *Ans.* $(-1/38,\ 148/38,\ 111/38)$

43. Find in symmetric form the equation of the locus of points equidistant from the points $(3, -1, 2),\ (4, -6, -5)$ and $(0, 0, -3)$.   *Ans.* $\dfrac{x}{16} = \dfrac{y + 175/32}{13} = \dfrac{z + 19/32}{-7}$

44. Find in symmetric form the equation of the locus of points which are equidistant from the points $(3, -2, 4),\ (5, 3, -2)$ and $(0, 4, 2)$.   *Ans.* $\dfrac{x - 18/11}{26} = \dfrac{y}{22} = \dfrac{z + 9/44}{27}$

CHAPTER 15

Surfaces

QUADRIC SURFACES. The surface defined by an equation of the second degree in three variables is called a *quadric surface* or *conicoid*. Any plane section of a quadric surface is a conic or a limiting form of a conic.

The most general equation of the second degree in three variables is of the form $Ax^2 + By^2 + Cz^2 + Dxy + Exz + Fyz + Gx + Hy + Iz + K = 0$.

By rotation or translation of axes, or both, the general equation can be converted into one of the two types:

$$(1) \quad Ax^2 + By^2 + Cz^2 = D$$
$$(2) \quad Ax^2 + By^2 + Iz = 0.$$

If none of the constants in (1) or (2) is zero, the equation may be written in one of the two forms:

$$(3) \quad \pm \frac{x^2}{a^2} \pm \frac{y^2}{b^2} \pm \frac{z^2}{c^2} = 1$$

$$(4) \quad \pm \frac{x^2}{a^2} \pm \frac{y^2}{b^2} = \frac{z}{c}.$$

Now (3) can represent only three fundamentally distinct surfaces, whose equations have the form

$$(5) \quad \frac{x^2}{a^2} + \frac{y^2}{b^2} + \frac{z^2}{c^2} = 1, \quad \frac{x^2}{a^2} + \frac{y^2}{b^2} - \frac{z^2}{c^2} = 1, \quad \frac{x^2}{a^2} - \frac{y^2}{b^2} - \frac{z^2}{c^2} = 1.$$

Since all surfaces in (5) are symmetrical with respect to the origin, they are called central quadrics.

The two surfaces represented by (4) are non-central quadrics.

THE SPHERE. If $a = b = c$ in $\frac{x^2}{a^2} + \frac{y^2}{b^2} + \frac{z^2}{c^2} = 1$, the equation can be written $x^2 + y^2 + z^2 = a^2$, the equation of a sphere with center at $(0,0,0)$ and radius a.

If the center of the sphere is (h,k,j) the equation has the form
$$(x-h)^2 + (y-k)^2 + (z-j)^2 = a^2.$$

THE ELLIPSOID. If a, b, c are unequal, the equation $\frac{x^2}{a^2} + \frac{y^2}{b^2} + \frac{z^2}{c^2} = 1$ represents the general case. If $a \neq b$, but $b = c$, the ellipsoid is an ellipsoid of revolution.

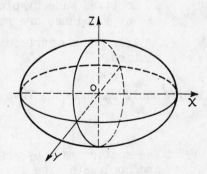

If the center of the ellipsoid is at (h,k,j) and its axes are parallel to the coordinate axes, the equation has the form
$$\frac{(x-h)^2}{a^2} + \frac{(y-k)^2}{b^2} + \frac{(z-j)^2}{c^2} = 1.$$

If the center is at the origin this equation becomes $\frac{x^2}{a^2} + \frac{y^2}{b^2} + \frac{z^2}{c^2} = 1.$

131

HYPERBOLOID OF ONE SHEET. If the equation has the sign of one variable changed,

as $\dfrac{x^2}{a^2} + \dfrac{y^2}{b^2} - \dfrac{z^2}{c^2} = 1$, the surface is called a *hyperboloid of one sheet*.

If $a = b$, the surface is the hyperboloid of revolution of one sheet.

Sections parallel to the xz- and yz-planes are hyperbolas. Sections parallel to the xy-plane are ellipses except in the hyperboloid of revolution, where the sections are circles.

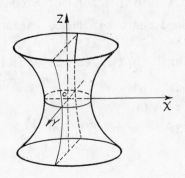
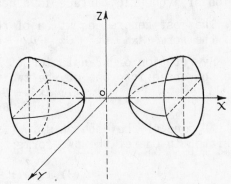

Hyperboloid of One Sheet.   *Hyperboloid of Two Sheets.*

HYPERBOLOID OF TWO SHEETS. The equation $\dfrac{x^2}{a^2} - \dfrac{y^2}{b^2} - \dfrac{z^2}{c^2} = 1$ represents a *hyper-*

boloid of two sheets. The equation is the same as that of the ellipsoid, with the signs of two variables changed. If $b = c$, the surface is a surface of revolution.

Sections parallel to the xy- and xz-planes are hyperbolas. Sections parallel to the yz-plane are ellipses except in the hyperboloid of revolution, where the sections are circles.

ELLIPTIC PARABOLOID. This is the locus of an equation of the form $\dfrac{x^2}{a^2} + \dfrac{y^2}{b^2} = 2cz$.

The section by a plane $z = k$ is an ellipse which increases in size as the cutting plane recedes from the xy-plane.

If $c > 0$, the surface lies wholly above the xy-plane. If $c < 0$, the surface lies wholly below the xy-plane.

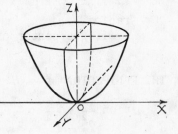

Sections made by planes parallel to the xz- or yz-planes are parabolas.

If $a = b$, the surface is a surface of revolution.

HYPERBOLIC PARABOLOID. This is the locus of an equation of the form

$$\frac{x^2}{a^2} - \frac{y^2}{b^2} = 2cz, \quad (c > 0).$$

The section made by a plane $z = k$ if $k > 0$ is a hyperbola whose transverse and conjugate axes are parallel to the x- and y-axes respectively and increase with k. If $k < 0$, the transverse and conjugate axes are parallel to the y- and x-axes respectively. If $k = 0$, the trace is the pair of straight lines $\dfrac{x^2}{a^2} - \dfrac{y^2}{b^2} = 0$.

The section made by a plane $y = k$ is a parabola open upward, and that made by a plane $x = k$ is a parabola open downward.

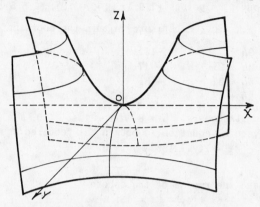

Hyperbolic Paraboloid.

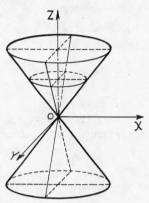

Right Circular Cone.

RIGHT CIRCULAR CONE, $x^2 + y^2 - c^2 z^2 = 0$.

This surface can be regarded as a surface of revolution generated by revolving a straight line $y = kx$ about the z-axis.

Any horizontal section made by a plane parallel to the xy-plane is a circle. Any section made by a plane parallel to the yz-plane or to the xz-plane is a hyperbola.

CYLINDRICAL SURFACE. A cylindrical surface is generated by a straight line which moves along a fixed curve and remains parallel to a fixed straight line. The fixed curve is called the *directrix* of the surface and the moving line is the *generatrix* of the surface.

A cylindrical surface whose generatrix is parallel to one of the coordinate axes and whose directrix is a curve in the coordinate plane that is perpendicular to the generatrix, has the same equation as the directrix.

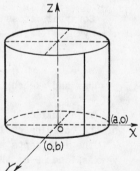

If the directrix is the ellipse $\dfrac{x^2}{a^2} + \dfrac{y^2}{b^2} = 1$,

the equation of the cylinder is $\dfrac{x^2}{a^2} + \dfrac{y^2}{b^2} = 1$.

SOLVED PROBLEMS

1. Find the equation of the sphere with its center at $(-2, 1, -3)$ and radius 4.

Substituting in $(x - h)^2 + (y - k)^2 + (z - j)^2 = a^2$, we have
$$(x + 2)^2 + (y - 1)^2 + (z + 3)^2 = 4^2.$$
Expanding and collecting terms, $x^2 + y^2 + z^2 + 4x - 2y + 6z - 2 = 0$.

2. Find the equation of the sphere with its center at $(3, 6, -4)$ and tangent to the plane
$$2x - 2y - z - 10 = 0.$$

The radius $a = \left| \dfrac{2(3) - 2(6) - 1(-4) - 10}{3} \right| = 4.$ Hence the required equation is

$(x-3)^2 + (y-6)^2 + (z+4)^2 = 16,$ or $x^2 + y^2 + z^2 - 6x - 12y + 8z + 45 = 0.$

3. Find the equation of the sphere through the points $(7,9,1)$, $(-2,-3,2)$, $(1,5,5)$, $(-6,2,5)$.

Use $x^2 + y^2 + z^2 + Gx + Hy + Iz + K = 0.$ Substitute in turn the coordinates of the four points.

$$7G + 9H + \ \ I + K = -131$$
$$-2G - 3H + 2I + K = -\ 17$$
$$G + 5H + 5I + K = -\ 51$$
$$-6G + 2H + 5I + K = -\ 65.$$

Solving these four equations simultaneously, $G = 8$, $H = -14$, $I = 18$, $K = -79$. Substitute these values in the general equation to obtain the required equation

$$x^2 + y^2 + z^2 + 8x - 14y + 18z - 79 = 0.$$

4. Find the coordinates of the center and the radius of the sphere
$$x^2 + y^2 + z^2 - 6x + 4y - 3z = 15.$$

Complete the squares of the terms in x, y, and z separately and compare with
$$(x-h)^2 + (y-k)^2 + (z-j)^2 = a^2.$$

$x^2 - 6x + 9 + y^2 + 4y + 4 + z^2 - 3z + \dfrac{9}{4} = \dfrac{121}{4},$ or $(x-3)^2 + (y+2)^2 + (z - \dfrac{3}{2})^2 = (\dfrac{11}{2})^2.$

The sphere has center $(3, -2, \dfrac{3}{2})$ and radius $\dfrac{11}{2}.$

5. Find the locus of a point whose distances from $(-2,2,-2)$ and $(3,-3,3)$ are in the ratio $2:3$ numerically.

$$\frac{\sqrt{(x+2)^2 + (y-2)^2 + (z+2)^2}}{\sqrt{(x-3)^2 + (y+3)^2 + (z-3)^2}} = \frac{2}{3}.$$

Squaring, cross multiplying, and collecting terms, we obtain
$x^2 + y^2 + z^2 + 12x - 12y + 12z = 0,$ a sphere with center $(-6,6,-6)$ and radius $6\sqrt{3}.$

6. Discuss and sketch the surface $\dfrac{x^2}{25} + \dfrac{y^2}{16} + \dfrac{z^2}{9} = 1.$

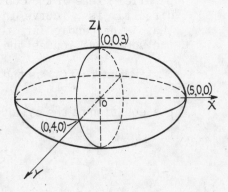

This surface is symmetric with respect to each of the coordinate planes and the origin.

The intercepts are ± 5 on the x-axis, ± 4 on the y-axis, and ± 3 on the z-axis.

Its trace in the xy-plane is the ellipse
$\dfrac{x^2}{25} + \dfrac{y^2}{16} = 1$ with semi-axes 5 and 4. Similarly,
its traces in the xz- and yz-planes are ellipses. This surface is an ellipsoid.

7. By completing squares in x, y, and z, show that the following equation represents an ellipsoid. Locate the center and determine the lengths of the semi-axes.
$$2x^2 + 3y^2 + z^2 - 8x + 6y - 4z - 3 = 0.$$

$$2(x^2 - 4x + 4) + 3(y^2 + 2y + 1) + (z^2 - 4z + 4) = 3 + 8 + 3 + 4 = 18,$$

or $\qquad\qquad 2(x-2)^2 + 3(y+1)^2 + (z-2)^2 = 18.$

Dividing the equation by 18, we obtain $\dfrac{(x-2)^2}{9} + \dfrac{(y+1)^2}{6} + \dfrac{(z-2)^2}{18} = 1$, an ellipsoid with center at $(2,-1,2)$ and semi-axes 3, $\sqrt{6}$, $3\sqrt{2}$.

8. Show that the locus of a point which moves so that the sum of its distances from $(2,3,4)$ and $(2,-3,4)$ is 8, is an ellipsoid. Find its center and the lengths of the semi-axes.

$$\sqrt{(x-2)^2 + (y-3)^2 + (z-4)^2} + \sqrt{(x-2)^2 + (y+3)^2 + (z-4)^2} = 8$$

or $\sqrt{(x-2)^2 + (y-3)^2 + (z-4)^2} = 8 - \sqrt{(x-2)^2 + (y+3)^2 + (z-4)^2}$.

Squaring and collecting terms, $3y + 16 = 4\sqrt{(x-2)^2 + (y+3)^2 + (z-4)^2}$.

Squaring and collecting terms, $16x^2 + 7y^2 + 16z^2 - 64x - 128z + 208 = 0$.

Completing the squares in x, y, z, this becomes $\dfrac{(x-2)^2}{7} + \dfrac{(y-0)^2}{16} + \dfrac{(z-4)^2}{7} = 1$, an ellipsoid of revolution with center at $(2,0,4)$ and semi-axes $\sqrt{7}$, 4, $\sqrt{7}$. Sections of the surface parallel to the xz-plane are circles.

9. Find the equation of the ellipsoid which passes through $(2,2,4)$, $(0,0,6)$, $(2,4,2)$ and has the coordinate planes as planes of symmetry.

Use $\dfrac{x^2}{a^2} + \dfrac{y^2}{b^2} + \dfrac{z^2}{c^2} = 1$ and substitute the coordinates of the given points for x, y, z.

Then $\dfrac{4}{a^2} + \dfrac{4}{b^2} + \dfrac{16}{c^2} = 1$, $\dfrac{0}{a^2} + \dfrac{0}{b^2} + \dfrac{36}{c^2} = 1$, and $\dfrac{4}{a^2} + \dfrac{16}{b^2} + \dfrac{4}{c^2} = 1$.

Solving for a^2, b^2 and c^2, we obtain $a^2 = 9$, $b^2 = 36$, $c^2 = 36$.

Substituting, $\dfrac{x^2}{9} + \dfrac{y^2}{36} + \dfrac{z^2}{36} = 1$, or $4x^2 + y^2 + z^2 = 36$.

10. Discuss and sketch the locus of $\dfrac{x^2}{9} + \dfrac{y^2}{4} - \dfrac{z^2}{16} = 1$.

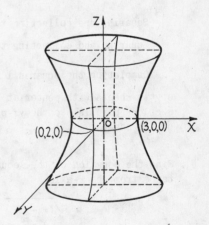

This surface is symmetric about each of the coordinate planes and the origin.

The x- and y-intercepts are ± 3 and ± 2 respectively. There is no z-intercept.

The sections by planes $z = k$ are ellipses with centers on the z-axis. These ellipses increase in size as k increases numerically.

The sections of this surface by planes parallel to the xz- or yz-planes are hyperbolas.

This surface is a hyperboloid of one sheet.

11. By completing the squares in x, y, z, determine the nature of the surface whose equation is $3x^2 + 4y^2 - 2z^2 + 6x - 16y + 8z = 13$.

$$3(x^2 + 2x + 1) + 4(y^2 - 4y + 4) - 2(z^2 - 4z + 4) = 13 + 11 = 24,$$

or $\dfrac{(x+1)^2}{8} + \dfrac{(y-2)^2}{6} - \dfrac{(z-2)^2}{12} = 1$.

This is a hyperboloid of one sheet with center at $(-1, 2, 2)$ and axis parallel to the z-axis. Sections by planes parallel to the xy-plane are ellipses. Sections by planes parallel to the xz- or yz-planes are hyperbolas.

12. Discuss and sketch the locus of $\dfrac{x^2}{9} - \dfrac{y^2}{4} - \dfrac{z^2}{16} = 1$.

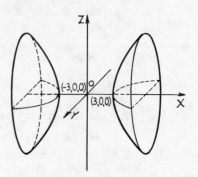

This surface is symmetric about each of the coordinate planes and the origin.

Its intercepts on the x-axis are ± 3. It has no y- or z-intercepts.

Sections of this surface by planes parallel to the xy- and xz-planes are hyperbolas. Sections by planes parallel to the yz-plane are ellipses.

This surface is a hyperboloid of two sheets.

13. By completing the squares in x, y, z, determine the nature of the surface whose equation is $2x^2 - 3y^2 - 2z^2 - 8x + 6y - 12z - 21 = 0$.

$$2(x^2 - 4x + 4) - 3(y^2 - 2y + 1) - 2(z^2 + 6z + 9) = 8, \quad \text{or} \quad \frac{(x-2)^2}{4} - \frac{(y-1)^2}{8/3} - \frac{(z+3)^2}{4} = 1,$$

a hyperboloid of two sheets with center at $(2, 1, -3)$ and transverse axis parallel to the x-axis.

14. Find the equation of the locus of a point, the difference of whose distances from $(-4, 3, 1)$ and $(4, 3, 1)$ is 6.

$$\sqrt{(x+4)^2 + (y-3)^2 + (z-1)^2} - \sqrt{(x-4)^2 + (y-3)^2 + (z-1)^2} = 6,$$

or $\quad \sqrt{(x+4)^2 + (y-3)^2 + (z-1)^2} = 6 + \sqrt{(x-4)^2 + (y-3)^2 + (z-1)^2}$.

Squaring and collecting terms, $\quad 4x - 9 = 3\sqrt{(x-4)^2 + (y-3)^2 + (z-1)^2}$.

Squaring and collecting terms, $\quad 7x^2 - 9y^2 - 9z^2 + 54y + 18z = 153$.

Completing the squares in x, y, z, $\quad \dfrac{(x-0)^2}{9} - \dfrac{(y-3)^2}{7} - \dfrac{(z-1)^2}{7} = 1$, a hyperboloid of two sheets with center at $(0, 3, 1)$ and transverse axis parallel to the x-axis. Since sections parallel to the yz-plane are circles, the surface is a hyperboloid of revolution of two sheets.

15. Find the equation of the locus of a point whose distance from $(2, -1, 3)$ is twice its distance from the x-axis.

$$\sqrt{(x-2)^2 + (y+1)^2 + (z-3)^2} = 2\sqrt{y^2 + z^2}.$$

Squaring and collecting terms, $\quad x^2 - 3y^2 - 3z^2 - 4x + 2y - 6z = -14$.

Completing the squares, $\quad (x-2)^2 - 3(y - 1/3)^2 - 3(z+1)^2 = -40/3$,

$$\text{or} \quad \frac{(y - 1/3)^2}{\dfrac{40}{9}} + \frac{(z+1)^2}{\dfrac{40}{9}} - \frac{(x-2)^2}{\dfrac{40}{3}} = 1, \qquad \text{a hyperboloid of}$$

revolution of one sheet, center at $(2, 1/3, -1)$, revolution about the x-axis.

16. Discuss and sketch the locus of $y^2 + z^2 = 4x$.

This surface is symmetric about the xz- and xy-planes, and about the x-axis.

The intercept on each axis is zero.

The traces are respectively the point circle $y^2 + z^2 = 0$, and the parabolas $z^2 = 4x$ and $y^2 = 4x$.

Since x can have no negative value, the surface lies wholly to the right of the yz-plane. Sections of this surface by planes parallel to the yz-plane are circles. Sections by planes parallel to the xy- or xz-planes are parabolas. This surface is a paraboloid of revolution.

17. Find the equation of the paraboloid with vertex at O, axis OZ, and passing through $(3, 0, 1)$ and $(3, 2, 2)$.

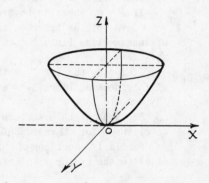

Use the equation $Ax^2 + By^2 = Cz$. Substituting the coordinates of the given points, we obtain

(1) $9A + 0B = C$, or $9A = C$
(2) $9A + 4B = 2C$.

Solving, $A = C/9$, $B = C/4$. Substituting these values for A and B in $Ax^2 + By^2 = Cz$, the required equation is $4x^2 + 9y^2 = 36z$ or $\dfrac{x^2}{9} + \dfrac{y^2}{4} = \dfrac{z}{1}$, an elliptic paraboloid.

18. Find the equation of the locus of a point the square of whose distance from the x-axis is always three times its distance from the yz-plane.

Let (x, y, z) represent the point. Then $y^2 + z^2 = 3x$.
This surface is a paraboloid of revolution, symmetric about the x-axis.

19. By completing the squares in x and y locate the vertex of the elliptic paraboloid
$$3x^2 + 2y^2 - 12z - 6x + 8y - 13 = 0.$$

$3(x^2 - 2x + 1) + 2(y^2 + 4y + 4) = 12z + 13 + 11 = 12z + 24$,

or $3(x - 1)^2 + 2(y + 2)^2 = 12(z + 2)$, or $\dfrac{(x - 1)^2}{4} + \dfrac{(y + 2)^2}{6} = \dfrac{z + 2}{1}$.

The vertex is at $(1, -2, -2)$.

20. Discuss and identify the following surface: $9x^2 - 4y^2 = 36z$.

The surface is symmetric about the xz-plane and yz-plane and the z-axis.
The intercept on each axis is zero.
If $z = 0$, the trace on the xy-plane is a pair of intersecting lines $9x^2 - 4y^2 = 0$, or $3x + 2y = 0$ and $3x - 2y = 0$.
If $y = 0$, the trace on the xz-plane is the parabola $9x^2 = 36z$, or $x^2 = 4z$. This parabola has its vertex at the origin and is open upward.
If $x = 0$, the trace on the yz-plane is the parabola $-4y^2 = 36z$, or $y^2 = -9z$. This parabola has its vertex at the origin but opens downward.
Sections of this surface by a plane $z = k$ are hyperbolas. If k is positive the transverse axis of the hyperbola is parallel to the x-axis. If k is negative the transverse axis of the hyperbola is parallel to the y-axis. Similarly, sections parallel to the xz-plane and parallel to the yz-plane are parabolas.
This surface is a hyperbolic paraboloid.

21. Find the equation of a paraboloid with vertex at $(0,0,0)$, axis OY, and passing through $(1,-2,1)$ and $(-3,-3,2)$.

Use the equation $Ax^2 + Cz^2 = By$. Substituting the coordinates of the two points,
$$A + C = -2B$$
$$9A + 4C = -3B.$$

Solving for A and C in terms of B, $A = B$, $C = -3B$.

Substitute these values for A and C, and divide the resulting equation through by B to obtain $x^2 - 3z^2 = y$, a hyperbolic paraboloid.

22. Discuss and sketch the locus of the cone $2y^2 + 3z^2 - x^2 = 0$.

This surface is symmetric about each of the coordinate planes and the origin.

The intercept on each of the axes is zero.

If $x = 0$, there is no trace on the yz-plane.

If $y = 0$, the trace on the xz-plane is the pair of intersecting lines $3z^2 - x^2 = 0$ or $\sqrt{3}z + x = 0$, $\sqrt{3}z - x = 0$.

If $z = 0$, the trace on the xy-plane is the pair of intersecting lines $2y^2 - x^2 = 0$ or $\sqrt{2}y + x = 0$, $\sqrt{2}y - x = 0$.

Sections of this surface by the plane $x = k$ will be ellipses for all values of k different from zero.

Similarly, sections of this surface by planes parallel to the xy- or xz-planes will be hyperbolas.

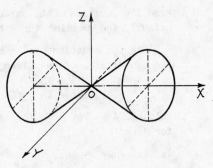

23. A point moves so that its distance from the y-axis is always 3 times its distance from the z-axis. Find the equation of its locus. Identify the surface.

$$\sqrt{x^2 + z^2} = 3\sqrt{x^2 + y^2}, \quad \text{or} \quad x^2 + z^2 = 9x^2 + 9y^2, \quad \text{or} \quad 8x^2 + 9y^2 - z^2 = 0.$$

The surface is a cone with vertex at the origin. The axis of the cone is the z-axis.

24. Sketch the locus of the surface $4x^2 + 9y^2 = 36$.

This surface is a cylinder with elements parallel to the z-axis, and having the ellipse $4x^2 + 9y^2 = 36$ as *directrix* curve.

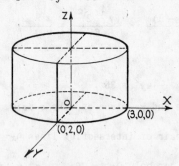

Problem 24.

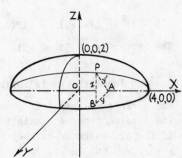

Problem 25.

25. Find the equation of the surface of revolution generated by revolving the following ellipse about the x-axis: $x^2 + 4z^2 - 16 = 0$.

Let $P(x,y,z)$ be any point on the surface. From P drop a perpendicular to the xy-plane. In the right triangle ABP, $AB = y$, $BP = z$.

Let $AP = y'$; then $y^2 + z^2 = y'^2$. But from the equation of the ellipse, $x^2 = 16 - 4y'^2$. Substituting, we have $x^2 = 16 - 4(y^2 + z^2)$, or $x^2 + 4y^2 + 4z^2 = 16$, an ellipsoid of revolution whose axis is the x-axis.

26. Find the equation of the surface of revolution generated by revolving the hyperbola $x^2 - 2z^2 = 1$ about the z-axis.

Let $P_1(x_1, 0, z_1)$ be any point on the hyperbola and $P'(0, 0, z_1)$ its projection on the z-axis. When the hyperbola is revolved about the z-axis, the point P_1 generates a circle having center P' and radius $P'P_1$. Let $P(x, y, z)$ be any point on this circle, and hence any point on the required surface.

Since $z_1 = z$ and $P'P_1 = P'P$, then $x_1 = \sqrt{(x-0)^2 + (y-0)^2 + (z-z_1)^2} = \sqrt{x^2 + y^2}$.

In the equation of the hyperbola, $x_1^2 - 2z_1^2 = 1$, put $x_1 = \sqrt{x^2 + y^2}$ and $z_1 = z$ and obtain $x^2 + y^2 - 2z^2 = 1$, a hyperboloid of one sheet.

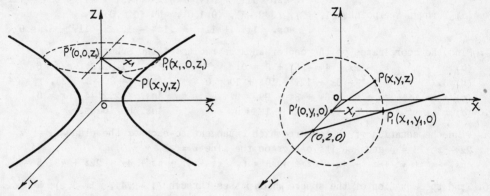

Problem 26.   Problem 27.

27. Find the equation of the surface of revolution generated by revolving the line $2x + 3y = 6$ about the y-axis.

Let $P_1(x_1, y_1, 0)$ be any point on the line and $P'(0, y_1, 0)$ its projection on the y-axis. When the line is revolved about the y-axis, the point P_1 generates a circle having center P' and radius $P'P_1$. Let $P(x, y, z)$ be any point on this circle, and hence any point on the required surface.

Since $y_1 = y$ and $P'P_1 = P'P$, then $x_1 = \sqrt{x^2 + z^2}$.

In the equation of the line, $2x_1 + 3y_1 = 6$, put $x_1 = \sqrt{x^2 + z^2}$ and $y_1 = y$ and obtain $2\sqrt{x^2 + z^2} + 3y = 6$. Simplifying, this equation reduces to $4x^2 - 9(y-2)^2 + 4z^2 = 0$, a cone with vertex at $(0, 2, 0)$.

SUPPLEMENTARY PROBLEMS

1. Find the equations of the following spheres.
   (*a*) Center $(2, -1, 3)$, radius 4.   *Ans.* $x^2 + y^2 + z^2 - 4x + 2y - 6z - 2 = 0$
   (*b*) Center $(-1, 2, 4)$, radius $\sqrt{13}$.   *Ans.* $x^2 + y^2 + z^2 + 2x - 4y - 8z + 8 = 0$
   (*c*) Having the line joining $(6, 2, -5)$ and $(-4, 0, 7)$ for a diameter.
                              Ans. $x^2 + y^2 + z^2 - 2x - 2y - 2z - 59 = 0$
   (*d*) Center $(-2, 2, 3)$ and passing through the point $(3, 4, -1)$.
                              Ans. $x^2 + y^2 + z^2 + 4x - 4y - 6z - 28 = 0$
   (*e*) Center $(6, 3, -4)$ and touching the x-axis.
                              Ans. $x^2 + y^2 + z^2 - 12x - 6y + 8z + 36 = 0$

2. Find the equations of each of the following spheres.

(a) Center $(-4, 2, 3)$ and tangent to the plane $2x - y - 2z + 7 = 0$.

$$Ans. \quad x^2 + y^2 + z^2 + 8x - 4y - 6z + 20 = 0$$

(b) Center $(2, -3, 2)$ and tangent to the plane $6x - 3y + 2z - 8 = 0$.

$$Ans. \quad 49x^2 + 49y^2 + 49z^2 - 196x + 294y - 196z + 544 = 0$$

(c) Center $(1, 2, 4)$ and tangent to the plane $3x - 2y + 4z - 7 = 0$.

$$Ans. \quad 29x^2 + 29y^2 + 29z^2 - 58x - 116y - 232z + 545 = 0$$

(d) Center $(-4, -2, 3)$ and tangent to the yz-plane.

$$Ans. \quad x^2 + y^2 + z^2 + 8x + 4y - 6z + 13 = 0$$

(e) Center $(0, 0, 0)$ and tangent to the plane $9x - 2y + 6z + 11 = 0$.

$$Ans. \quad x^2 + y^2 + z^2 = 1$$

3. Find the equation of each of the following spheres.

(a) Through the points $(1, 1, 1)$, $(1, 2, 1)$, $(1, 1, 2)$, and $(2, 1, 1)$.

$$Ans. \quad x^2 + y^2 + z^2 - 3x - 3y - 3z + 6 = 0$$

(b) Through the points $(2, 1, 3)$, $(3, -2, 1)$, $(-4, 1, 1)$, and $(1, 1, -3)$.

$$Ans. \quad 51x^2 + 51y^2 + 51z^2 + 45x + 37y - 33z - 742 = 0$$

(c) Through the points $(1, 3, 2)$, $(3, 2, -5)$, $(0, 1, 0)$, and $(0, 0, 0)$.

$$Ans. \quad 11x^2 + 11y^2 + 11z^2 - 127x - 11y + 3z = 0$$

4. Find the coordinates of the center and the radius of the sphere.

(a) $x^2 + y^2 + z^2 - 2x + 4y - 6z + 8 = 0$. $Ans.$ $(1, -2, 3)$, $r = \sqrt{6}$

(b) $3x^2 + 3y^2 + 3z^2 - 8x + 12y - 10z + 10 = 0$. $Ans.$ $(4/3, -2, 5/3)$, $r = \sqrt{47}/3$

(c) $x^2 + y^2 + z^2 + 4x - 6y + 8z + 29 = 0$. $Ans.$ $(-2, 3, -4)$, $r = 0$

(d) $x^2 + y^2 + z^2 - 6x + 2y - 2z + 18 = 0$. $Ans.$ No locus.

5. Find the equation of the sphere which is tangent to each of the planes $x - 2z - 8 = 0$ and $2x - z + 5 = 0$ and has its center on the line $x = -2$, $y = 0$.

$Ans.$ $x^2 + y^2 + z^2 + 4x + 6z + 49/5 = 0$, $x^2 + y^2 + z^2 + 4x + 22z + 481/5 = 0$

6. Find the equation of the sphere which passes through $(1, -3, 4)$, $(1, -5, 2)$, and $(1, -3, 0)$ and has its center in the plane $x + y + z = 0$.

$Ans.$ $x^2 + y^2 + z^2 - 2x + 6y - 4z + 10 = 0$

7. A point moves so that the sum of the squares of its distances from the three planes $x + 4y + 2z = 0$, $2x - y + z = 0$, and $2x + y - 3z = 0$ is 10. Find the equation of its locus.

$Ans.$ $x^2 + y^2 + z^2 = 10$

8. Find the equation of the locus of a point whose distances from $(1, 1, -2)$ and $(-2, 3, 2)$ are in the ratio $3 : 4$ numerically.

$Ans.$ $7x^2 + 7y^2 + 7z^2 - 68x + 22y + 100z - 57 = 0$

9. Discuss and sketch each of the following ellipsoids.

(a) $25x^2 + 16y^2 + 4z^2 = 100$.

(b) $4x^2 + y^2 + 9z^2 = 144$.

(c) $8x^2 + 2y^2 + 9z^2 = 144$.

(d) $x^2 + 4y^2 + 4z^2 - 12x = 0$.

(e) $x^2 + 4y^2 + 9z^2 = 36$.

(f) $\dfrac{(x - 1)^2}{36} + \dfrac{(y - 2)^2}{16} + \dfrac{(z - 3)^2}{9} = 1$.

10. In each of the following equations determine the coordinates of the center and the lengths of the semi-axes.

(a) $x^2 + 16y^2 + z^2 - 4x + 32y = 5$. $Ans.$ $(2, -1, 0)$, 5, $5/4$, 5

(b) $3x^2 + y^2 + 2z^2 + 3x + 3y + 4z = 0$. $Ans.$ $(-1/2, -3/2, -1)$, $\sqrt{15}/3$, $\sqrt{5}$, $\sqrt{10}/2$

(c) $x^2 + 4y^2 + z^2 - 4x - 8y + 8z + 15 = 0$. $Ans.$ $(2, 1, -4)$, 3, $3/2$, 3

(d) $3x^2 + 4y^2 + z^2 - 12x - 16y + 4z = 4$. $Ans.$ $(2, 2, -2)$, $2\sqrt{3}$, 3, 6

(e) $4x^2 + 5y^2 + 3z^2 + 12x - 20y + 24z + 77 = 0$. $Ans.$ Point $(-3/2, 2, -4)$

11. Find the equation of the standard ellipsoid (center at origin, axes parallel to coordinate axes) passing through the points given. Use the form $Ax^2 + By^2 + Cz^2 = D$.

(a) $(2,-1,1)$, $(-3,0,0)$, $(1,-1,-2)$. *Ans.* $x^2 + 4y^2 + z^2 = 9$

(b) $(\sqrt{3}, 1, 1)$, $(1, \sqrt{3}, -1)$, $(-1, -1, \sqrt{5})$. *Ans.* $2x^2 + 2y^2 + z^2 = 9$

(c) $(2,2,2)$, $(3, 1, \sqrt{3})$, $(-2,0,4)$. *Ans.* $2x^2 + 3y^2 + z^2 = 24$

(d) $(1,3,4)$, $(3,1, -2\sqrt{2})$, the axis of revolution being the x-axis. *Ans.* $2x^2 + y^2 + z^2 = 27$

12. A point moves so that the sum of its distances from $(0,3,0)$ and $(0,-3,0)$ is 8. Find the equation of the locus. *Ans.* $16x^2 + 7y^2 + 16z^2 = 112$

13. A point moves so that the sum of its distances from $(3,2,-4)$ and $(3,2,4)$ is 10. Find the equation of its locus. *Ans.* $\dfrac{(x-3)^2}{9} + \dfrac{(y-2)^2}{9} + \dfrac{(z-0)^2}{25} = 1$

14. A point moves so that the sum of its distances from $(-5,0,2)$ and $(5,0,2)$ is 12. Find the equation of its locus. *Ans.* $\dfrac{x^2}{36} + \dfrac{y^2}{11} + \dfrac{(z-2)^2}{11} = 1$

15. Find the locus of a point whose distance from the yz-plane equals twice its distance from the point $(1,-2,2)$. *Ans.* $3x^2 + 4y^2 + 4z^2 - 8x + 16y - 16z + 36 = 0$

16. Find the locus of a point whose distance from the point $(2,-3,1)$ is 1/4 its distance from the plane $y + 4 = 0$. *Ans.* $16x^2 + 15y^2 + 16z^2 - 64x + 88y - 32z + 208 = 0$

17. Find the locus of a point whose distance from the x-axis is 3 times its distance from $(2,3,-3)$. *Ans.* $9x^2 + 8y^2 + 8z^2 - 36x - 54y + 54z + 198 = 0$

18. Discuss and sketch the locus of the hyperboloid of one sheet.

(a) $\dfrac{x^2}{16} + \dfrac{y^2}{9} - \dfrac{z^2}{36} = 1.$

(d) $16y^2 - 36x^2 + 9z^2 = 144$

(b) $\dfrac{x^2}{4} - \dfrac{y^2}{36} + \dfrac{z^2}{16} = 1.$

(e) $\dfrac{x^2}{16} + \dfrac{y^2}{4} - \dfrac{(z-1)^2}{25} = 1.$

(c) $4x^2 - 25y^2 + 16z^2 = 100.$

(f) $9y^2 - x^2 + 4z^2 = 36.$

19. Discuss and sketch the locus of the hyperboloid of two sheets.

(a) $\dfrac{x^2}{16} - \dfrac{y^2}{9} - \dfrac{z^2}{36} = 1.$

(d) $\dfrac{(x-1)^2}{16} - \dfrac{y^2}{4} - \dfrac{z^2}{25} = 1.$

(b) $36x^2 - 4y^2 - 9z^2 = 144.$

(e) $36y^2 - 9x^2 - 16z^2 = 144.$

(c) $25x^2 - 16y^2 - 4z^2 = 100.$

(f) $4z^2 - x^2 - 9y^2 = 36.$

20. Determine the coordinates of the center and discuss the nature of each of the following.

(a) $2x^2 - 3y^2 + 4z^2 - 8x - 6y + 12z - 10 = 0.$

Ans. $(2, -1, -\frac{3}{2})$. Hyperboloid of one sheet. Axis parallel to y-axis.

(b) $x^2 + 2y^2 - 3z^2 + 4x - 4y - 6z - 9 = 0.$

Ans. $(-2,1,-1)$. Hyperboloid of one sheet. Axis parallel to z-axis.

(c) $2x^2 - 3y^2 - 4z^2 - 12x - 6y - 21 = 0.$

Ans. $(3,-1,0)$. Hyperboloid of two sheets. Axis parallel to x-axis.

(d) $4y^2 - 3x^2 - 6z^2 - 16y - 6x + 36z - 77 = 0.$

Ans. $(-1,2,3)$. Hyperboloid of two sheets. Axis parallel to y-axis.

(e) $16y^2 - 9x^2 + 4z^2 - 36x - 64y - 24z = 80.$

Ans. $(-2,2,3)$. Hyperboloid of one sheet. Axis parallel to x-axis.

(f) $5z^2 - 9x^2 - 15y^2 + 54x + 60y + 20z = 166.$

Ans. $(3,2,-2)$. Hyperboloid of two sheets. Axis parallel to z-axis.

(g) $2x^2 - y^2 - 3z^2 - 8x - 6y + 24z - 49 = 0.$ *Ans.* Point $(2,-3,4)$.

21. Find the equation of the locus of a point, the difference of whose distances from $(0,0,3)$ and $(0,0,-3)$ is 4.

 Ans. $5z^2 - 4x^2 - 4y^2 = 20$. Hyperboloid of two sheets. Center at the origin.

22. Find the equation of the locus of a point the difference of whose distances from $(2,-3,4)$ and $(2,3,4)$ is 5.

 Ans. $44y^2 - 100x^2 - 100z^2 + 400x + 800z = 2275$. Hyperboloid of two sheets. Center $(2,0,4)$.

23. Find the equation of the hyperboloid of one sheet which passes through $(4, 2\sqrt{3}, 0)$ and $(-1, 3, 3\sqrt{6}/2)$, center at $(0,0,0)$, the axis of revolution being the y-axis.

 Ans. $2x^2 - y^2 + 2z^2 = 20$. Hyperboloid of revolution of one sheet.

24. Find the equation of the hyperboloid of two sheets, center at the origin, axes on the coordinate axes, and passing through $(3,1,2)$, $(2, \sqrt{11}, 3)$, and $(6, 2, \sqrt{15})$.

 Ans. $3z^2 - x^2 - 2y^2 = 1$. Hyperboloid of two sheets, transverse axis along the z-axis.

25. Discuss and sketch each of the following surfaces.

 (a) $3x^2 + z^2 - 4y = 0$.   (e) $4x^2 + 3y^2 - 12z = 0$.

 (b) $x^2 + 2y^2 - 6z = 0$.   (f) $4x^2 - y^2 - 4z = 0$.

 (c) $y^2 - 4z^2 + 4x = 0$.   (g) $4x^2 + y^2 + z = 0$.

 (d) $x^2 + 4z^2 - 16y = 0$.   (h) $x^2 + 2y^2 = 8 - 4z$.

26. Find the equation of the paraboloid with its vertex at $(0,0,0)$, axis along the z-axis, and passing through the points $(2,0,3)$ and $(1,2,3)$.

 Ans. $12x^2 + 9y^2 - 16z = 0$. Elliptic paraboloid.

27. Find the equation of the paraboloid with vertex at $(0,0,0)$, axis along the z-axis, and passing through $(1,0,1)$ and $(0,2,1)$.

 Ans. $4x^2 + y^2 - 4z = 0$. Elliptic paraboloid.

28. Find the equation of the paraboloid with vertex at $(0,0,0)$, axis along the z-axis, and passing through $(1,2,1)$ and $(2,1,1)$.

 Ans. $x^2 + y^2 - 5z = 0$. Paraboloid of revolution.

29. Find the equation of the paraboloid with vertex at $(0,0,0)$, axis along the y-axis, and passing through $(1,1,1)$ and $(3/2, 7/12, 1/2)$.

 Ans. $x^2 + 5z^2 - 6y = 0$. Elliptic paraboloid.

30. Find the equation of the paraboloid containing the origin, symmetric about the x-axis, and passing through the points $(1,2,2)$ and $(2,6,8)$.

 Ans. $z^2 - 2y^2 + 4x = 0$, hyperbolic paraboloid; $2x^2 = z$, parabolic cylinder.

31. Find the equation of the locus of a point the square of whose distance from the z-axis is always twice its distance from the xy-plane.

 Ans. $x^2 + y^2 - 2z = 0$. Paraboloid of revolution about the z-axis.

32. By completing the squares locate the vertex of the paraboloid.

 (a) $2x^2 + 3y^2 - 8x + 12y + 3z + 23 = 0$.   *Ans.* $(2,-2,-1)$

 (b) $2x^2 + 4z^2 - 4x - 24z - y + 36 = 0$.   *Ans.* $(1,-2,3)$

 (c) $3z^2 + 5y^2 - 2x + 10y - 12z + 21 = 0$.   *Ans.* $(2,-1,2)$

 (d) $y^2 - 4x^2 + 2z - 6y - 12x + 6 = 0$.   *Ans.* $(-3/2, 3, -3)$

 (e) $4x^2 + 3z^2 - 4y + 12z + 12 = 0$.   *Ans.* $(0,0,-2)$

33. Discuss and sketch the locus of each of the following cones.

 (a) $x^2 + 2y^2 = 4z^2$.   (e) $2x^2 + 3y^2 - 6(z - 4)^2 = 0$.

 (b) $3x^2 + 2y^2 = 6z^2$.

 (c) $z^2 + y^2 = 2x^2$.   (f) $z^2 + 2y^2 - 4(x + 3)^2 = 0$.

 (d) $3x^2 + 4z^2 = 12y^2$.   (g) $3x^2 + 4z^2 - 12(y - 4)^2 = 0$.

34. Discuss and sketch the locus of each of the following cylinders.

(a) $x^2 + y^2 = 9.$

(b) $4x^2 + 9y^2 = 36.$

(c) $y^2 = 4x.$

(d) $16y^2 + 9z^2 = 144.$

(e) $x^2 - 9y^2 = 36.$

(f) $z = 4 - x^2.$

(g) $x^{2/3} + y^{2/3} = a^{2/3}$ (first quadrant).

35. Find the equation of each of the surfaces obtained by revolving the following curves about the axis indicated. Name the surface.

(a) $x^2 - 2z^2 = 1$, about the x-axis.

Ans. $x^2 - 2y^2 - 2z^2 = 1.$ Hyperboloid of two sheets.

(b) $x^2 - 2z^2 = 1$, about the z-axis.

Ans. $x^2 + y^2 - 2z^2 = 1.$ Hyperboloid of one sheet.

(c) $x = 4 - y^2$, about the x-axis.

Ans. $x = 4 - y^2 - z^2.$ Paraboloid.

(d) $2x - y = 10$, about the x-axis.

Ans. $4(x - 5)^2 = y^2 + z^2.$ Cone.

(e) $x^2 + z^2 = a^2$, about the z-axis.

Ans. $x^2 + y^2 + z^2 = a^2.$ Sphere.

(f) $x^2 + 4z^2 = 16$, about the z-axis.

Ans. $x^2 + y^2 + 4z^2 = 16.$ Ellipsoid.

(g) 1. $2x + 3y = 6$, about the y-axis.   *Ans.* $4x^2 - 9(y - 2)^2 + 4z^2 = 0.$ Cone.

2. What are the coordinates of the vertex of the cone?   *Ans.* $(0, 2, 0).$

3. What is the curve common to the cone and the plane $y = 0$?

Ans. $x^2 + z^2 = 9$, a circle of radius 3.

4. What is the curve common to the cone and the plane $y = 2$?

Ans. $x^2 + z^2 = 0$, a point circle, or the vertex of the cone.

5. What is the curve common to the cone and the plane $x = 0$?

Ans. $3(y - 2) = \pm 2z$, two straight lines in the yz-plane and intersecting in the point $(0, 2, 0)$, the vertex of the cone.

CHAPTER 16

Other Systems of Coordinates

POLAR, CYLINDRICAL, AND SPHERICAL COORDINATES. In addition to rectangular coordinates in space, the three other systems of coordinates frequently used are polar, cylindrical, and spherical coordinates.

POLAR COORDINATES. The polar coordinates in space of the point P in the adjacent figure are $(\rho, \alpha, \beta, \gamma)$, where ρ is the distance OP, and α, β, and γ are the direction angles of OP. The relations connecting the polar and rectangular coordinates of the point P are

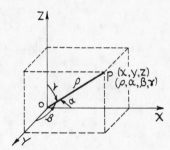

$$x = \rho \cos\alpha, \quad y = \rho \cos\beta, \quad \text{and} \quad z = \rho \cos\gamma.$$

$$\rho = \pm\sqrt{x^2 + y^2 + z^2},$$

$$\cos\alpha = \frac{x}{\rho} = \frac{x}{\pm\sqrt{x^2+y^2+z^2}}, \quad \cos\beta = \frac{y}{\rho} = \frac{y}{\pm\sqrt{x^2+y^2+z^2}}, \quad \cos\gamma = \frac{z}{\rho} = \frac{z}{\pm\sqrt{x^2+y^2+z^2}}.$$

Since $\cos^2\alpha + \cos^2\beta + \cos^2\gamma = 1$, the four coordinates are not independent. For example, if $\alpha = 60°$ and $\beta = 45°$, then $\cos^2\gamma = 1 - \cos^2\alpha - \cos^2\beta = 1 - \frac{1}{4} - \frac{1}{2} = \frac{1}{4}$. Since $\gamma \leq 180°$, $\gamma = 60°$ or $120°$.

CYLINDRICAL COORDINATES. In the cylindrical coordinate system the point $P(x,y,z)$ is located by coordinates ρ, θ, z, where ρ and θ are the polar coordinates of the projection Q, of the point P upon the xy-plane. These coordinates are written (ρ, θ, z). The relations between cylindrical and rectangular coordinates are

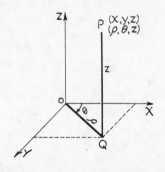

$$x = \rho \cos\theta, \quad y = \rho \sin\theta, \quad z = z.$$

$$\rho = \pm\sqrt{x^2 + y^2}, \quad \theta = \arctan\frac{y}{x}.$$

Note that the angle θ is not restricted in value, so that ρ may have negative values, as in the case of polar coordinates.

SPHERICAL COORDINATES. Let $P(x,y,z)$ be any point in space, and Q the projection of the point P upon the xy-plane. Denote the distance OP by ρ as in polar coordinates. Denote the angle ZOP by ϕ. Consider the angle ϕ as positive and $0° \leq \phi \leq 180°$. Denote the angle XOQ by θ. The symbols ρ, θ, and ϕ are called the spherical coordinates of the point P, represented by $P(\rho, \theta, \phi)$, ρ is called the radius vector, θ the longitude, and ϕ the co-latitude of P. The angle θ may have any value.

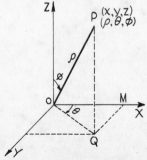

From the right triangle OPQ, we have

$$OQ = \rho \sin\phi, \quad QP = \rho \cos\phi.$$

144

From the right triangle OMQ, we have $OM = OQ \cos \theta$, $MQ = OQ \sin \theta$. Hence,

$$x = OM = \rho \sin \phi \cos \theta, \quad y = MQ = \rho \sin \phi \sin \theta, \quad z = QP = \rho \cos \phi.$$

$$\rho = \pm \sqrt{x^2 + y^2 + z^2}, \quad \theta = \arctan \frac{y}{x}, \quad \phi = \arccos \frac{z}{\pm \sqrt{x^2 + y^2 + z^2}}.$$

In many problems involving the determination of areas of surfaces or of volumes under a surface by methods employed in Calculus, the work is much simplified by the use of Cylindrical or Spherical Coordinates. Cylindrical coordinates are especially useful when a bounding surface is a surface of revolution.

SOLVED PROBLEMS

1. Find the polar, cylindrical and spherical coordinates for the point whose rectangular coordinates are (1,–2,2).

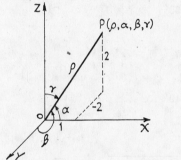   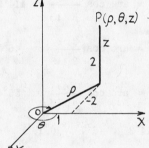   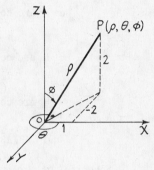

Polar Coordinates.   *Cylindrical Coordinates.*   *Spherical Coordinates.*

Polar Coordinates. $\rho = \sqrt{x^2 + y^2 + z^2} = \sqrt{1^2 + (-2)^2 + 2^2} = \sqrt{9} = 3.$

$\alpha = \arccos \dfrac{x}{\rho} = \arccos \dfrac{1}{3} = 70°32',$ $\beta = \arccos \dfrac{y}{\rho} = \arccos \left(-\dfrac{2}{3}\right) = 131°49',$

$\gamma = \arccos \dfrac{z}{\rho} = \arccos \dfrac{2}{3} = 48°11'.$ *Ans.* (3, 70°32′, 131°49′, 48°11′)

Cylindrical Coordinates. $\rho = \sqrt{x^2 + y^2} = \sqrt{1^2 + (-2)^2} = \sqrt{5}.$

$\theta = \arctan \dfrac{y}{x} = \arctan (-2) = 296°34',$ $z = 2.$ *Ans.* $(\sqrt{5}, 296°34', 2)$

Spherical Coordinates. $\rho = \sqrt{x^2 + y^2 + z^2} = \sqrt{1^2 + (-2)^2 + (2)^2} = 3.$

$\theta = \arctan \dfrac{y}{x} = \arctan (-2) = 296°34',$ $\phi = \arccos \dfrac{z}{\rho} = \arccos \dfrac{2}{3} = 48°11'.$

Ans. (3, 296°34′, 48°11′)

2. Find rectangular coordinates for the point whose cylindrical coordinates are (6, 120°, –2).

$x = \rho \cos \theta = 6 \cos 120° = -3,$ $y = \rho \sin \theta = 6 \sin 120° = 3\sqrt{3},$ $z = -2.$

Ans. (–3, 3√3, –2)

3. Find the rectangular coordinates of the point whose spherical coordinates are $(4, -45°, 30°)$.

$$x = \rho \sin \phi \cos \theta = 4 \sin 30° \cos (-45°) = \sqrt{2},$$
$$y = \rho \sin \phi \sin \theta = 4 \sin 30° \sin (-45°) = -\sqrt{2},$$
$$z = \rho \cos \phi \qquad\quad = 4 \cos 30° \qquad\qquad = 2\sqrt{3}. \qquad Ans. \; (\sqrt{2}, -\sqrt{2}, 2\sqrt{3})$$

4. Find the rectangular coordinates of the point having polar coordinates $(3, 120°, 120°, 135°)$.

$$x = \rho \cos \alpha = 3 \cos 120° = -3/2,$$
$$y = \rho \cos \beta = 3 \cos 120° = -3/2,$$
$$z = \rho \cos \gamma = 3 \cos 135° = -3\sqrt{2}/2. \qquad Ans. \; (-\frac{3}{2}, -\frac{3}{2}, -\frac{3\sqrt{2}}{2})$$

5. Find the rectangular, polar, and spherical coordinates for a point whose cylindrical coordinates are $(6, 120°, 4)$.

Rectangular. $\quad x = \rho \cos \theta = 6 \cos 120° = -3,$
$$y = \rho \sin \theta = 6 \sin 120° = 3\sqrt{3}, \; z = 4. \quad Ans. \; (-3, 3\sqrt{3}, 4)$$

Polar. $\quad \rho = \sqrt{x^2 + y^2 + z^2} = \sqrt{(-3)^2 + (3\sqrt{3})^2 + 4^2} = 2\sqrt{13},$

$$\alpha = \text{arc cos} \frac{x}{\rho} = \text{arc cos} \frac{-3}{2\sqrt{13}} = 114°35',$$

$$\beta = \text{arc cos} \frac{y}{\rho} = \text{arc cos} \frac{3\sqrt{3}}{2\sqrt{13}} = 46°7',$$

$$\gamma = \text{arc cos} \frac{z}{\rho} = \text{arc cos} \frac{4}{2\sqrt{13}} = 56°19'.$$

$$Ans. \; (2\sqrt{13}, 114°35', 46°7', 56°19')$$

Spherical. $\quad \rho = \sqrt{x^2 + y^2 + z^2} = \sqrt{(-3)^2 + (3\sqrt{3})^2 + 4^2} = 2\sqrt{13},$

$$\theta = \text{arc tan} \frac{y}{x} = \text{arc tan} \frac{3\sqrt{3}}{-3} = 120°,$$

$$\phi = \text{arc cos} \frac{z}{\rho} = \text{arc cos} \frac{4}{2\sqrt{13}} = 56°19'. \quad Ans. \; (2\sqrt{13}, 120°, 56°19')$$

6. Transform the equation $x^2 + y^2 + 2z^2 - 2x - 3y - z + 2 = 0$ to cylindrical coordinates.

Use $x = \rho \cos \theta, \; y = \rho \sin \theta, \; z = z.$

Substituting, $\rho^2 \cos^2\theta + \rho^2 \sin^2\theta + 2z^2 - 2\rho \cos \theta - 3\rho \sin \theta - z + 2 = 0.$

Simplifying, $\rho^2 - \rho(2 \cos \theta + 3 \sin \theta) + 2z^2 - z + 2 = 0.$

7. Transform the equation $2x^2 + 3y^2 - 6z = 0$ to spherical coordinates.

Use $x = \rho \sin \phi \cos \theta, \; y = \rho \sin \phi \sin \theta, \; z = \rho \cos \phi.$

Substituting, $2\rho^2 \sin^2\phi \cos^2\theta + 3\rho^2 \sin^2\phi \sin^2\theta - 6\rho \cos \phi = 0,$

or $\quad 2\rho \sin^2\phi \cos^2\theta + 3\rho \sin^2\phi \sin^2\theta - 6 \cos \phi = 0.$

8. Change the equation $\rho + 6 \sin \phi \cos \theta + 4 \sin \phi \sin \theta - 8 \cos \phi = 0$ to rectangular coordinates.

This equation is expressed in spherical coordinates. Multiply through by ρ. Using values for x, y, z in Problem 7, we have

$$\rho^2 + 6\rho \sin \phi \cos \theta + 4\rho \sin \phi \sin \theta - 8\rho \cos \phi = 0, \quad \text{or}$$

$$x^2 + y^2 + z^2 + 6x + 4y - 8z = 0.$$

This is the equation of a sphere with its center at $(-3,-2,4)$ and radius $r = \sqrt{29}$.

9. Change the equation $z = \rho^2 \cos 2\theta$ from cylindrical to rectangular coordinates.

Write $\cos 2\theta = \cos^2\theta - \sin^2\theta$. Then $z = \rho^2(\cos^2\theta - \sin^2\theta) = \rho^2 \cos^2\theta - \rho^2 \sin^2\theta$.

Since $\rho \cos \theta = x$ and $\rho \sin \theta = y$, the required equation is $z = x^2 - y^2$.

10. Change the equation $x^2 + y^2 - z^2 = 25$ into an equation in polar coordinates.

In polar coordinates, $x = \rho \cos \alpha$, $y = \rho \cos \beta$, $z = \rho \cos \gamma$.

Hence the given equation becomes $\rho^2 \cos^2\alpha + \rho^2 \cos^2\beta - \rho^2 \cos^2\gamma = 25$,

$$\text{or} \quad \rho^2(\cos^2\alpha + \cos^2\beta - \cos^2\gamma) = 25.$$

Since $\cos^2\alpha + \cos^2\beta + \cos^2\gamma = 1$, the required equation is $\rho^2(1 - 2\cos^2\gamma) = 25$.

11. Change the polar coordinate equation $\cos \gamma = \rho \cos \alpha \cos \beta$ into rectangular coordinates.

Multiply both sides of the given equation by ρ.

Then $\rho \cos \gamma = \rho^2 \cos \alpha \cos \beta$. Since $\rho \cos \gamma = z$, $\rho \cos \alpha = x$, $\rho \cos \beta = y$, the required equation is $z = xy$.

SUPPLEMENTARY PROBLEMS

1. Find polar coordinates for the following points:
   (a) $(0,1,1)$; (b) $(0,-2,-2)$; (c) $(1,-2,2)$; (d) $(6,3,2)$; (e) $(8,-4,1)$.
   Ans. (a) $(\sqrt{2}, 90°, 45°, 45°)$; (b) $(2\sqrt{2}, 90°, 135°, 135°)$;
   (c) $(3, \text{arc cos } 1/3, \text{arc cos } (-2/3), \text{arc cos } 2/3)$;
   (d) $(7, \text{arc cos } 6/7, \text{arc cos } 3/7, \text{arc cos } 2/7)$;
   (e) $(9, \text{arc cos } 8/9, \text{arc cos } (-4/9), \text{arc cos } 1/9)$.

2. Find cylindrical coordinates for the points in Problem 1.
   Ans. (a) $(1, 90°, 1)$; (b) $(2, 270°, -2)$; (c) $(\sqrt{5}, 2\pi - \text{arc tan} \frac{1}{2}, 2)$;
   (d) $(3\sqrt{5}, \text{arc tan} \frac{1}{2}, 2)$; (e) $(4\sqrt{5}, 2\pi - \text{arc tan} 2, 1)$.

3. Find spherical coordinates for the points in Problem 1.
   Ans. (a) $(\sqrt{2}, 90°, 45°)$; (b) $(2\sqrt{2}, 270°, 135°)$; (c) $(3, 2\pi - \text{arc tan} 2, \text{arc cos } 2/3)$;
   (d) $(7, \text{arc tan } 1/2, \text{arc cos } 2/7)$; (e) $(9, 2\pi - \text{arc tan} \frac{1}{2}, \text{arc cos } 1/9)$.

4. Find rectangular coordinates for the points whose polar coordinates are:
   (a) $(2, 90°, 30°, 60°)$; (b) $(3, 60°, -45°, 120°)$; (c) $(4, 120°, 120°, 135°)$;
   (d) $(3, 150°, 60°, 90°)$; (e) $(2, 45°, 120°, -60°)$.
   Ans. (a) $(0, \sqrt{3}, 1)$; (b) $(3/2, 3\sqrt{2}/2, -3/2)$; (c) $(-2, -2, -2\sqrt{2})$;
   (d) $(-3\sqrt{3}/2, 3/2, 0)$; (e) $(\sqrt{2}, -1, 1)$.

5. Find rectangular coordinates for the points whose cylindrical coordinates are:
   (a) $(6, 120°, -2)$; (b) $(1, 330°, -2)$; (c) $(4, 45°, 2)$; (d) $(8, 120°, 3)$; (e) $(6, 30°, -3)$.
   Ans. (a) $(-3, 3\sqrt{3}, -2)$; (b) $(\sqrt{3}/2, -1/2, -2)$; (c) $(2\sqrt{2}, 2\sqrt{2}, 2)$;
   (d) $(-4, 4\sqrt{3}, 3)$; (e) $(3\sqrt{3}, 3, -3)$.

6. Find rectangular coordinates for the points whose spherical coordinates are:
   (a) $(4, 210°, 30°)$; (b) $(3, 120°, 240°)$; (c) $(6, 330°, 60°)$;
   (d) $(5, 150°, 210°)$; (e) $(2, 180°, 270°)$.

Ans. (a) $(-\sqrt{3}, -1, 2\sqrt{3})$; (b) $(\frac{3\sqrt{3}}{4}, -\frac{9}{4}, -\frac{3}{2})$; (c) $(\frac{9}{2}, -\frac{3\sqrt{3}}{2}, 3)$;

(d) $(\frac{5\sqrt{3}}{4}, -\frac{5}{4}, -\frac{5\sqrt{3}}{2})$; (e) $(2,0,0)$.

7. Find the spherical coordinates of the points whose cylindrical coordinates are:
(a) $(8, 120°, 6)$; (b) $(4, 30°, -3)$; (c) $(6, 135°, 2)$; (d) $(3, 150°, 4)$;
(e) $(12, -90°, 5)$.

Ans. (a) $(10, 120°, \text{arc } \cos \frac{3}{5})$; (b) $(5, 30°, \text{arc } \cos(-\frac{3}{5}))$; (c) $(2\sqrt{10}, 135°, \frac{\sqrt{10}}{10})$;

(d) $(5, 150°, \text{arc } \cos \frac{4}{5})$; (e) $(13, -90°, \text{arc } \cos \frac{5}{13})$.

8. Transform the following equations into equations in spherical coordinates:
(a) $3x^2 - 3y^2 = 8z$; (b) $x^2 - y^2 - z^2 = a^2$; (c) $3x + 5y - 2z = 6$.
Ans. (a) $3\rho \sin^2\phi \cos 2\theta = 8 \cos \theta$;   (b) $\rho^2(\sin^2\phi \cos 2\theta - \cos^2\phi) = a^2$;
(c) $\rho(3 \sin \phi \cos \theta + 5 \sin \phi \sin \theta - 2 \cos \phi) = 6$.

9. Transform the following equations from rectangular into cylindrical coordinates:
(a) $5x + 4y = 0$; (b) $5x^2 - 4y^2 + 2x + 3y = 0$; (c) $x^2 + y^2 - 8x = 0$;
(d) $x^2 - y^2 + 2y - 6 = 0$; (e) $x^2 + y^2 - z^2 = a^2$.
Ans. (a) $\theta = \text{arc } \tan(-5/4)$; (b) $5\rho \cos^2\theta - 4\rho \sin^2\theta + 2 \cos \theta + 3 \sin \theta = 0$;
(c) $\rho - 8 \cos \theta = 0$; (d) $\rho^2 \cos 2\theta + 2\rho \sin \theta - 6 = 0$; (e) $\rho^2 - z^2 = a^2$.

10. The following surfaces are given in cylindrical coordinates. Find the equations in rectangular coordinates and name the surfaces.
(a) $\rho^2 + 3z^2 = 36$; (b) $\rho = a \sin \theta$; (c) $\rho^2 + z^2 = 16$; (d) $\theta = 45°$; (e) $\rho^2 - z^2 = 1$.
Ans. (a) $x^2 + y^2 + 3z^2 = 36$. Ellipsoid of revolution.
(b) $x^2 + y^2 = ay$. Right circular cylinder.
(c) $x^2 + y^2 + z^2 = 16$. Sphere.
(d) $y = x$. Plane.
(e) $x^2 + y^2 - z^2 = 1$. Hyperboloid of one sheet.

11. Transform the following equations from rectangular into polar coordinates:
(a) $x^2 + y^2 + 4z = 0$; (b) $x^2 + y^2 - z^2 = a^2$; (c) $2x^2 + 3y^2 + 2z^2 - 6x + 2y = 0$;
(d) $z = 2xy$.
Ans. (a) $\rho(\cos^2\alpha + \cos^2\beta) + 4 \cos \gamma = 0$, or $\rho(1 - \cos^2\gamma) + 4 \cos \gamma = 0$;
(b) $\rho^2(1 - 2\cos^2\gamma) = a^2$; (c) $\rho(2 + \cos^2\beta) - 6 \cos \alpha + 2 \cos \beta = 0$;
(d) $\cos \gamma = 2\rho \cos \alpha \cos \beta$.

12. Transform the following equations in spherical coordinates into equations in rectangular coordinates: (a) $\rho = 5a \cos \phi$; (b) $\theta = 60°$; (c) $\rho \sin \phi = a$; (d) $\rho = 4$.
Ans. (a) $x^2 + y^2 + z^2 = 5az$; (b) $y = \sqrt{3}x$; (c) $x^2 + y^2 = a^2$; (d) $x^2 + y^2 + z^2 = 16$.

13. Transform the following equations in polar coordinates into equations in rectangular coordinates: (a) $\rho(\cos \alpha + \cos \beta + \cos \gamma) = 5$; (b) $\rho^2(2 \cos^2\alpha - 1) = 25$;
(c) $\cos \gamma = \rho(\cos^2\alpha - \cos^2\beta)$; (d) $\rho^2 - \rho^2 \cos^2\gamma - 4\rho \cos \gamma - 2 = 0$.
Ans. (a) $x + y + z = 5$; (b) $x^2 - y^2 - z^2 = 25$; (c) $z = x^2 - y^2$;
(d) $x^2 + y^2 - 4z - 2 = 0$.

14. Develop a formula for the distance between two points $P_1(\rho_1, \theta_1, \phi_1)$, $P_2(\rho_2, \theta_2, \phi_2)$ when spherical coordinates are used. Hint: Use the formula for the distance between two points expressed in rectangular coordinates and transform into spherical coordinates.

Ans. $\sqrt{\rho_1^2 + \rho_2^2 - 2\rho_1\rho_2[\cos(\theta_2 - \theta_1) \sin \phi_1 \sin \phi_2 + \cos \phi_1 \cos \phi_2]} = d$.

INDEX

149

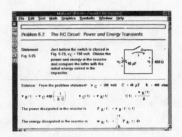

SCHAUM'S SOLVED PROBLEMS SERIES

- ■ Learn the best strategies for solving tough problems in step-by-step detail
- ■ Prepare effectively for exams and save time in doing homework problems
- ■ Use the indexes to quickly locate the types of problems you need the most help solving
- ■ Save these books for reference in other courses and even for your professional library

To order, please check the appropriate box(es) and complete the following coupon.

❑ **3000 SOLVED PROBLEMS IN BIOLOGY**
ORDER CODE 005022-8/**$16.95** 406 pp.

❑ **3000 SOLVED PROBLEMS IN CALCULUS**
ORDER CODE 041523-4/**$19.95** 442 pp.

❑ **3000 SOLVED PROBLEMS IN CHEMISTRY**
ORDER CODE 023684-4/**$20.95** 624 pp.

❑ **2500 SOLVED PROBLEMS IN COLLEGE ALGEBRA & TRIGONOMETRY**
ORDER CODE 055373-4/**$14.95** 608 pp.

❑ **2500 SOLVED PROBLEMS IN DIFFERENTIAL EQUATIONS**
ORDER CODE 007979-x/**$19.95** 448 pp.

❑ **2000 SOLVED PROBLEMS IN DISCRETE MATHEMATICS**
ORDER CODE 038031-7/**$16.95** 412 pp.

❑ **3000 SOLVED PROBLEMS IN ELECTRIC CIRCUITS**
ORDER CODE 045936-3/**$21.95** 746 pp.

❑ **2000 SOLVED PROBLEMS IN ELECTROMAGNETICS**
ORDER CODE 045902-9/**$18.95** 480 pp.

❑ **2000 SOLVED PROBLEMS IN ELECTRONICS**
ORDER CODE 010284-8/**$19.95** 640 pp.

❑ **2500 SOLVED PROBLEMS IN FLUID MECHANICS & HYDRAULICS**
ORDER CODE 019784-9/**$21.95** 800 pp.

❑ **1000 SOLVED PROBLEMS IN HEAT TRANSFER**
ORDER CODE 050204-8/**$19.95** 750 pp.

❑ **3000 SOLVED PROBLEMS IN LINEAR ALGEBRA**
ORDER CODE 038023-6/**$19.95** 750 pp.

❑ **2000 SOLVED PROBLEMS IN Mechanical Engineering THERMODYNAMICS**
ORDER CODE 037863-0/**$19.95** 406 pp.

❑ **2000 SOLVED PROBLEMS IN NUMERICAL ANALYSIS**
ORDER CODE 055233-9/**$20.95** 704 pp.

❑ **3000 SOLVED PROBLEMS IN ORGANIC CHEMISTRY**
ORDER CODE 056424-8/**$22.95** 688 pp.

❑ **2000 SOLVED PROBLEMS IN PHYSICAL CHEMISTRY**
ORDER CODE 041716-4/**$21.95** 448 pp.

❑ **3000 SOLVED PROBLEMS IN PHYSICS**
ORDER CODE 025734-5/**$20.95** 752 pp.

❑ **3000 SOLVED PROBLEMS IN PRECALCULUS**
ORDER CODE 055365-3/**$16.95** 385 pp.

❑ **800 SOLVED PROBLEMS IN VECTOR MECHANICS FOR ENGINEERS
Vol I: STATICS**
ORDER CODE 056582-1/**$20.95** 800 pp.

❑ **700 SOLVED PROBLEMS IN VECTOR MECHANICS FOR ENGINEERS
Vol II: DYNAMICS**
ORDER CODE 056687-9/**$20.95** 672 pp.

DAILY
Word Ladders

Grades 2–3

by Timothy Rasinski
Kent State University

New York • Toronto • London • Auckland • Sydney
Mexico City • New Delhi • Hong Kong • Buenos Aires

Teaching
Resources

To my children—Mike, Emily, Mary, and Jenny—
Word Wizards in their own right.

A father couldn't ask for better kids.

Cover design by Maria Lilja

Interior design by Ellen Matlach for Boultinghouse & Boultinghouse, Inc.

Interior illustrations by Teresa Anderko

ISBN: 0-439-51383-9

Copyright © 2005 by Timothy Rasinski

Contents

Welcome to Word Ladders!

In this book you'll find 100 mini-word-study lessons that are also kid-pleasing games! To complete each Word Ladder takes just ten minutes but actively involves each learner in analyzing the structure and meaning of words. To play, students begin with one word and then make a series of other words by changing or rearranging the letters in the word before. With regular use, Word Ladders can go a long way toward developing your students' decoding and vocabulary skills.

How do Word Ladders work?

Let's say our first Word Ladder begins with the word *walk*. The directions will tell students to change one letter in *walk* to make a word that means "to speak." The word students will make, of course, is *talk*. The next word will then ask students to make a change in *talk* to form another word—perhaps *chalk*, or *tall*. At the top of the ladder, students will have a final word that is in some way related to the first word—for example, *run*. If students get stuck on a rung along the way, they can come back to it, because the words before and after will give them the clues they need to go on.

How do Word Ladders benefit students?

Word Ladders are great for building students' decoding, phonics, spelling, and vocabulary skills. When students add or rearrange letters to make a new word from one they have just made, they must examine sound-symbol relationships closely. This is just the kind of analysis that all children need to do in order to learn how to decode and spell accurately. And when the puzzle adds a bit of meaning in the form of a definition (for example, "make a word that means to say something"), it helps extend students' understanding of words and concepts. All of these skills are key to students' success in learning to read and write. So even though Word Ladders will feel like a game your students will be practicing essential literacy skills at the same time!

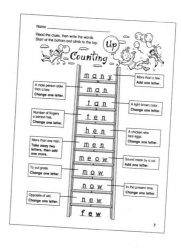

How do I teach a Word Ladder lesson?

Word Ladders are incredibly easy and quick to implement. Here are four simple steps:

1. Choose a Word Ladder to try. (The first five pages feature easier ladders; you may want to start with those.)

2. Make a copy of the Word Ladder for each student.

3. Choose whether you want to do your Word Ladders with the class as a whole, or by having students work alone, in pairs, or in groups. (You might do the

first few together, until students are ready to work more independently.)

4. At each new word, students will see two clues: the kinds of changes they need to make to the previous word ("rearrange letters" or "add two letters"), and a definition of or clue to the meaning of the word. Sometimes this clue will be a sentence in which the word is used in context but is left out for children to fill in. Move from word to word this way, up the whole Word Ladder.

Look for the **Bonus Boxes** with stars. These are particularly difficult words you may want to preteach. Or you can do these ladders as a group so that children will not get stuck on this rung.

That's the lesson in a nutshell! It should take no longer than ten minutes to do. Once you're done, you may wish to extend the lesson by having students sort the words into various categories. This can help them deepen their understanding of word relationships. For instance, they could sort them into:

- Grammatical categories. (Which words are nouns? Verbs?)

- Word structure. (Which words have a long vowel and which don't? Which contain a consonant blend?)

- Word meaning. (Which words express what a person can do or feel? Which do not?)

Tips for Working With Word Ladders

To give students extra help, mix up and write on the board all the "answers" for the ladder (that is, the words for each rung) for them to choose from as they go through the puzzle. In addition:

- Add your own clues to give students extra help as they work through each rung of a ladder. A recent event in your classroom or community could even inspire clues for words.

- If students are having difficulty with a particular word, you might simply say the word aloud and see if students can spell it correctly by making appropriate changes in the previous word. Elaborate on the meanings of the words as students move their way up the ladder.

- If students are stuck on a particular rung of the Word Ladder, tell them to skip it and come back to it later.

- Challenge students to come up with alternative definitions for the same words. Many words, like *lock, fall,* and *stock,* have multiple meanings.

Timothy Rasinski is a professor of literacy education at Kent State University, with a special focus on young and struggling readers. Dr. Rasinski has served on the board of directors at the International Reading Association and as president of the College Reading Association. He is the author of numerous books and professional articles on effective reading instruction.

Name _____

Read the clues, then write the words.
Start at the bottom and climb to the top.

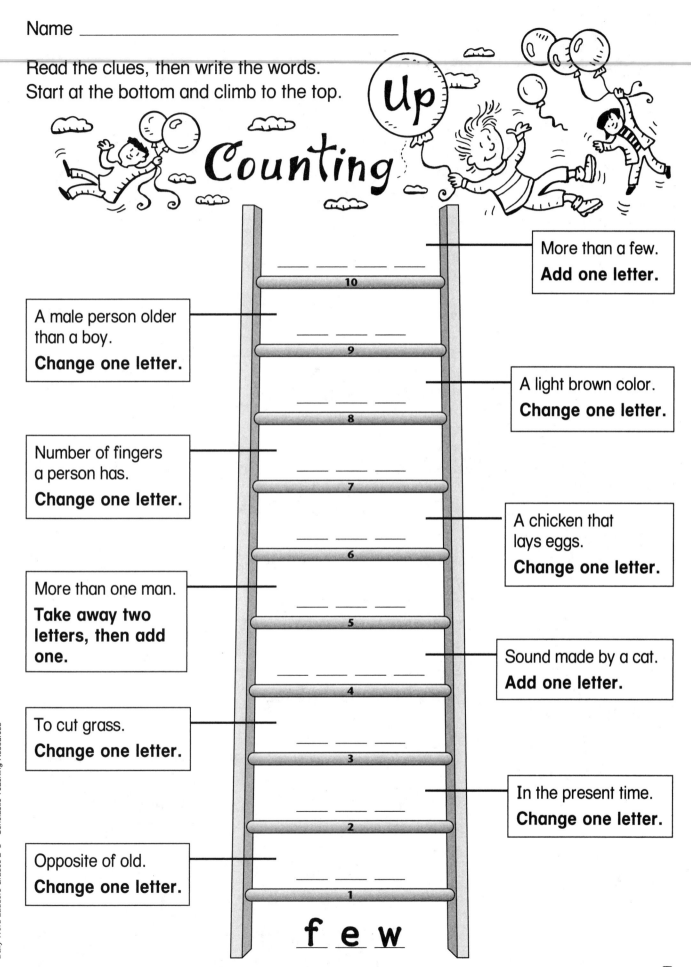

Counting **Up**

More than a few.
Add one letter.

10 _ _ _ _

A male person older than a boy.
Change one letter.

9 _ _ _

A light brown color.
Change one letter.

8 _ _ _

Number of fingers a person has.
Change one letter.

7 _ _ _

A chicken that lays eggs.
Change one letter.

6 _ _ _

More than one man.
Take away two letters, then add one.

5 _ _ _

Sound made by a cat.
Add one letter.

4 _ _ _

To cut grass.
Change one letter.

3 _ _ _

In the present time.
Change one letter.

2 _ _ _

Opposite of old.
Change one letter.

1 _ _ _

<u>f</u> <u>e</u> <u>w</u>

Name _____

Read the clues, then write the words.
Start at the bottom and climb to the top.

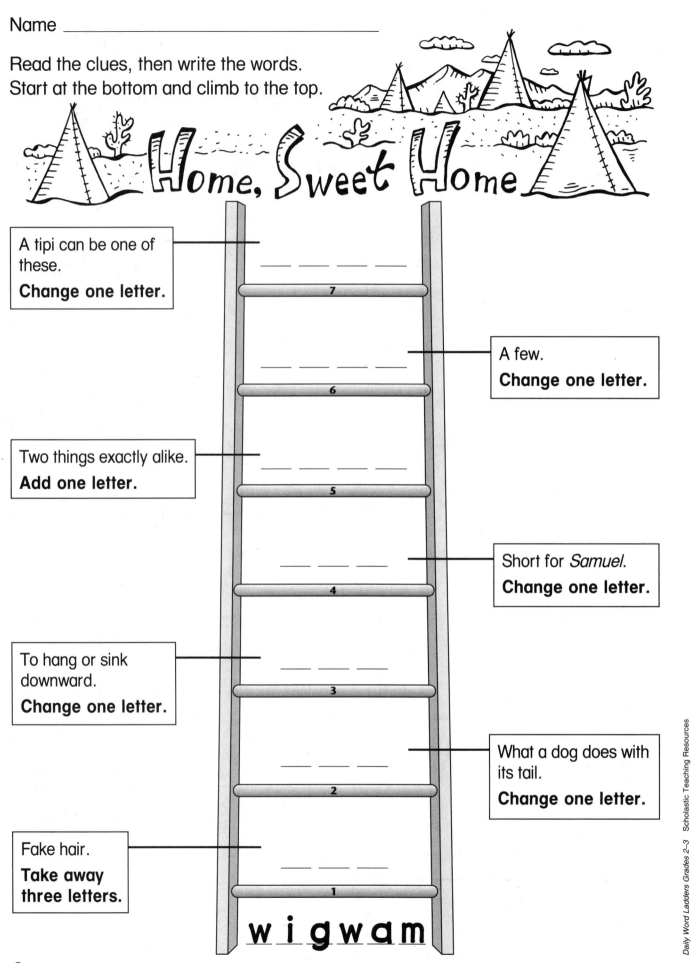

Home, Sweet Home

A tipi can be one of these.
Change one letter.

A few.
Change one letter.

Two things exactly alike.
Add one letter.

Short for *Samuel*.
Change one letter.

To hang or sink downward.
Change one letter.

What a dog does with its tail.
Change one letter.

Fake hair.
Take away three letters.

w i g w a m

7

6

5

4

3

2

1

Daily Word Ladders Grades 2–3 Scholastic Teaching Resources

Name _____

Read the clues, then write the words.
Start at the bottom and climb to the top.

In the Doghouse

What a dog might like to chew on.
Change one letter.

Something you put ice cream in.
Change one letter.

A kind of cola soda.
Change one letter.

A secret way of writing.
Add one letter.

A kind of fish.
Change one letter.

To say "yes," you can ___ your head.
Rearrange the letters.

Short for *Donald*.
Change one letter.

A small round spot.
Change one letter.

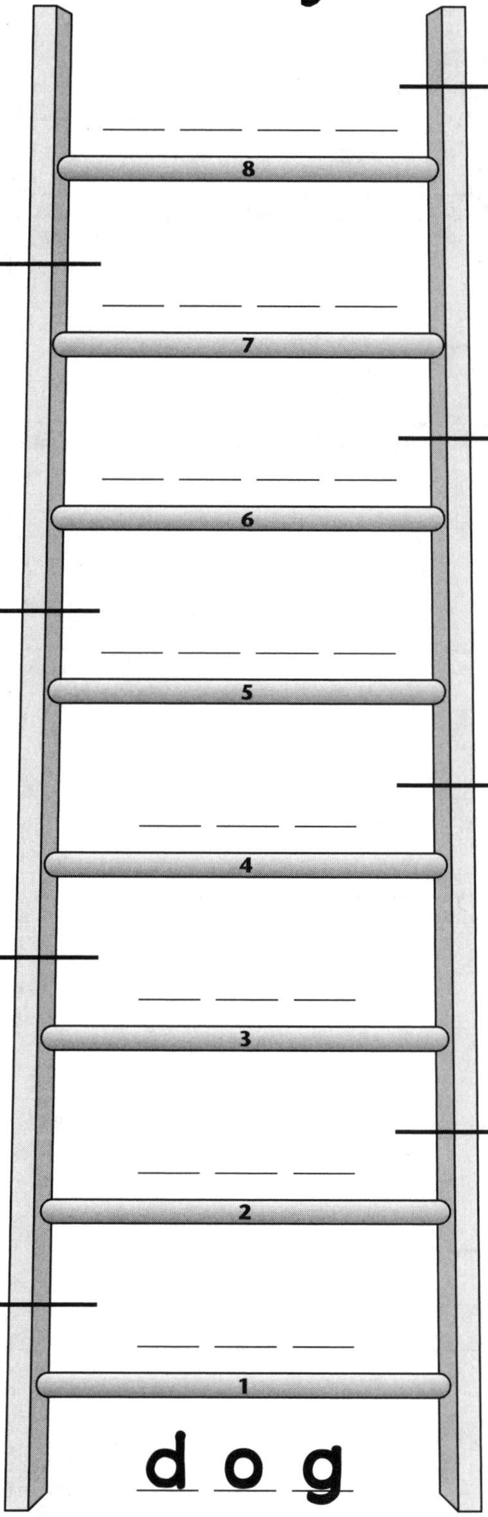

8

7

6

5

4

3

2

1

d o g

Name _____

Read the clues, then write the words.
Start at the bottom and climb to the top.

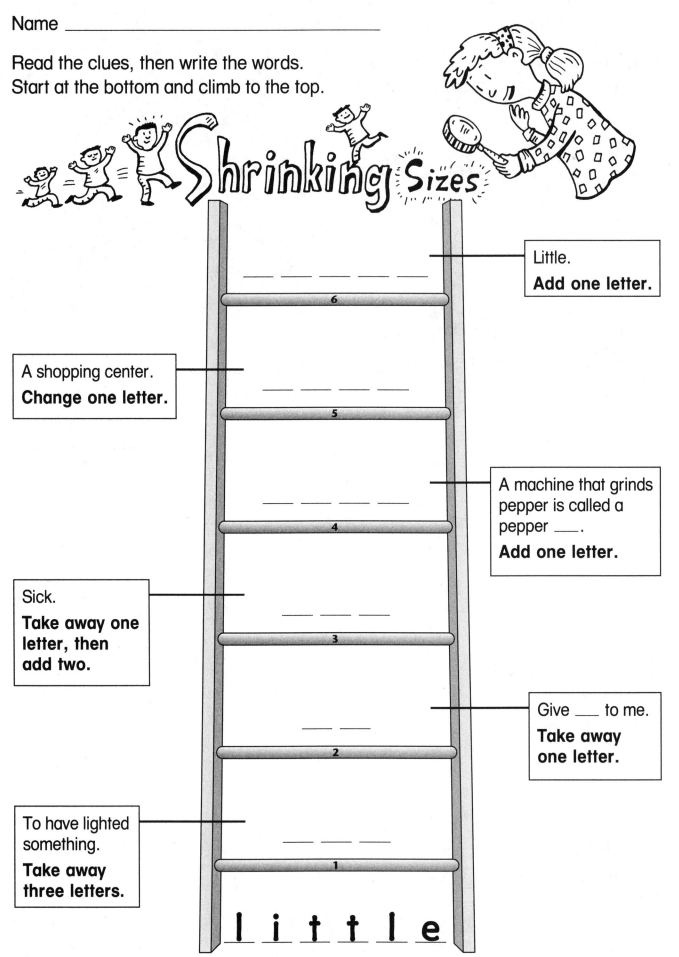

Shrinking Sizes

Little.
Add one letter.

A shopping center.
Change one letter.

A machine that grinds pepper is called a pepper ___.
Add one letter.

Sick.
Take away one letter, then add two.

Give ___ to me.
Take away one letter.

To have lighted something.
Take away three letters.

l i t t l e

6

5

4

3

2

1

Daily Word Ladders Grades 2–3   Scholastic Teaching Resources

Read the clues, then write the words.
Start at the bottom and climb to the top.

All Wet

It falls from the sky and makes you wet.
Add one letter.

Walked very fast.
Take away one letter.

Part of the wheat plant.
Change one letter.

Short for Bradley.
Add one letter.

The opposite of good.
Change one letter.

A stick used for hitting balls.
Change one letter.

A wager or guess that something will happen.
Take away one letter.

You wear it to hold up your pants.
Change one letter.

Something that makes a ringing noise.
Change one letter.

Healthy.
Add two letters.

All of us.
Take away one letter.

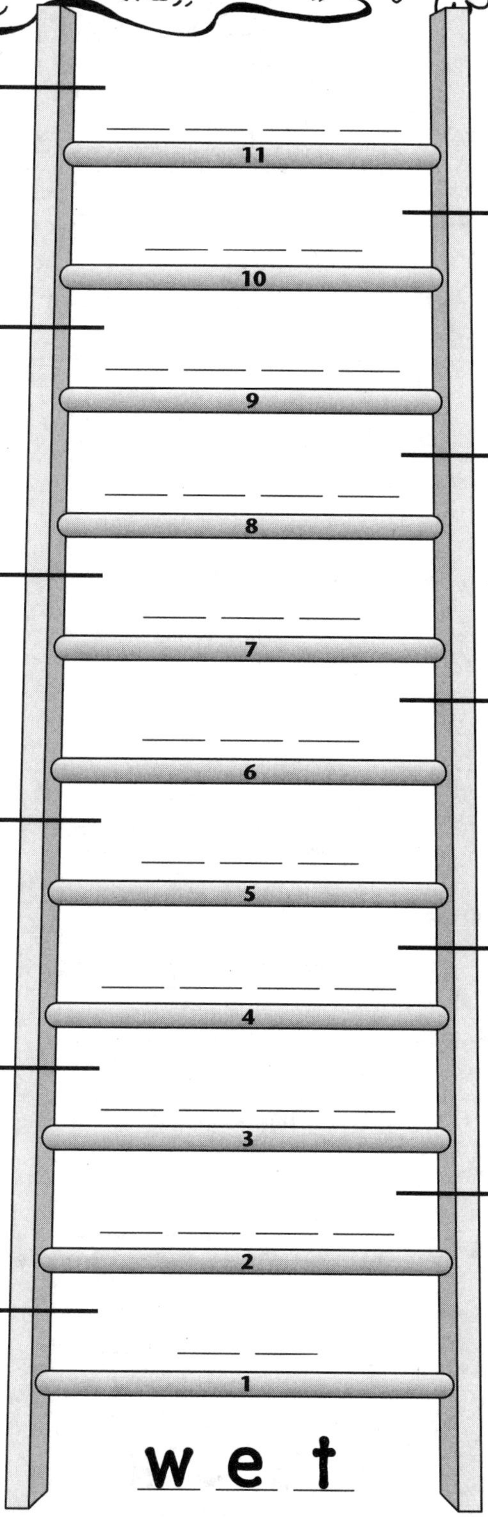

11
10
9
8
7
6
5
4
3
2
1

w e t

Name _____

Read the clues, then write the words.
Start at the bottom and climb to the top.

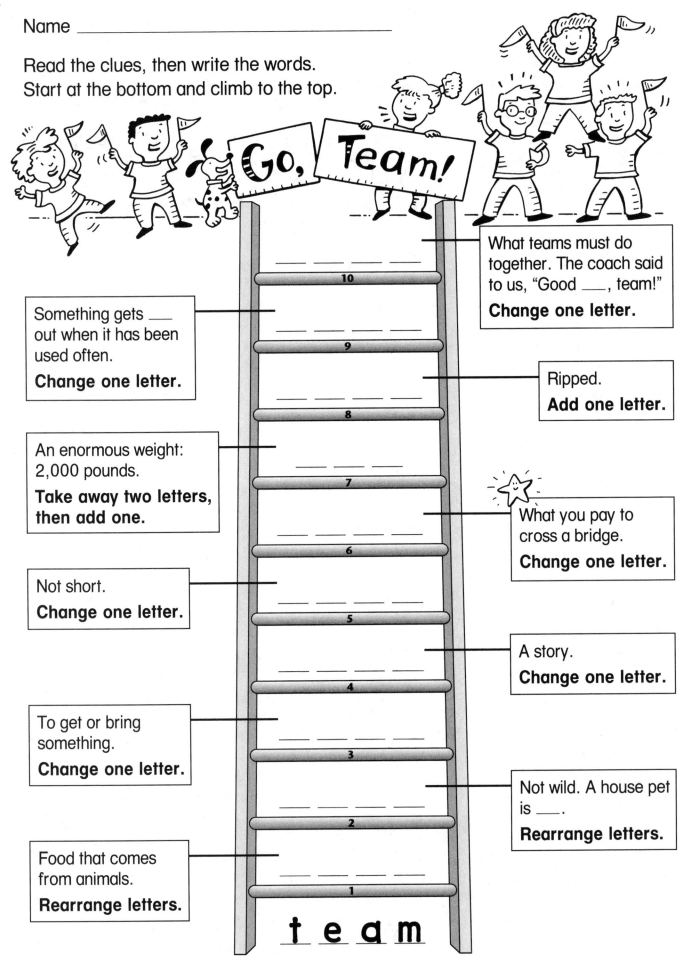

Go, Team!

10 _ _ _ _

What teams must do together. The coach said to us, "Good ___, team!"
Change one letter.

Something gets ___ out when it has been used often.
Change one letter.

9 _ _ _ _

8 _ _ _ _

Ripped.
Add one letter.

An enormous weight: 2,000 pounds.
Take away two letters, then add one.

7 _ _ _

6 _ _ _ _

What you pay to cross a bridge.
Change one letter.

Not short.
Change one letter.

5 _ _ _ _

A story.
Change one letter.

4 _ _ _ _

To get or bring something.
Change one letter.

3 _ _ _ _

Not wild. A house pet is ___.
Rearrange letters.

2 _ _ _ _

Food that comes from animals.
Rearrange letters.

1 _ _ _ _

t e a m

12

Daily Word Ladders, Grades 2–3   Scholastic Professional Books

Name _____

Read the clues, then write the words.
Start at the bottom and climb to the top.

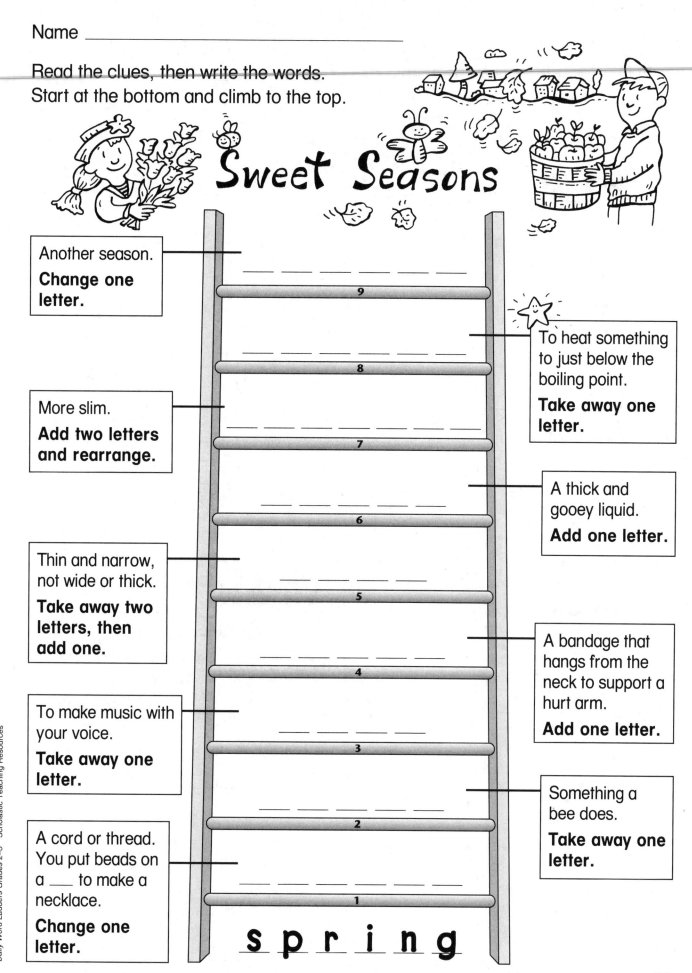

Sweet Seasons

Another season.
Change one letter.

More slim.
Add two letters and rearrange.

Thin and narrow, not wide or thick.
Take away two letters, then add one.

To make music with your voice.
Take away one letter.

A cord or thread. You put beads on a ___ to make a necklace.
Change one letter.

To heat something to just below the boiling point.
Take away one letter.

A thick and gooey liquid.
Add one letter.

A bandage that hangs from the neck to support a hurt arm.
Add one letter.

Something a bee does.
Take away one letter.

9

8

7

6

5

4

3

2

1

s p r i n g

Name _____

Read the clues, then write the words.
Start at the bottom and climb to the top.

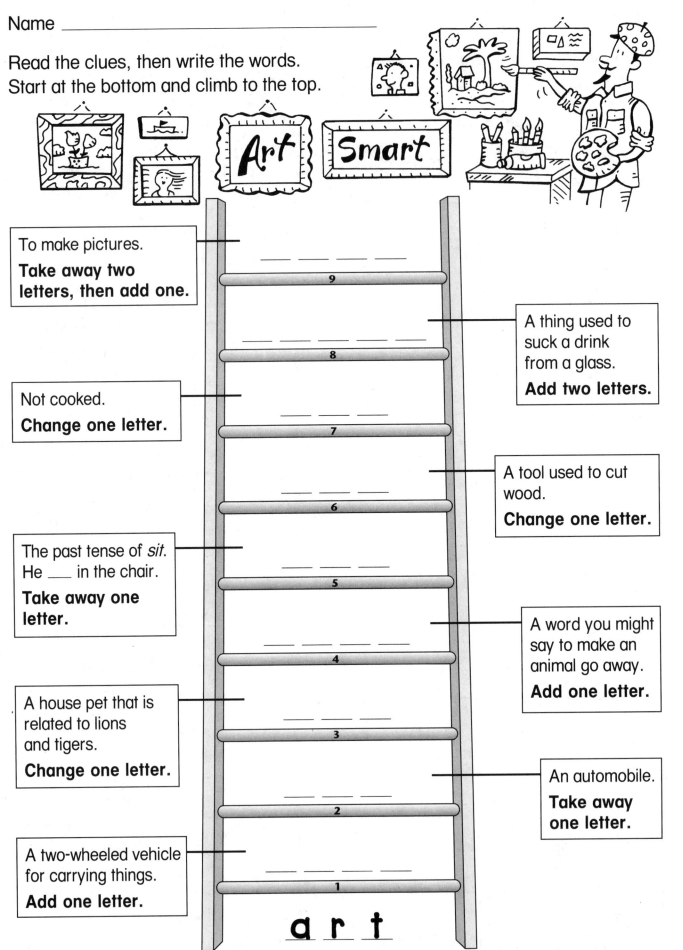

To make pictures.
**Take away two
letters, then add one.**

A thing used to
suck a drink
from a glass.
Add two letters.

Not cooked.
Change one letter.

A tool used to cut
wood.
Change one letter.

The past tense of *sit*.
He ___ in the chair.
**Take away one
letter.**

A word you might
say to make an
animal go away.
Add one letter.

A house pet that is
related to lions
and tigers.
Change one letter.

An automobile.
**Take away
one letter.**

A two-wheeled vehicle
for carrying things.
Add one letter.

9

8

7

6

5

4

3

2

1

a r t

Daily Word Ladders Grades 2–3 Scholastic Teaching Resources

Read the clues, then write the words.
Start at the bottom and climb to the top.

Sleepytime

A short rest.
Change one letter.

A liquid that comes from trees.
Take away one letter.

To hit.
Change one letter.

To make applause.
Add one letter.

Children often sit on their parents' ___s.
Rearrange the letters.

A friend.
Take away one letter.

Another name for a bucket.
Change two letters.

To take off the skin of an orange.
Change one letter.

Baby birds make this sound.
Take away two letters, then add one.

An animal that gives us wool.
Add one letter.

To flow or trickle slowly.
Take away one letter.

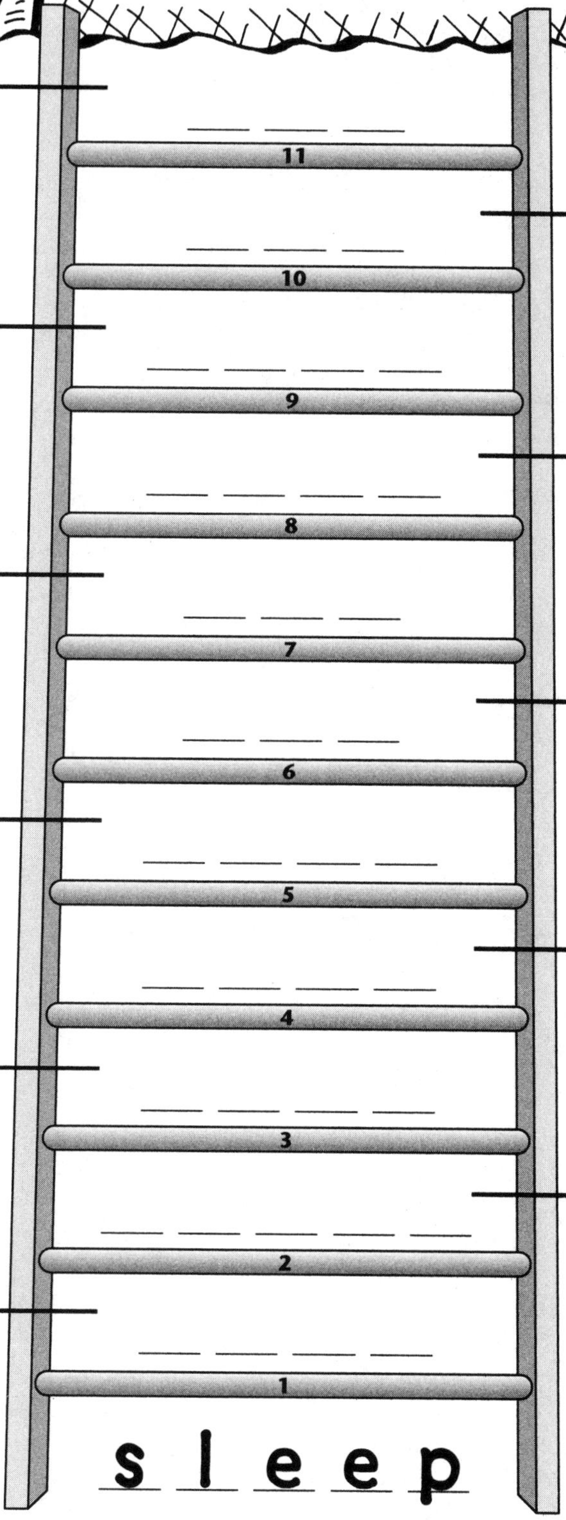

11
10
9
8
7
6
5
4
3
2
1

s   l   e   e   p

Name _____

Read the clues, then write the words.
Start at the bottom and climb to the top.

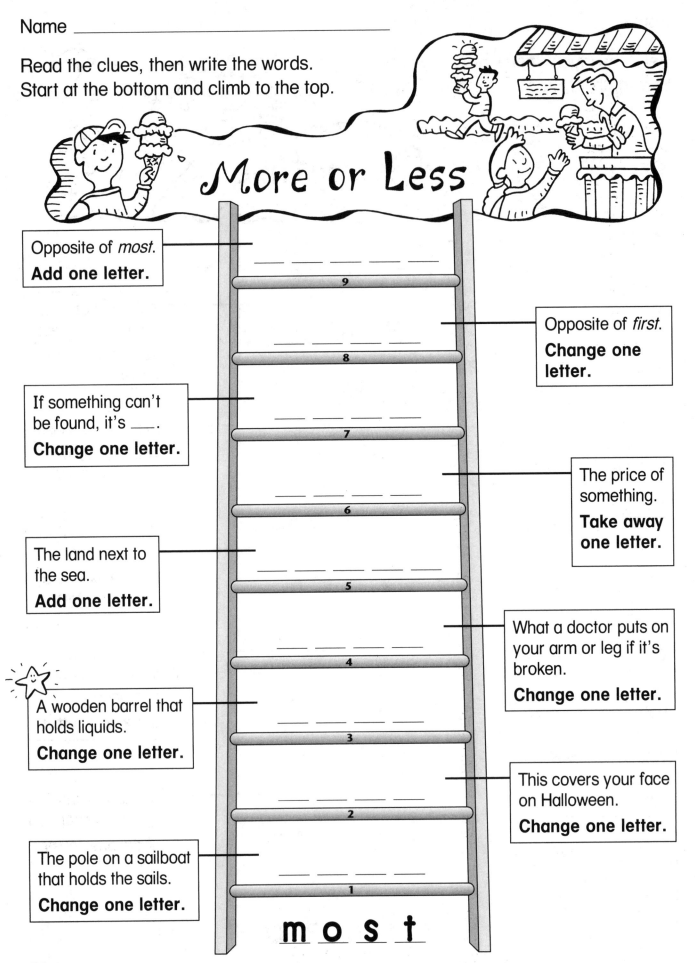

More or Less

Opposite of *most*.
Add one letter.

9 ___ ___ ___ ___ ___

Opposite of *first*.
**Change one
letter.**

8 ___ ___ ___ ___

If something can't
be found, it's ___.
Change one letter.

7 ___ ___ ___ ___

The price of
something.
**Take away
one letter.**

6 ___ ___ ___ ___ ___

The land next to
the sea.
Add one letter.

5 ___ ___ ___ ___ ___

What a doctor puts on
your arm or leg if it's
broken.
Change one letter.

4 ___ ___ ___ ___

A wooden barrel that
holds liquids.
Change one letter.

3 ___ ___ ___ ___

This covers your face
on Halloween.
Change one letter.

2 ___ ___ ___ ___

The pole on a sailboat
that holds the sails.
Change one letter.

1 ___ ___ ___ ___

m o s t

16

Name _____

Read the clues, then write the words.
Start at the bottom and climb to the top.

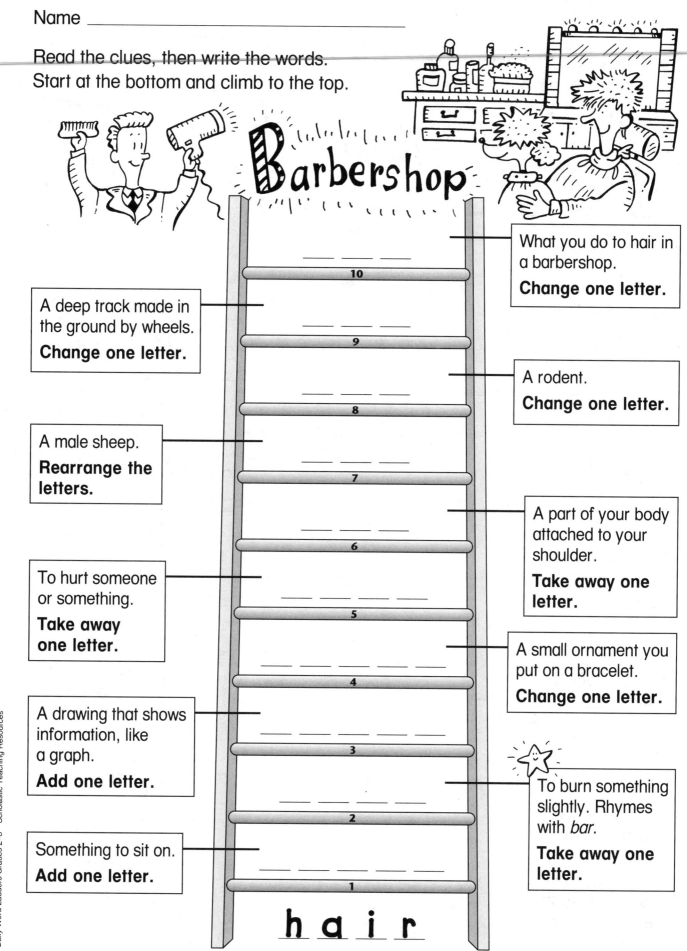

Barbershop

What you do to hair in a barbershop.
Change one letter.

A deep track made in the ground by wheels.
Change one letter.

A rodent.
Change one letter.

A male sheep.
Rearrange the letters.

A part of your body attached to your shoulder.
Take away one letter.

To hurt someone or something.
Take away one letter.

A small ornament you put on a bracelet.
Change one letter.

A drawing that shows information, like a graph.
Add one letter.

To burn something slightly. Rhymes with *bar*.
Take away one letter.

Something to sit on.
Add one letter.

h a i r

Name _____

Read the clues, then write the words.
Start at the bottom and climb to the top.

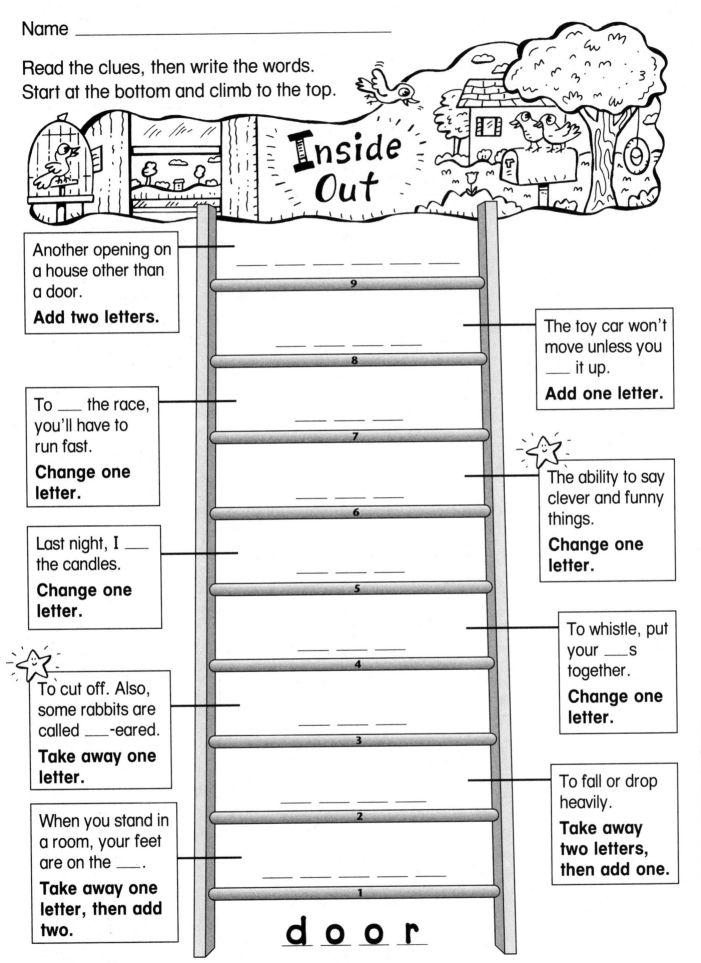

Inside Out

Another opening on a house other than a door.
Add two letters.

The toy car won't move unless you ___ it up.
Add one letter.

To ___ the race, you'll have to run fast.
Change one letter.

The ability to say clever and funny things.
Change one letter.

Last night, I ___ the candles.
Change one letter.

To whistle, put your ___s together.
Change one letter.

To cut off. Also, some rabbits are called ___-eared.
Take away one letter.

To fall or drop heavily.
Take away two letters, then add one.

When you stand in a room, your feet are on the ___.
Take away one letter, then add two.

9
8
7
6
5
4
3
2
1

d o o r

Daily Word Ladders Grades 2–3 Scholastic Teaching Resources

Name _____

Read the clues, then write the words.
Start at the bottom and climb to the top.

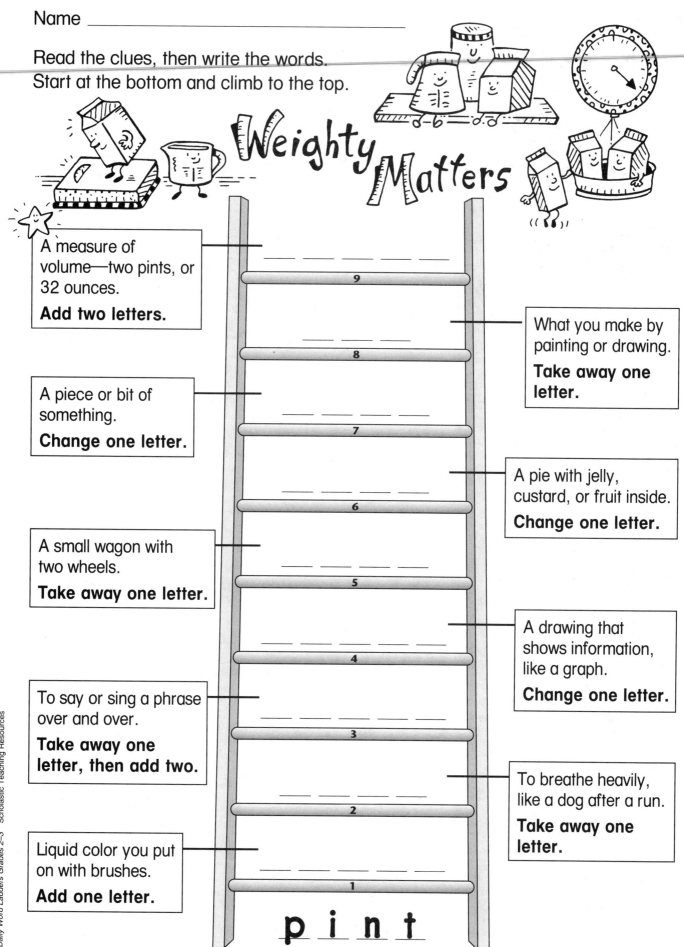

Weighty Matters

A measure of volume—two pints, or 32 ounces.
Add two letters.

A piece or bit of something.
Change one letter.

A small wagon with two wheels.
Take away one letter.

To say or sing a phrase over and over.
Take away one letter, then add two.

Liquid color you put on with brushes.
Add one letter.

What you make by painting or drawing.
Take away one letter.

A pie with jelly, custard, or fruit inside.
Change one letter.

A drawing that shows information, like a graph.
Change one letter.

To breathe heavily, like a dog after a run.
Take away one letter.

9

8

7

6

5

4

3

2

1

p i n t

Name _____

Read the clues, then write the words.
Start at the bottom and climb to the top.

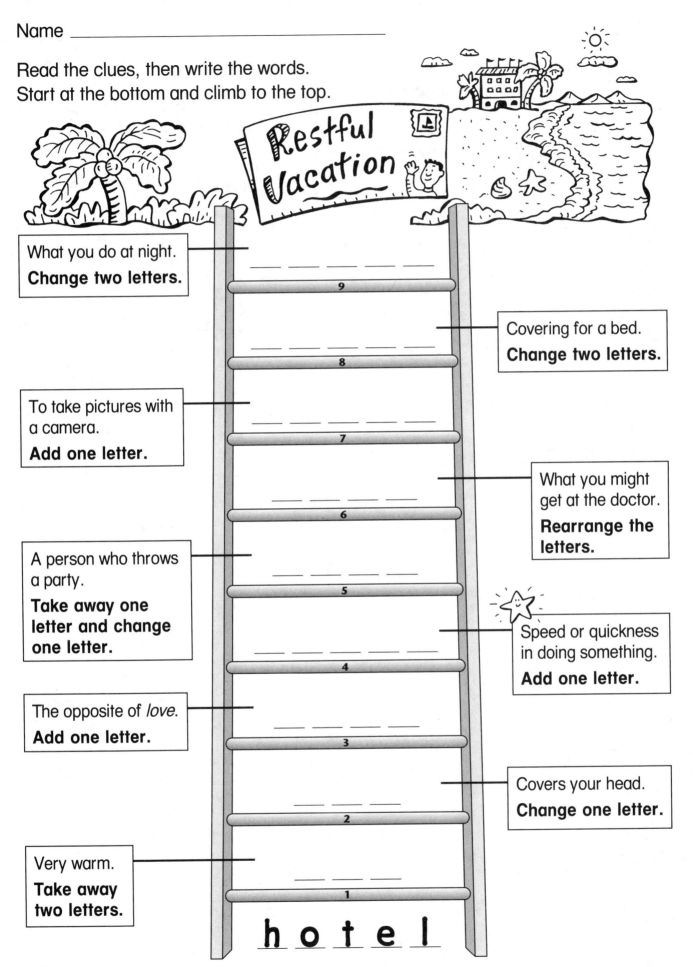

Restful Vacation

What you do at night.
Change two letters.

— 9 —

Covering for a bed.
Change two letters.

— 8 —

To take pictures with
a camera.
Add one letter.

— 7 —

What you might
get at the doctor.
**Rearrange the
letters.**

— 6 —

A person who throws
a party.
**Take away one
letter and change
one letter.**

— 5 —

Speed or quickness
in doing something.
Add one letter.

— 4 —

The opposite of *love*.
Add one letter.

— 3 —

Covers your head.
Change one letter.

— 2 —

Very warm.
**Take away
two letters.**

— 1 —

h o t e l

Daily Word Ladders Grades 2–3   Scholastic Teaching Resources

Read the clues, then write the words.
Start at the bottom and climb to the top.

Gardening

What a flower might be planted in.
Rearrange the letters.

9 _ _ _

The highest point of something.
Take away one letter.

8 _ _ _

To come to a halt.
Change one letter.

7 _ _ _ _

A store.
Take away three letters.

6 _ _ _ _

Someone who buys things.
Take away one letter, then add two.

5 _ _ _ _ _

What you might take to get clean.
Add two letters.

4 _ _ _ _ _

To display something.
Change one letter.

3 _ _ _ _

Not fast.
Change one letter.

2 _ _ _ _

To move along slowly, like a river.
Take away two letters.

1 _ _ _ _

f l o w e r

Name _____

Read the clues, then write the words.
Start at the bottom and climb to the top.

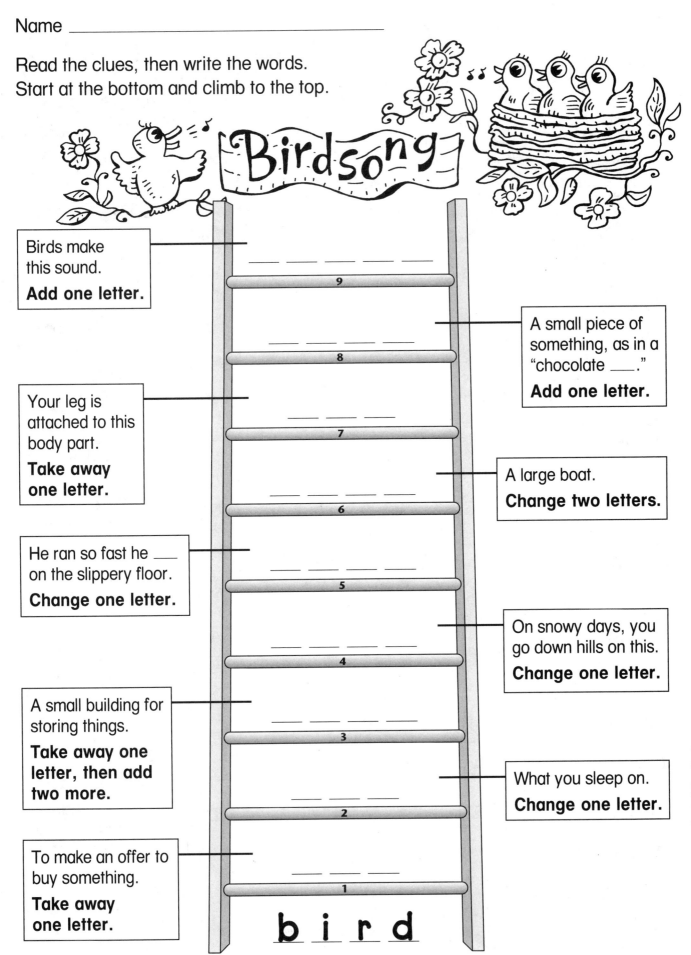

Birdsong

Birds make this sound.
Add one letter.

A small piece of something, as in a "chocolate ___."
Add one letter.

Your leg is attached to this body part.
Take away one letter.

A large boat.
Change two letters.

He ran so fast he ___ on the slippery floor.
Change one letter.

On snowy days, you go down hills on this.
Change one letter.

A small building for storing things.
Take away one letter, then add two more.

What you sleep on.
Change one letter.

To make an offer to buy something.
Take away one letter.

9
8
7
6
5
4
3
2
1

b i r d

Daily Word Ladders Grades 2–3 Scholastic Teaching Resources

Name _____

Read the clues, then write the words.
Start at the bottom and climb to the top.

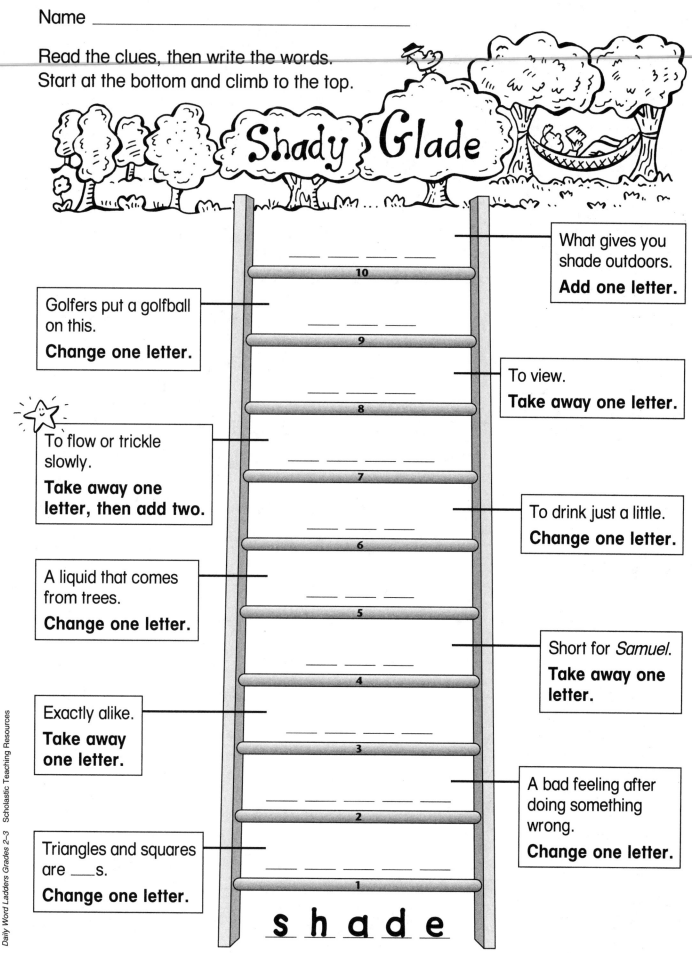

Shady Glade

What gives you shade outdoors.
Add one letter.

Golfers put a golfball on this.
Change one letter.

To view.
Take away one letter.

To flow or trickle slowly.
Take away one letter, then add two.

To drink just a little.
Change one letter.

A liquid that comes from trees.
Change one letter.

Short for *Samuel*.
Take away one letter.

Exactly alike.
Take away one letter.

A bad feeling after doing something wrong.
Change one letter.

Triangles and squares are ___s.
Change one letter.

10
9
8
7
6
5
4
3
2
1

s h a d e

Name _____

Read the clues, then write the words.
Start at the bottom and climb to the top.

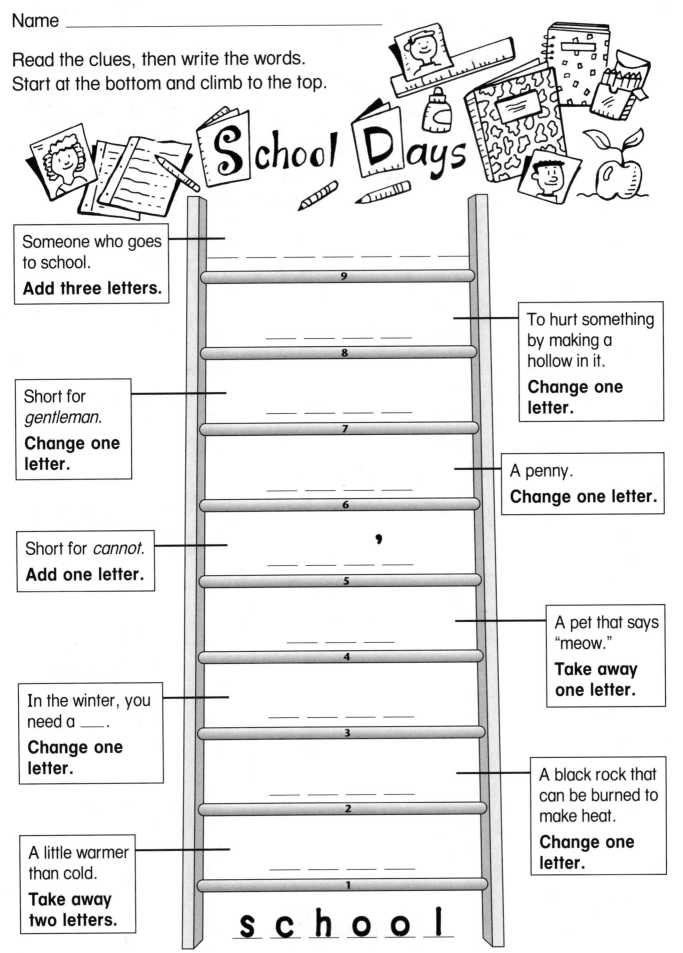

School Days

Someone who goes to school.
Add three letters.

To hurt something by making a hollow in it.
Change one letter.

Short for *gentleman*.
Change one letter.

A penny.
Change one letter.

Short for *cannot*.
Add one letter.

A pet that says "meow."
Take away one letter.

In the winter, you need a ___.
Change one letter.

A black rock that can be burned to make heat.
Change one letter.

A little warmer than cold.
Take away two letters.

9

8

7

6

,

5

4

3

2

1

s c h o o l

24

Name _____

Read the clues, then write the words.
Start at the bottom and climb to the top.

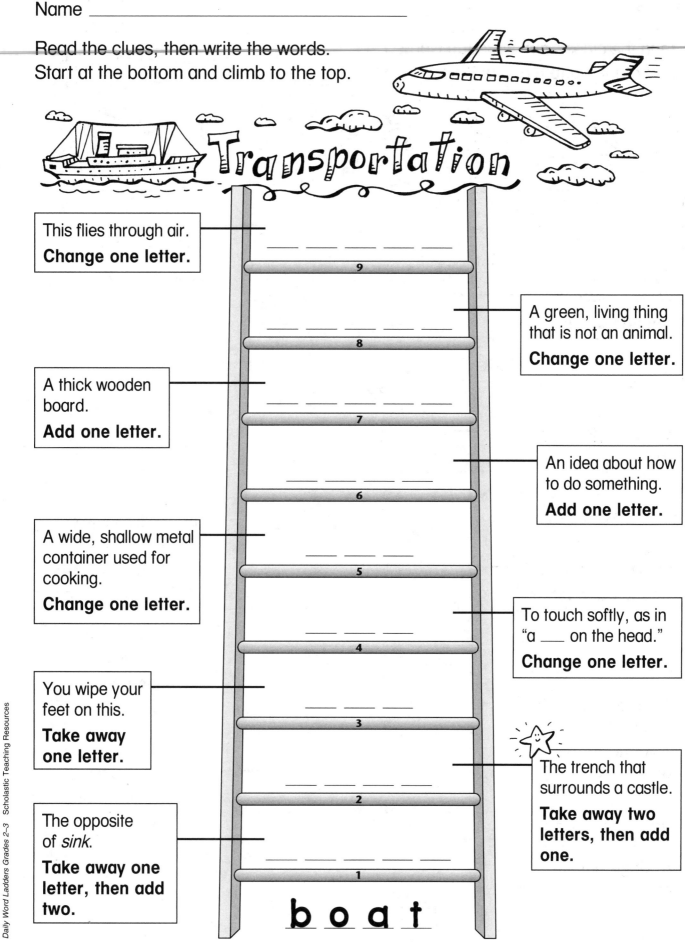

Transportation

This flies through air.
Change one letter.

9

A green, living thing that is not an animal.
Change one letter.

8

A thick wooden board.
Add one letter.

7

An idea about how to do something.
Add one letter.

6

A wide, shallow metal container used for cooking.
Change one letter.

5

To touch softly, as in "a ___ on the head."
Change one letter.

4

You wipe your feet on this.
Take away one letter.

3

The trench that surrounds a castle.
Take away two letters, then add one.

2

The opposite of *sink*.
Take away one letter, then add two.

1

b o a t

Name _____

Read the clues, then write the words.
Start at the bottom and climb to the top.

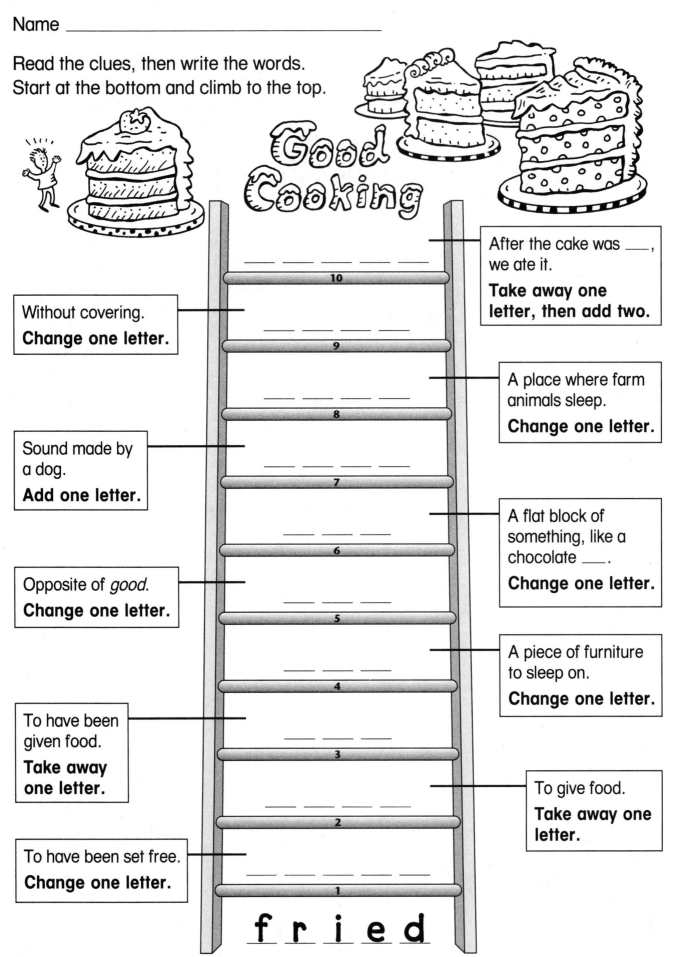

Good Cooking

10. _ _ _ _

After the cake was ___, we ate it.
Take away one letter, then add two.

9. _ _ _ _

Without covering.
Change one letter.

8. _ _ _ _

A place where farm animals sleep.
Change one letter.

7. _ _ _ _

Sound made by a dog.
Add one letter.

6. _ _ _ _

A flat block of something, like a chocolate ___.
Change one letter.

5. _ _ _ _

Opposite of *good*.
Change one letter.

4. _ _ _ _

A piece of furniture to sleep on.
Change one letter.

3. _ _ _

To have been given food.
Take away one letter.

2. _ _ _ _

To give food.
Take away one letter.

1. _ _ _ _ _

To have been set free.
Change one letter.

f r i e d

Daily Word Ladders Grades 2–3   Scholastic Teaching Resources

Name _____

Read the clues, then write the words.
Start at the bottom and climb to the top.

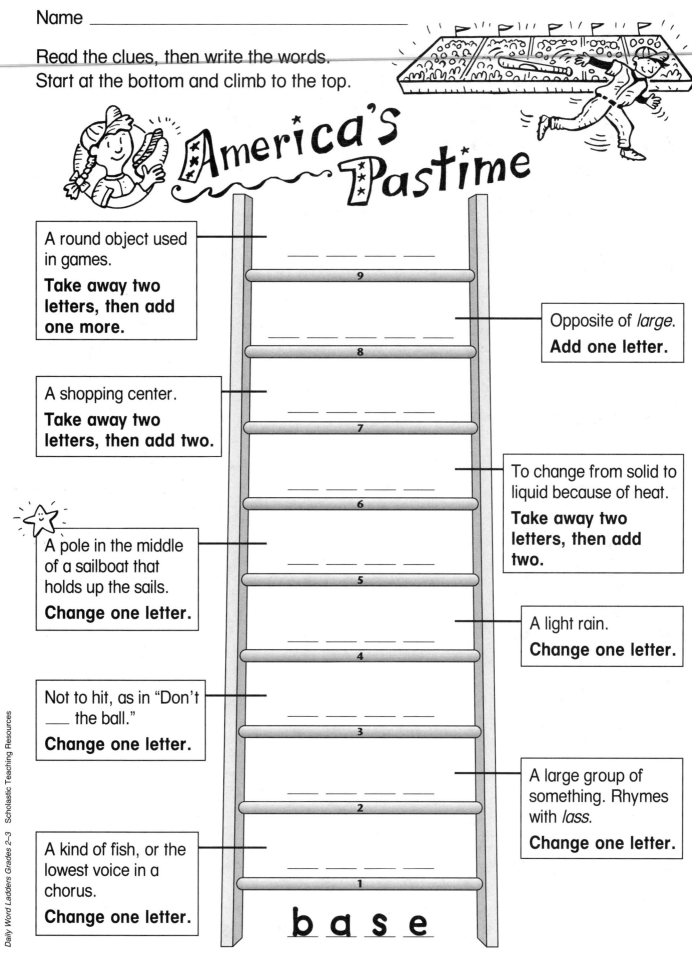

America's Pastime

A round object used in games.
Take away two letters, then add one more.

Opposite of *large*.
Add one letter.

A shopping center.
Take away two letters, then add two.

To change from solid to liquid because of heat.
Take away two letters, then add two.

A pole in the middle of a sailboat that holds up the sails.
Change one letter.

A light rain.
Change one letter.

Not to hit, as in "Don't ___ the ball."
Change one letter.

A large group of something. Rhymes with *lass*.
Change one letter.

A kind of fish, or the lowest voice in a chorus.
Change one letter.

9

8

7

6

5

4

3

2

1

b a s e

Name _____

Read the clues, then write the words.
Start at the bottom and climb to the top.

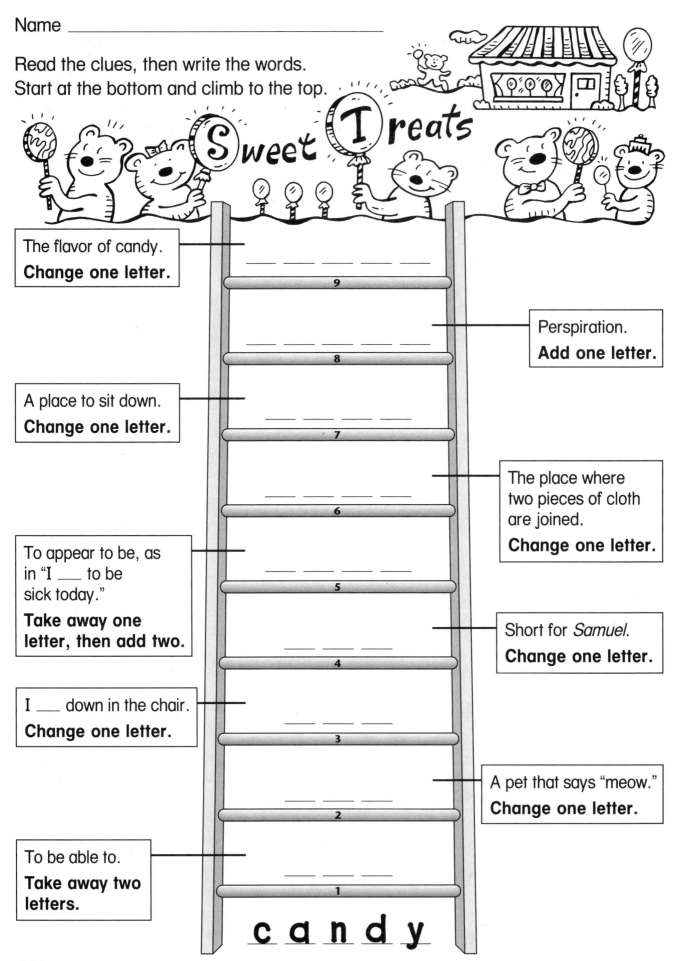

Sweet Treats

The flavor of candy.
Change one letter.

Perspiration.
Add one letter.

A place to sit down.
Change one letter.

The place where
two pieces of cloth
are joined.
Change one letter.

To appear to be, as
in "I ___ to be
sick today."
**Take away one
letter, then add two.**

Short for *Samuel*.
Change one letter.

I ___ down in the chair.
Change one letter.

A pet that says "meow."
Change one letter.

To be able to.
**Take away two
letters.**

9
8
7
6
5
4
3
2
1

c a n d y

Daily Word Ladders Grades 2–3 Scholastic Teaching Resources

Name _____

Read the clues, then write the words.
Start at the bottom and climb to the top.

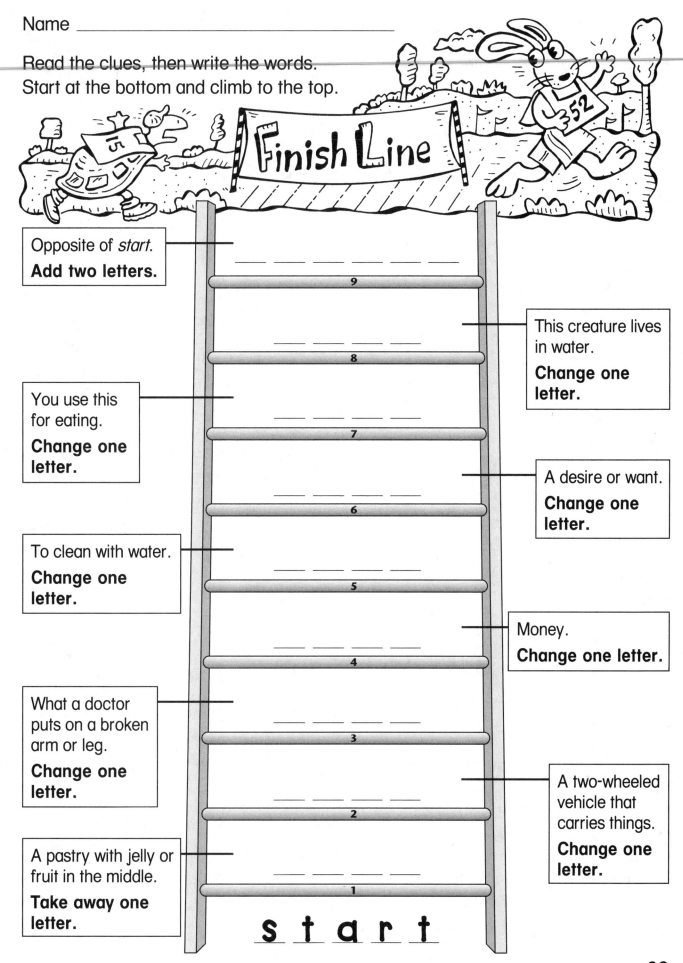

Finish Line

Opposite of *start*.
Add two letters.

- - - - - -   9

This creature lives in water.
Change one letter.

- - - - -   8

You use this for eating.
Change one letter.

- - - - -   7

A desire or want.
Change one letter.

- - - -   6

To clean with water.
Change one letter.

- - - -   5

Money.
Change one letter.

- - - -   4

What a doctor puts on a broken arm or leg.
Change one letter.

- - - -   3

A two-wheeled vehicle that carries things.
Change one letter.

- - - -   2

A pastry with jelly or fruit in the middle.
Take away one letter.

- - - -   1

s t a r t

Name _____

Read the clues, then write the words.
Start at the bottom and climb to the top.

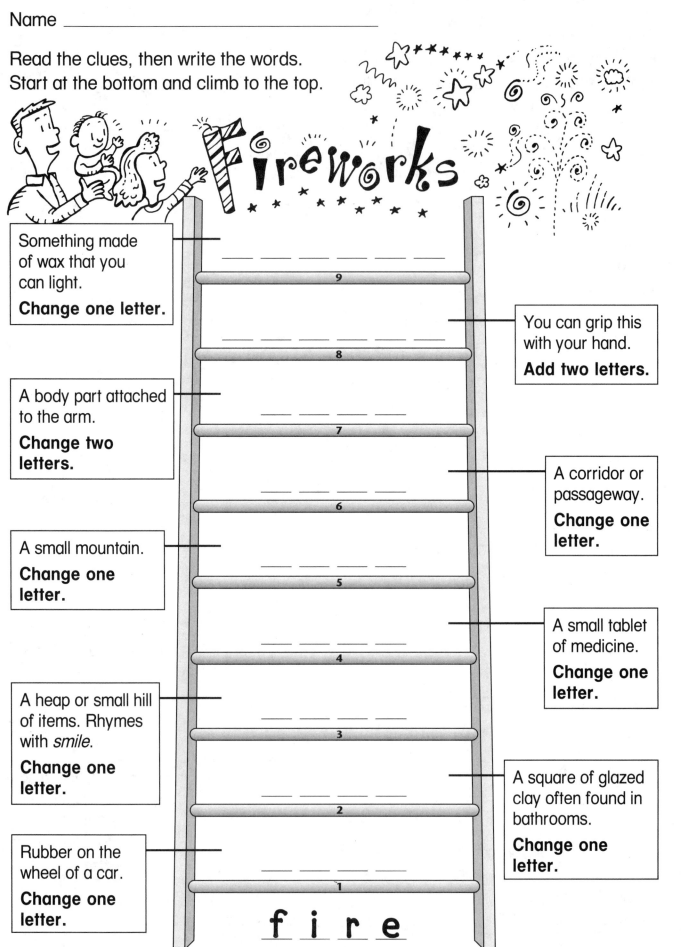

Something made of wax that you can light.
Change one letter.

You can grip this with your hand.
Add two letters.

A body part attached to the arm.
Change two letters.

A corridor or passageway.
Change one letter.

A small mountain.
Change one letter.

A small tablet of medicine.
Change one letter.

A heap or small hill of items. Rhymes with *smile*.
Change one letter.

A square of glazed clay often found in bathrooms.
Change one letter.

Rubber on the wheel of a car.
Change one letter.

f i r e

9
8
7
6
5
4
3
2
1

Daily Word Ladders Grades 2–3 Scholastic Teaching Resources

Name _____

Read the clues, then write the words.
Start at the bottom and climb to the top.

Opposites Attract

Opposite of *black*.
Add two letters.

9 _ _ _ _ _

To strike.
Change one letter.

8 _ _ _ _

Very warm.
Change one letter.

7 _ _ _ _

Plenty of something.
Take away one letter.

6 _ _ _ _

To steal or rob.
Rhymes with *boot*.
Change one letter.

5 _ _ _ _

To use your eyes
to see things.
Change one letter.

4 _ _ _ _

Part of a door you
can open or shut
with a key.
Take away one letter.

3 _ _ _ _

Used for telling time.
Change one letter.

2 _ _ _ _ _

A cube of something
hard, like a ___ of
wood.
Change one letter.

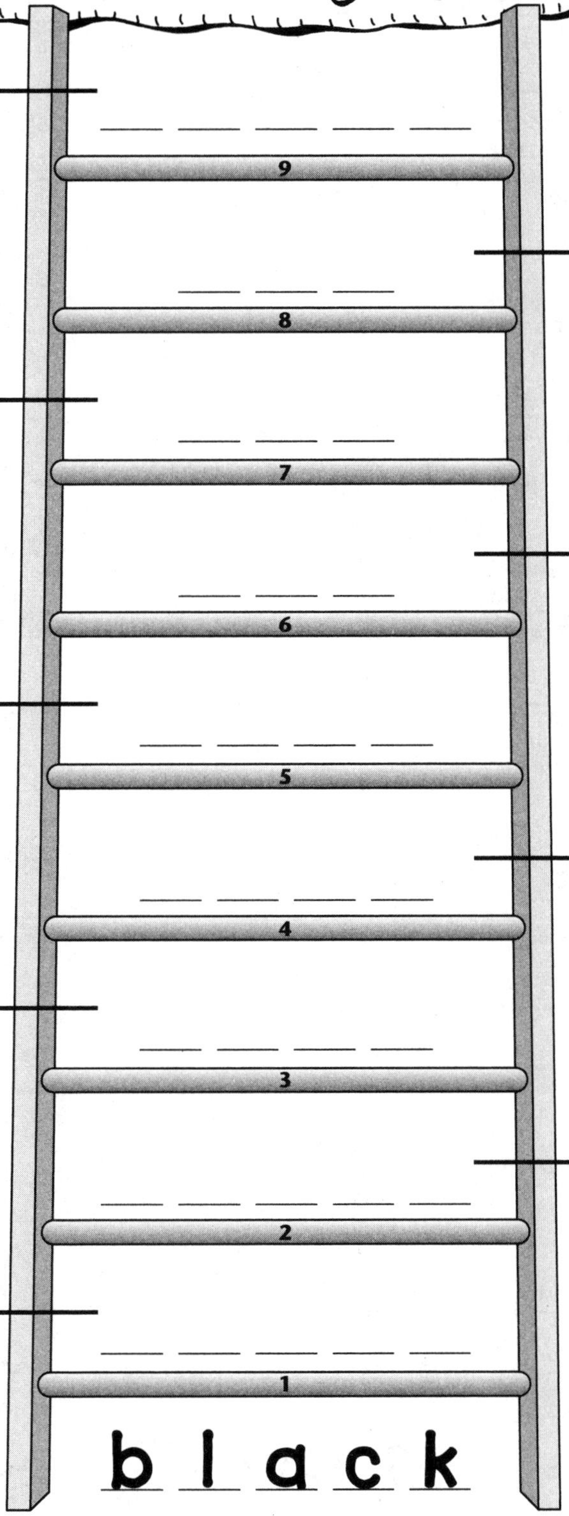

1 _ _ _ _ _

b l a c k

Name _____

Read the clues, then write the words.
Start at the bottom and climb to the top.

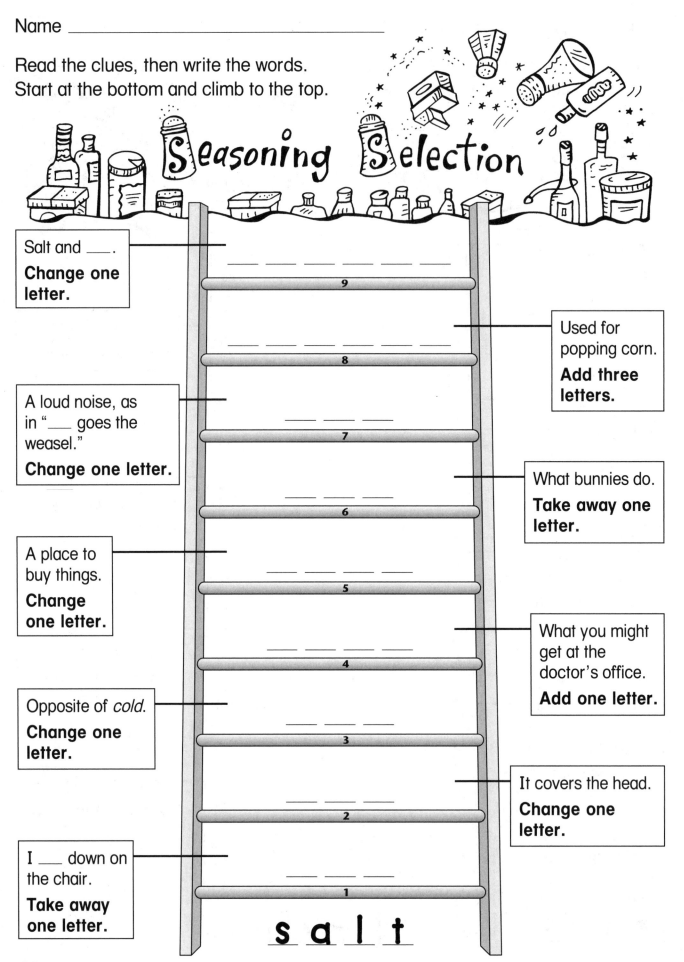

Seasoning Selection

Salt and ___.
Change one letter.
_ _ _ _ _ _ 9

_ _ _ _ _ _ 8
Used for popping corn.
Add three letters.

A loud noise, as in "___ goes the weasel."
Change one letter.
_ _ _ _ 7

_ _ _ _ 6
What bunnies do.
Take away one letter.

A place to buy things.
Change one letter.
_ _ _ _ _ 5

_ _ _ _ _ 4
What you might get at the doctor's office.
Add one letter.

Opposite of *cold*.
Change one letter.
_ _ _ _ 3

_ _ _ _ 2
It covers the head.
Change one letter.

I ___ down on the chair.
Take away one letter.
_ _ _ _ 1

s a l t

Daily Word Ladders Grades 2–3 Scholastic Teaching Resources

Name _____

Read the clues, then write the words.
Start at the bottom and climb to the top.

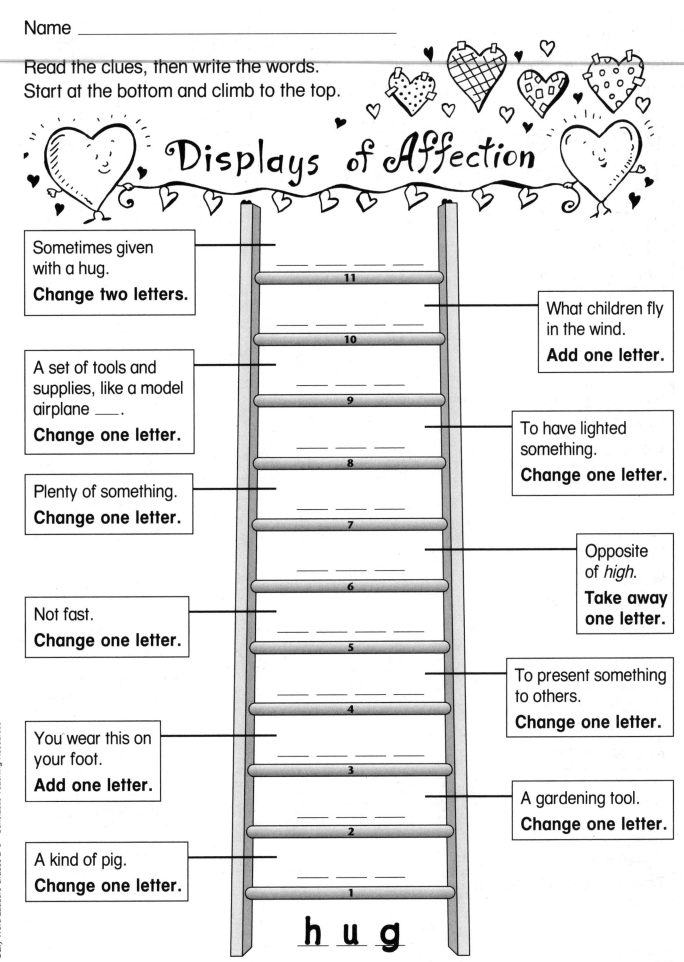

Displays of Affection

Sometimes given with a hug.
Change two letters.
— 11

A set of tools and supplies, like a model airplane ___.
Change one letter.
— 9

Plenty of something.
Change one letter.
— 7

Not fast.
Change one letter.
— 5

You wear this on your foot.
Add one letter.
— 3

A kind of pig.
Change one letter.
— 1

What children fly in the wind.
Add one letter.
— 10

To have lighted something.
Change one letter.
— 8

Opposite of *high*.
Take away one letter.
— 6

To present something to others.
Change one letter.
— 4

A gardening tool.
Change one letter.
— 2

h u g

Name _____

Read the clues, then write the words.
Start at the bottom and climb to the top.

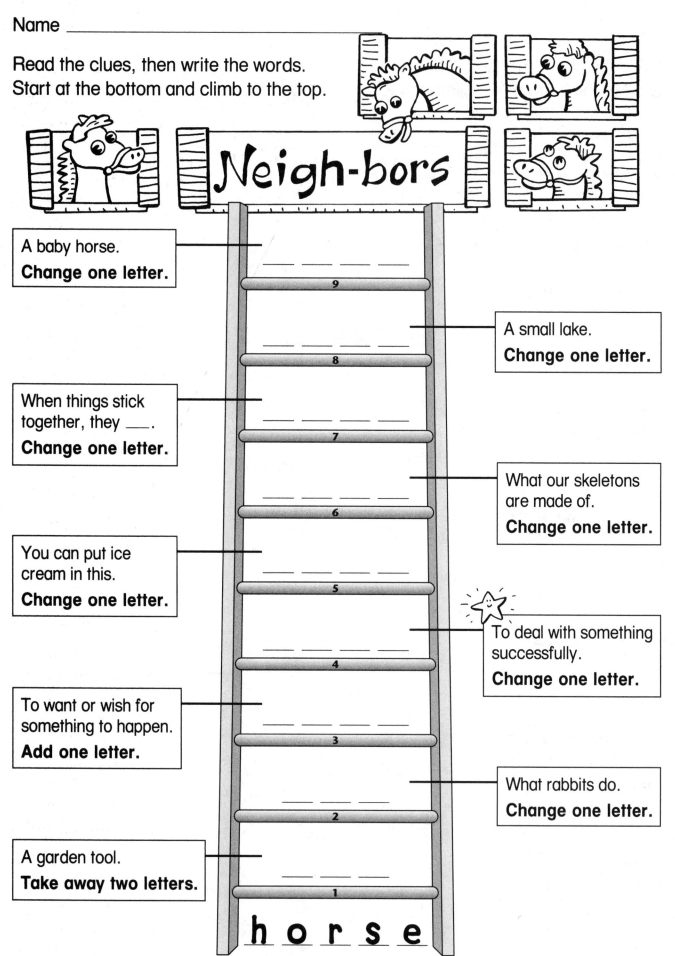

Neigh-bors

A baby horse.
Change one letter.

9 _ _ _ _ _

A small lake.
Change one letter.

8 _ _ _ _ _

When things stick
together, they ___.
Change one letter.

7 _ _ _ _ _

What our skeletons
are made of.
Change one letter.

6 _ _ _ _ _

You can put ice
cream in this.
Change one letter.

5 _ _ _ _ _

To deal with something
successfully.
Change one letter.

4 _ _ _ _ _

To want or wish for
something to happen.
Add one letter.

3 _ _ _ _ _

What rabbits do.
Change one letter.

2 _ _ _ _

A garden tool.
Take away two letters.

1 _ _ _ _

h o r s e

34

Name _____

Read the clues, then write the words.
Start at the bottom and climb to the top.

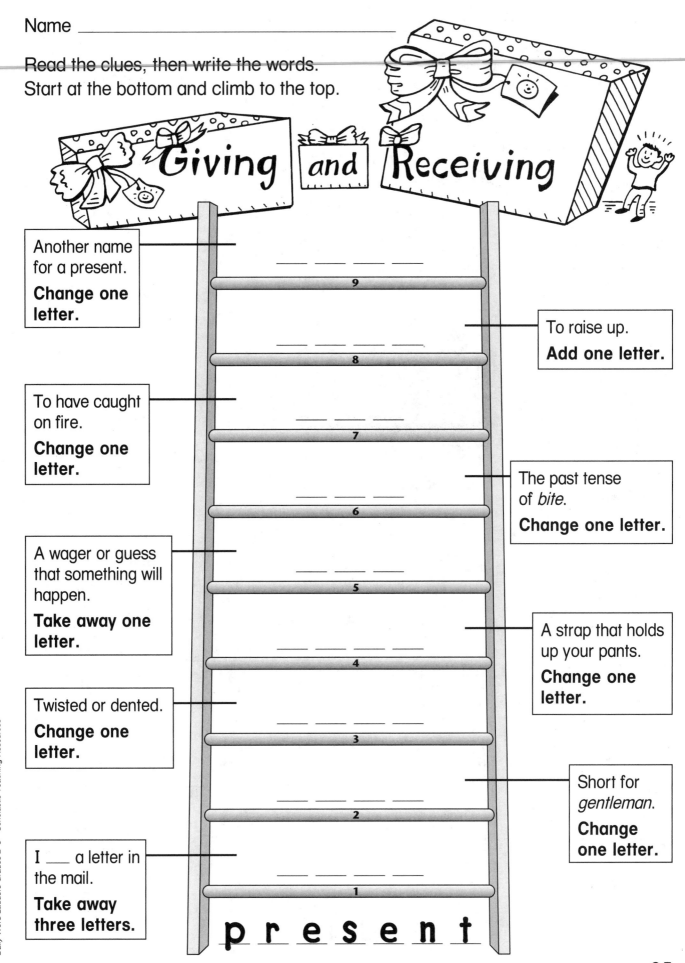

Giving and Receiving

Another name for a present.
Change one letter.

To raise up.
Add one letter.

To have caught on fire.
Change one letter.

The past tense of *bite*.
Change one letter.

A wager or guess that something will happen.
Take away one letter.

A strap that holds up your pants.
Change one letter.

Twisted or dented.
Change one letter.

Short for *gentleman*.
Change one letter.

I ___ a letter in the mail.
Take away three letters.

9
8
7
6
5
4
3
2
1

p r e s e n t

Name _____

Read the clues, then write the words.
Start at the bottom and climb to the top.

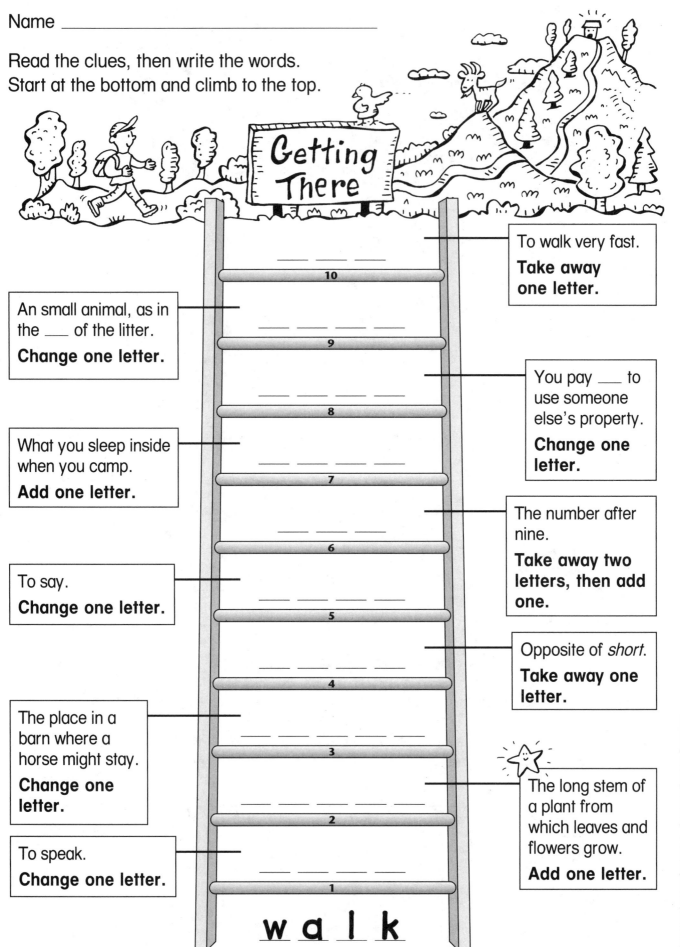

Getting There

To walk very fast.
Take away one letter.

_ _ _ _ _ 10

An small animal, as in the ___ of the litter.
Change one letter.

_ _ _ _ 9

_ _ _ _ 8

You pay ___ to use someone else's property.
Change one letter.

What you sleep inside when you camp.
Add one letter.

_ _ _ _ 7

_ _ _ 6

The number after nine.
Take away two letters, then add one.

To say.
Change one letter.

_ _ _ _ 5

_ _ _ _ 4

Opposite of *short*.
Take away one letter.

The place in a barn where a horse might stay.
Change one letter.

_ _ _ _ _ 3

_ _ _ _ _ 2

The long stem of a plant from which leaves and flowers grow.
Add one letter.

To speak.
Change one letter.

_ _ _ _ 1

w a l k

Daily Word Ladders Grades 2–3   Scholastic Teaching Resources

Name _____

Read the clues, then write the words.
Start at the bottom and climb to the top.

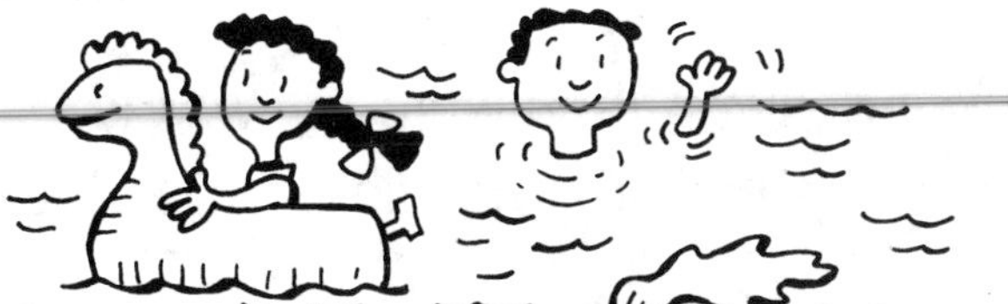

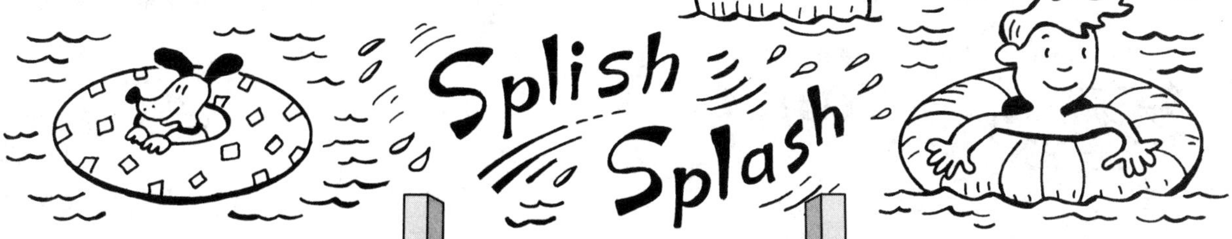

Splish
Splash

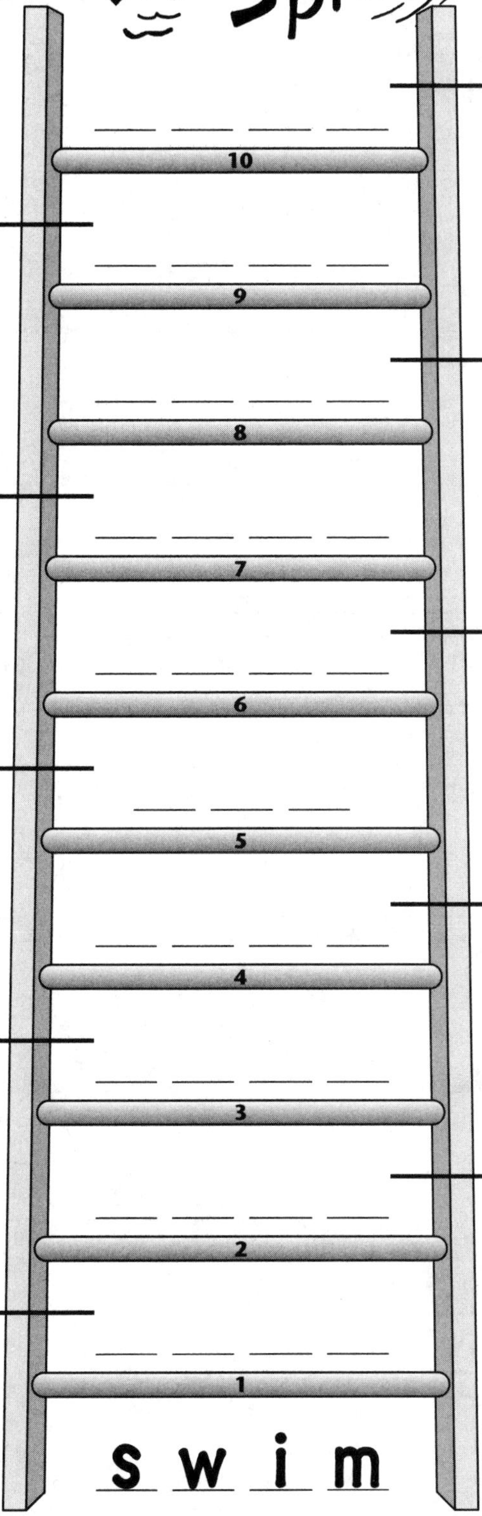

A place to swim.
Change one letter.

The hair of sheep.
Change one letter.

What trees are made of.
Change two letters.

Opposite of tame.
Change one letter.

You have to ___ up some toys to make them go.
Add one letter.

To beat someone in a game.
Take away one letter.

The blowing air.
Change one letter.

A stick used by magicians.
Take away one letter, then add another.

A large white bird with a curved neck.
Change one letter.

Today I swim; yesterday I ___.
Change one letter.

10

9

8

7

6

5

4

3

2

1

s w i m

Name _____

Read the clues, then write the words.
Start at the bottom and climb to the top.

Fancy Footwear

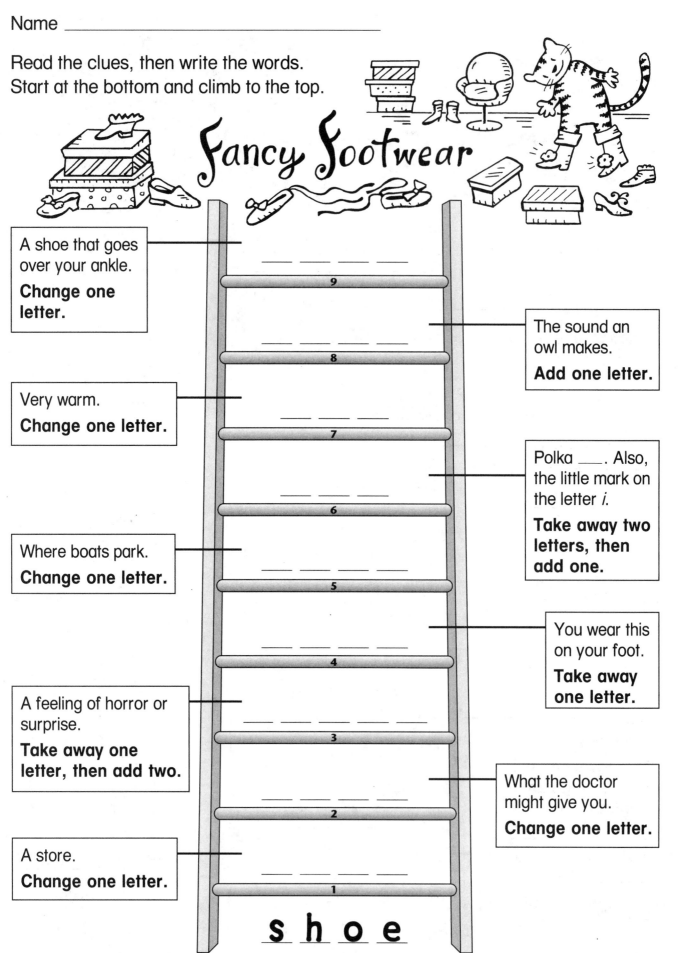

A shoe that goes over your ankle.
Change one letter.

The sound an owl makes.
Add one letter.

Very warm.
Change one letter.

Polka ___. Also, the little mark on the letter *i*.
Take away two letters, then add one.

Where boats park.
Change one letter.

You wear this on your foot.
Take away one letter.

A feeling of horror or surprise.
Take away one letter, then add two.

What the doctor might give you.
Change one letter.

A store.
Change one letter.

9

8

7

6

5

4

3

2

1

s h o e

Daily Word Ladders Grades 2–3 Scholastic Teaching Resources

Name _____

Read the clues, then write the words.
Start at the bottom and climb to the top.

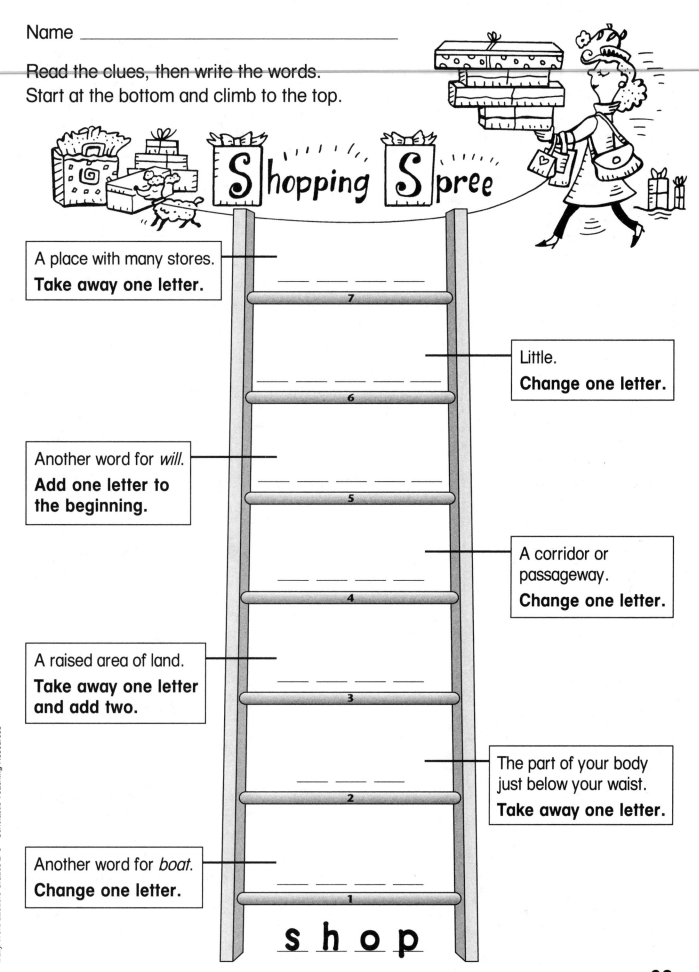

Shopping Spree

A place with many stores.
Take away one letter.

7 _ _ _ _ _ _

Little.
Change one letter.

6 _ _ _ _ _

Another word for *will*.
Add one letter to the beginning.

5 _ _ _ _ _

A corridor or passageway.
Change one letter.

4 _ _ _ _

A raised area of land.
Take away one letter and add two.

3 _ _ _ _

The part of your body just below your waist.
Take away one letter.

2 _ _ _ _

Another word for *boat*.
Change one letter.

1 _ _ _ _

s h o p

Name _____

Read the clues, then write the words.
Start at the bottom and climb to the top.

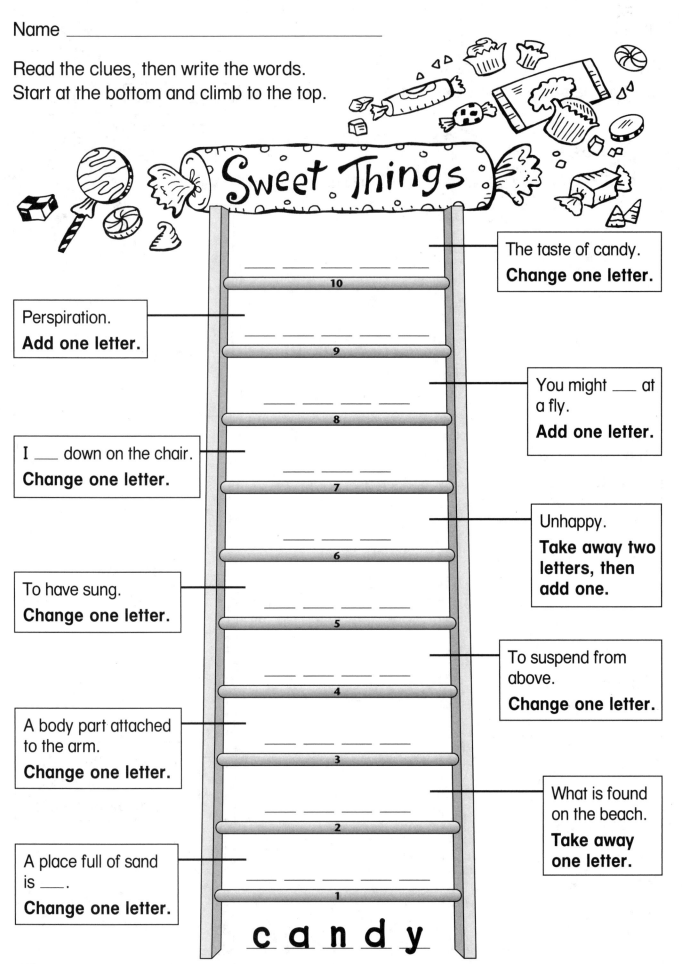

Sweet Things

The taste of candy.
Change one letter.

Perspiration.
Add one letter.

You might ___ at a fly.
Add one letter.

I ___ down on the chair.
Change one letter.

Unhappy.
Take away two letters, then add one.

To have sung.
Change one letter.

To suspend from above.
Change one letter.

A body part attached to the arm.
Change one letter.

What is found on the beach.
Take away one letter.

A place full of sand is ___.
Change one letter.

c a n d y

Daily Word Ladders Grades 2–3   Scholastic Teaching Resources

Name _____

Read the clues, then write the words.
Start at the bottom and climb to the top.

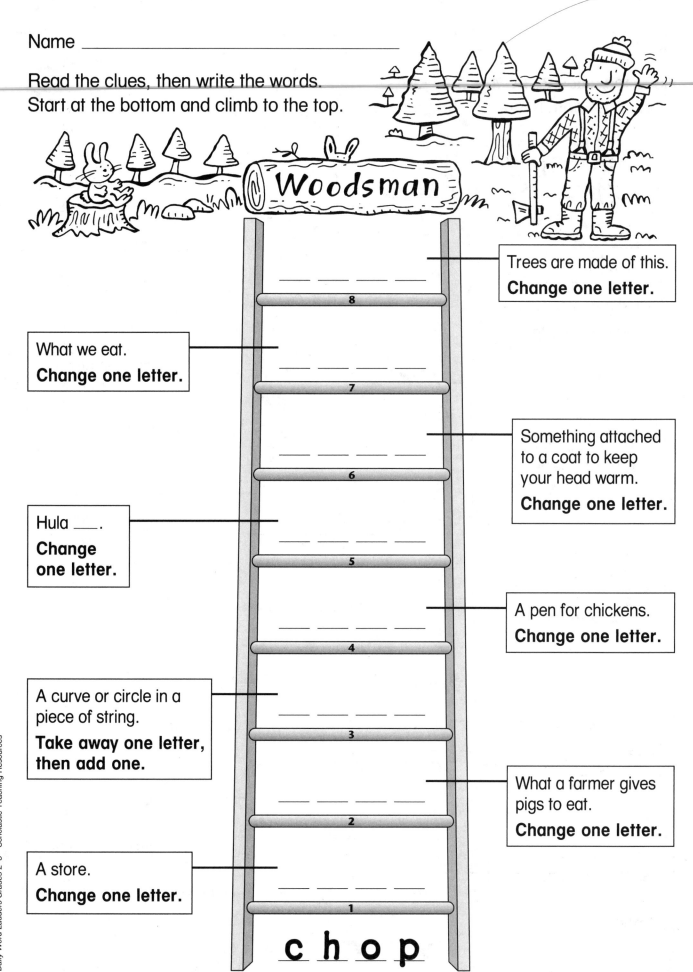

Woodsman

Trees are made of this.
Change one letter.

8

What we eat.
Change one letter.

7

Something attached
to a coat to keep
your head warm.
Change one letter.

6

Hula ___.
**Change
one letter.**

5

A pen for chickens.
Change one letter.

4

A curve or circle in a
piece of string.
**Take away one letter,
then add one.**

3

What a farmer gives
pigs to eat.
Change one letter.

2

A store.
Change one letter.

1

c h o p

Name _____

Read the clues, then write the words.
Start at the bottom and climb to the top.

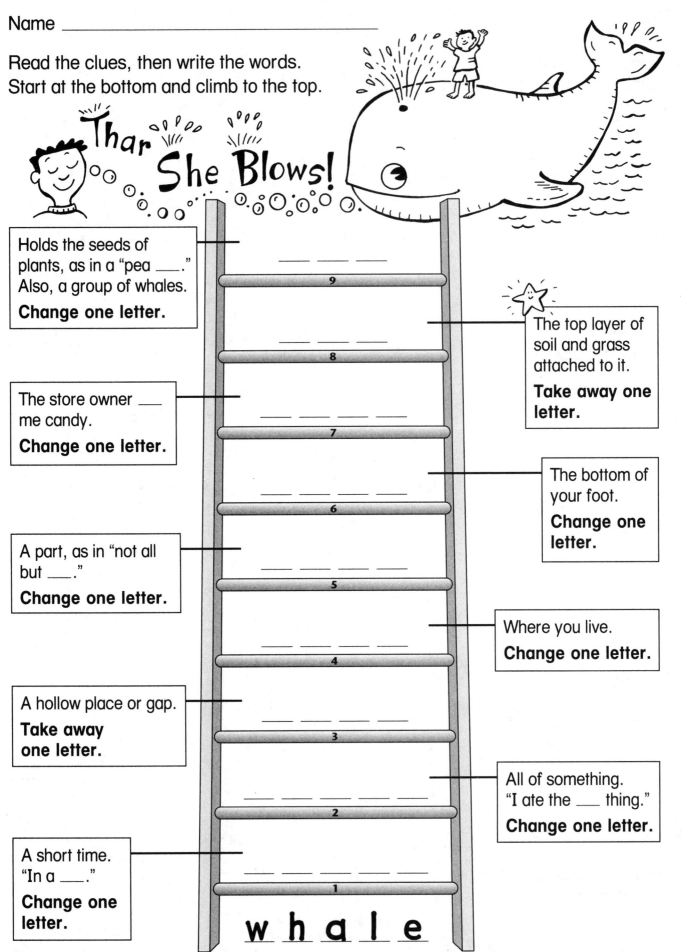

Thar She Blows!

Holds the seeds of
plants, as in a "pea ___."
Also, a group of whales.
Change one letter.

The store owner ___
me candy.
Change one letter.

A part, as in "not all
but ___."
Change one letter.

A hollow place or gap.
**Take away
one letter.**

A short time.
"In a ___."
**Change one
letter.**

The top layer of
soil and grass
attached to it.
**Take away one
letter.**

The bottom of
your foot.
**Change one
letter.**

Where you live.
Change one letter.

All of something.
"I ate the ___ thing."
Change one letter.

9

8

7

6

5

4

3

2

1

w h a l e

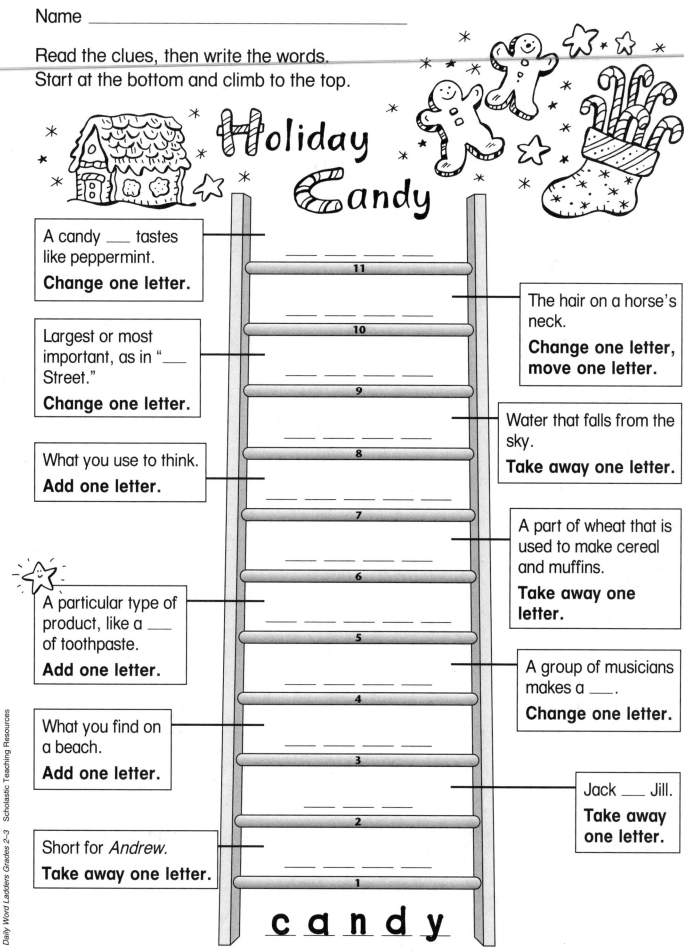

Name _____

Read the clues, then write the words.
Start at the bottom and climb to the top.

Holiday Candy

A candy ___ tastes like peppermint.
Change one letter.

Largest or most important, as in "___ Street."
Change one letter.

What you use to think.
Add one letter.

A particular type of product, like a ___ of toothpaste.
Add one letter.

What you find on a beach.
Add one letter.

Short for *Andrew*.
Take away one letter.

The hair on a horse's neck.
Change one letter, move one letter.

Water that falls from the sky.
Take away one letter.

A part of wheat that is used to make cereal and muffins.
Take away one letter.

A group of musicians makes a ___.
Change one letter.

Jack ___ Jill.
Take away one letter.

11
10
9
8
7
6
5
4
3
2
1

c a n d y

Name _____

Read the clues, then write the words.
Start at the bottom and climb to the top.

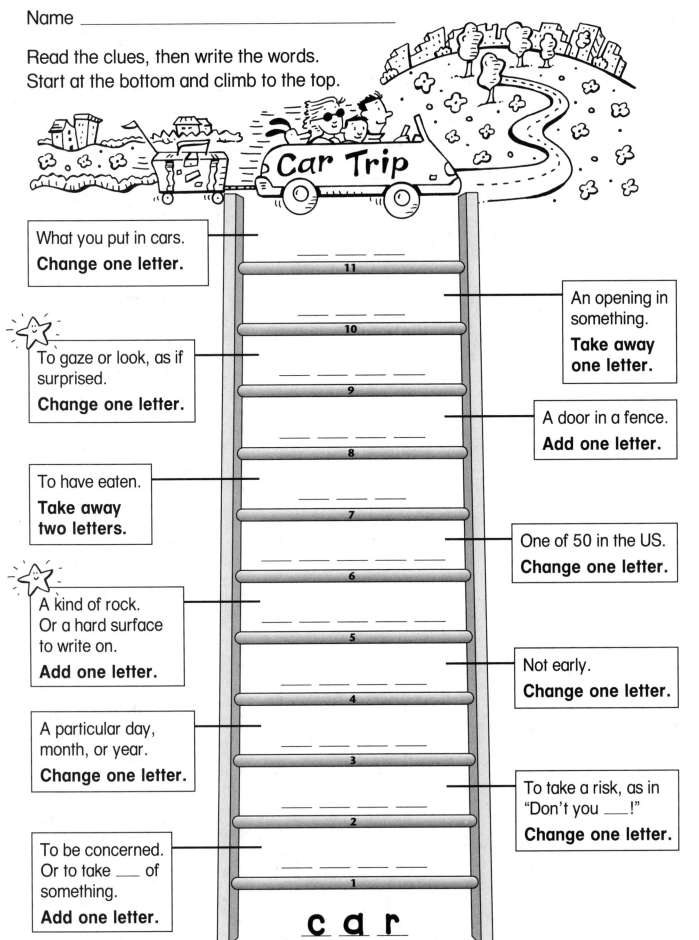

Car Trip

What you put in cars.
Change one letter.

11 _ _ _ _ _

An opening in
something.
**Take away
one letter.**

10 _ _ _ _

To gaze or look, as if
surprised.
Change one letter.

9 _ _ _ _

A door in a fence.
Add one letter.

8 _ _ _ _

To have eaten.
**Take away
two letters.**

7 _ _ _

One of 50 in the US.
Change one letter.

6 _ _ _ _ _

A kind of rock.
Or a hard surface
to write on.
Add one letter.

5 _ _ _ _ _

Not early.
Change one letter.

4 _ _ _ _

A particular day,
month, or year.
Change one letter.

3 _ _ _ _

To take a risk, as in
"Don't you ___!"
Change one letter.

2 _ _ _ _

To be concerned.
Or to take ___ of
something.
Add one letter.

1 _ _ _

c a r

44

Name _____

Read the clues, then write the words.
Start at the bottom and climb to the top.

Friendship

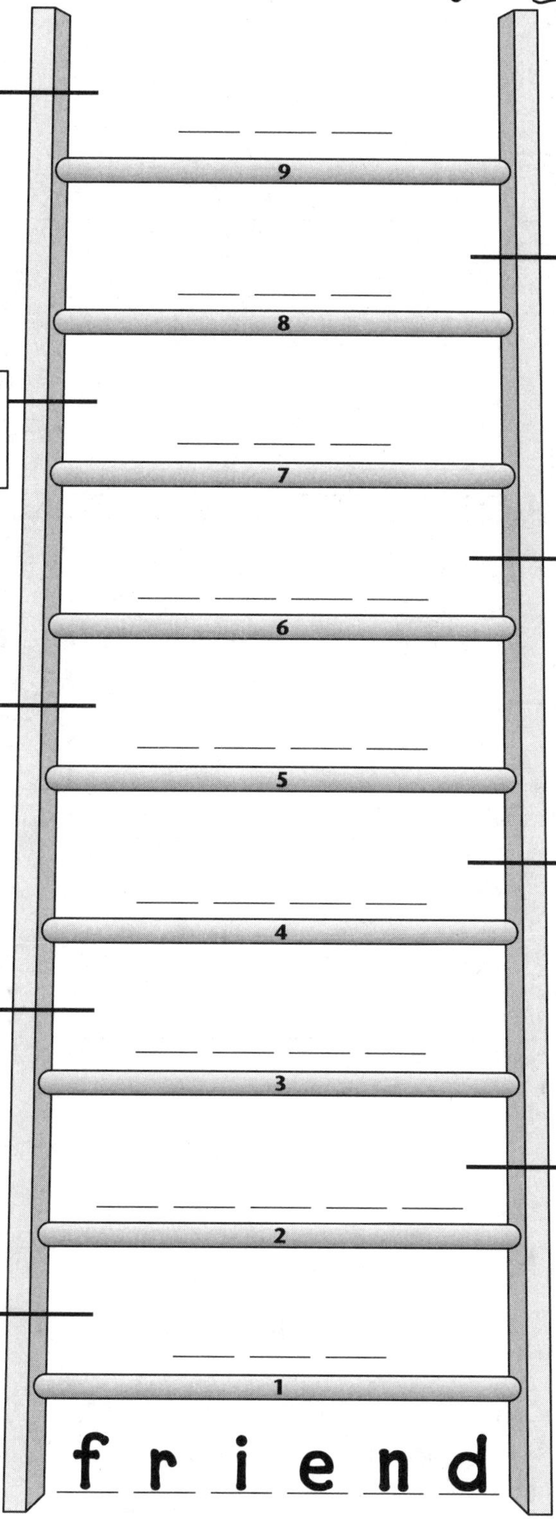

Another name for a friend.
Change one letter.

A frog sits on a lily ___.
Change one letter.

Not good.
Take away one letter.

A group of musicians makes a ___.
Change one letter.

Found at the beach.
Change one letter.

To make someone or something go somewhere.
Change one letter.

To take care of something. "I like to ___ to the plants in my house."
Take away one letter.

Style. The way things are changing.
Add two letters.

Finish.
Take away three letters.

9

8

7

6

5

4

3

2

1

f r i e n d

Name _____

Read the clues, then write the words.
Start at the bottom and climb to the top.

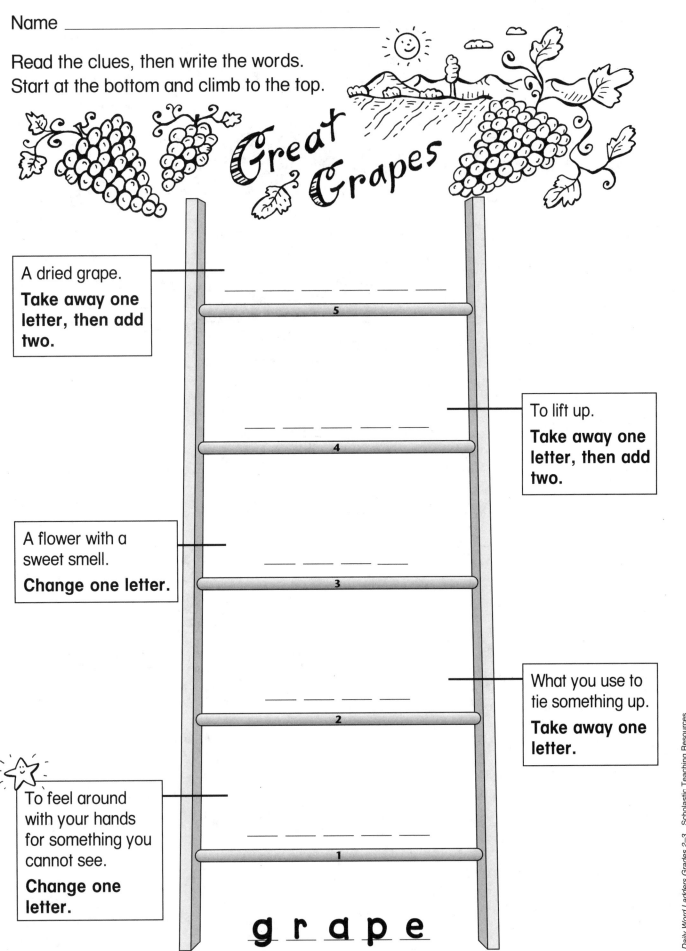

Great Grapes

A dried grape.
Take away one letter, then add two.

_ _ _ _ _ _ _
5

To lift up.
Take away one letter, then add two.

_ _ _ _ _ _
4

A flower with a sweet smell.
Change one letter.

_ _ _ _ _
3

What you use to tie something up.
Take away one letter.

_ _ _ _ _
2

To feel around with your hands for something you cannot see.
Change one letter.

_ _ _ _ _
1

g r a p e

Daily Word Ladders Grades 2–3   Scholastic Teaching Resources

Name _____

Read the clues, then write the words.
Start at the bottom and climb to the top.

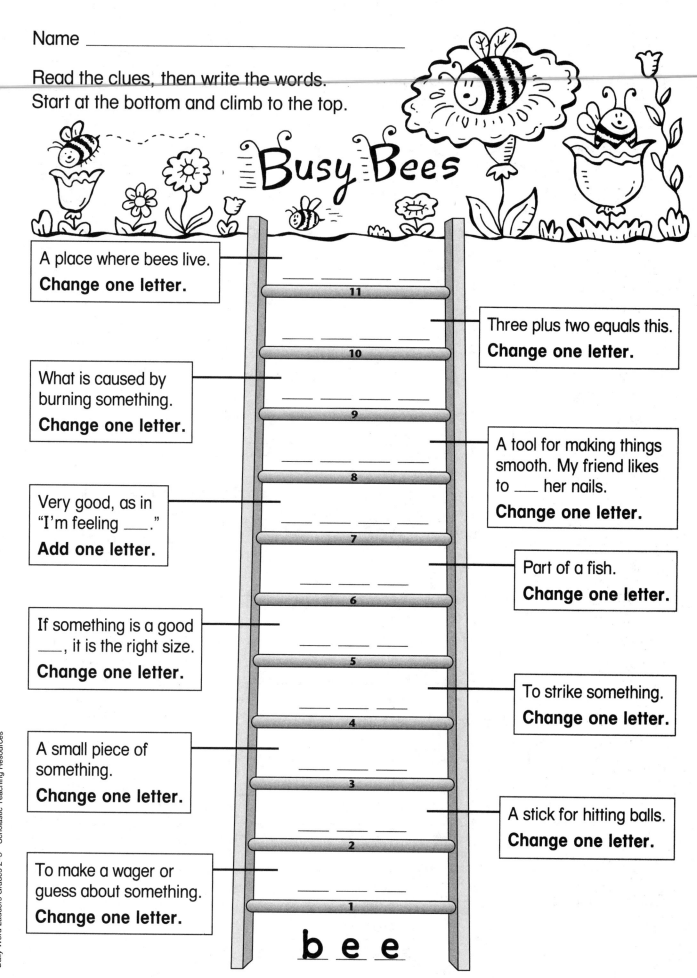

Busy Bees

A place where bees live.
Change one letter.

— — — —
11

Three plus two equals this.
Change one letter.

— — — —
10

What is caused by burning something.
Change one letter.

— — — —
9

A tool for making things smooth. My friend likes to ___ her nails.
Change one letter.

— — — —
8

Very good, as in "I'm feeling ___."
Add one letter.

— — — —
7

Part of a fish.
Change one letter.

— — —
6

If something is a good ___, it is the right size.
Change one letter.

— — —
5

To strike something.
Change one letter.

— — —
4

A small piece of something.
Change one letter.

— — —
3

A stick for hitting balls.
Change one letter.

— — —
2

To make a wager or guess about something.
Change one letter.

— — —
1

b e e

Name _____

Read the clues, then write the words.
Start at the bottom and climb to the top.

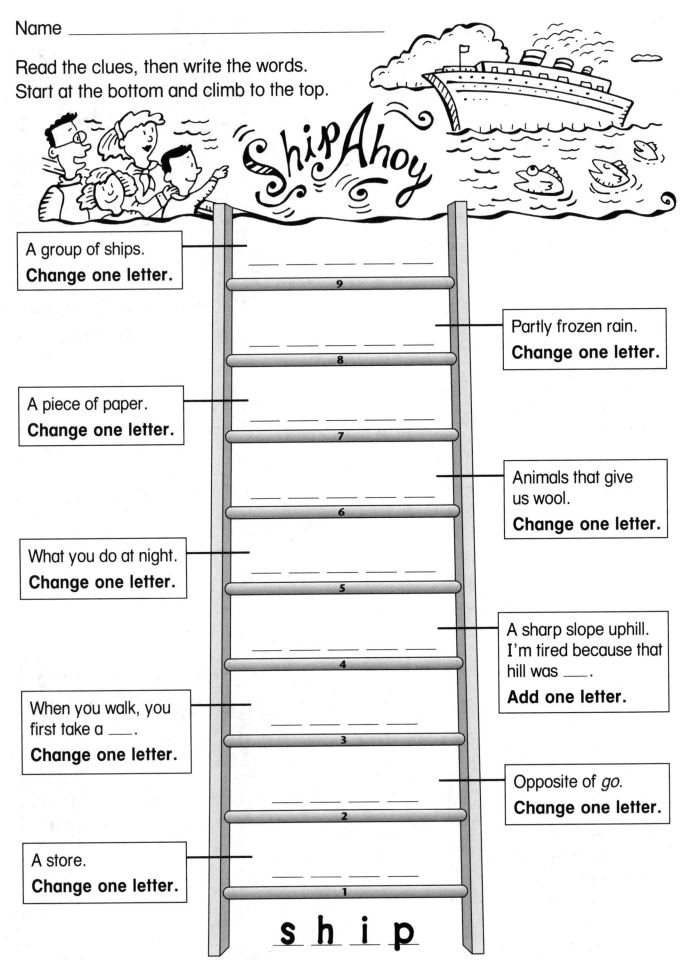

Ship Ahoy

A group of ships.
Change one letter.

Partly frozen rain.
Change one letter.

A piece of paper.
Change one letter.

Animals that give
us wool.
Change one letter.

What you do at night.
Change one letter.

A sharp slope uphill.
I'm tired because that
hill was ___.
Add one letter.

When you walk, you
first take a ___.
Change one letter.

Opposite of *go.*
Change one letter.

A store.
Change one letter.

s h i p

9
8
7
6
5
4
3
2
1

48

Daily Word Ladders Grades 2–3 Scholastic Teaching Resources

Read the clues, then write the words.
Start at the bottom and climb to the top.

Score!

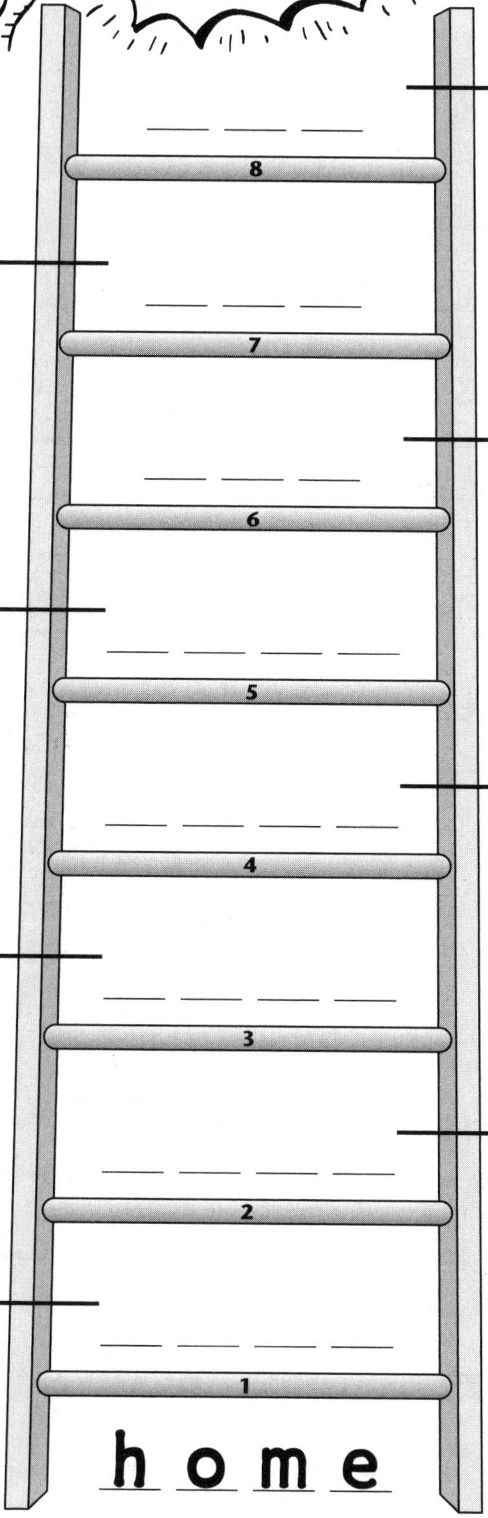

Not walk, but ___.
Change one letter.

8 _ _ _ _

A groove worn in the ground by a wheel.
Change one letter.

7 _ _ _ _

A rodent that looks like a large mouse.
Take away one letter.

6 _ _ _ _

To judge the quality of a person or thing.
Change one letter.

5 _ _ _ _

Not common, hard to find.
Change one letter.

4 _ _ _ _

A female horse.
Add one letter.

3 _ _ _ _

Not less, but ___.
Rearrange the letters.

2 _ _ _ _

A large city in Italy.
Change one letter.

1 _ _ _ _

h o m e

Name _____

Read the clues, then write the words.
Start at the bottom and climb to the top.

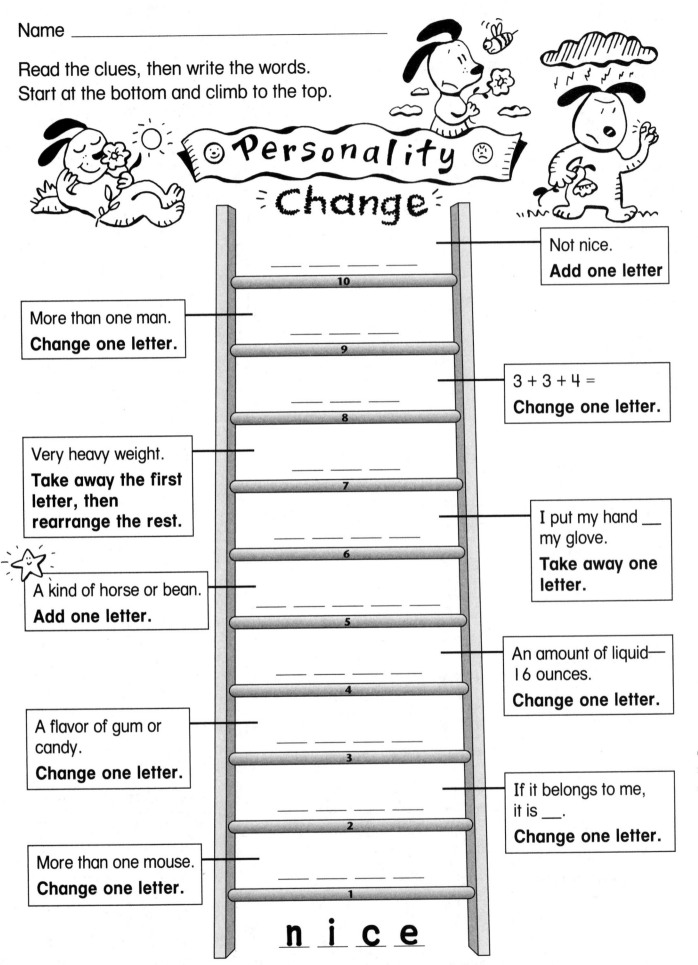

Personality Change

Not nice.
Add one letter

More than one man.
Change one letter.

$3 + 3 + 4 =$
Change one letter.

Very heavy weight.
Take away the first letter, then rearrange the rest.

I put my hand __ my glove.
Take away one letter.

A kind of horse or bean.
Add one letter.

An amount of liquid—16 ounces.
Change one letter.

A flavor of gum or candy.
Change one letter.

If it belongs to me, it is __.
Change one letter.

More than one mouse.
Change one letter.

10

9

8

7

6

5

4

3

2

1

n i c e

50

Daily Word Ladders Grades 2–3 Scholastic Teaching Resources

Name _____

Read the clues, then write the words.
Start at the bottom and climb to the top.

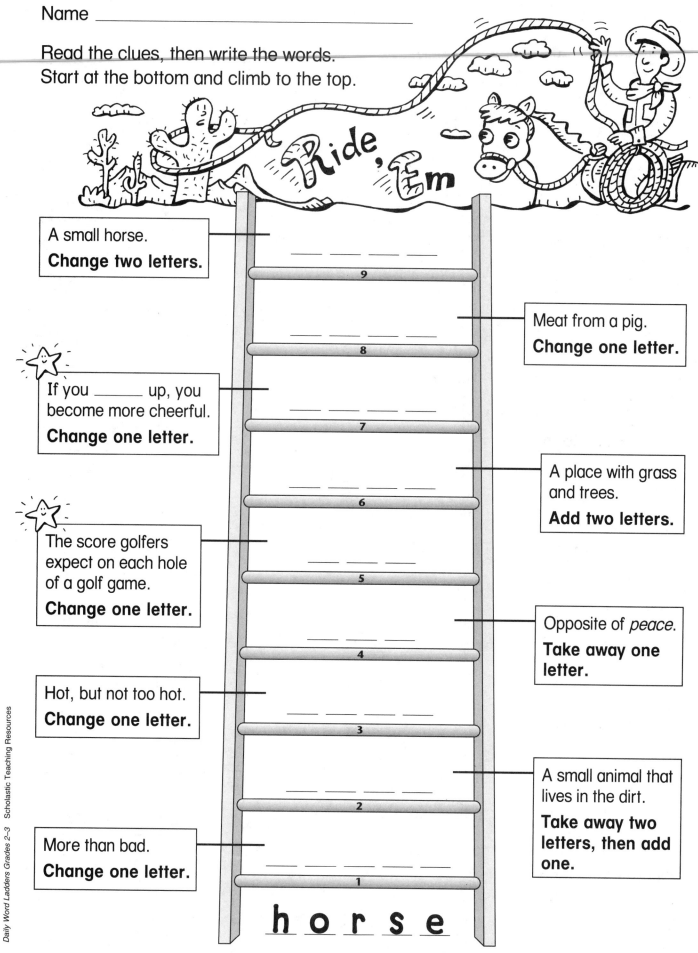

Ride, 'Em

A small horse.
Change two letters.

If you _____ up, you become more cheerful.
Change one letter.

The score golfers expect on each hole of a golf game.
Change one letter.

Hot, but not too hot.
Change one letter.

More than bad.
Change one letter.

Meat from a pig.
Change one letter.

A place with grass and trees.
Add two letters.

Opposite of *peace*.
Take away one letter.

A small animal that lives in the dirt.
Take away two letters, then add one.

9
8
7
6
5
4
3
2
1

h o r s e

Name _____

Read the clues, then write the words.
Start at the bottom and climb to the top.

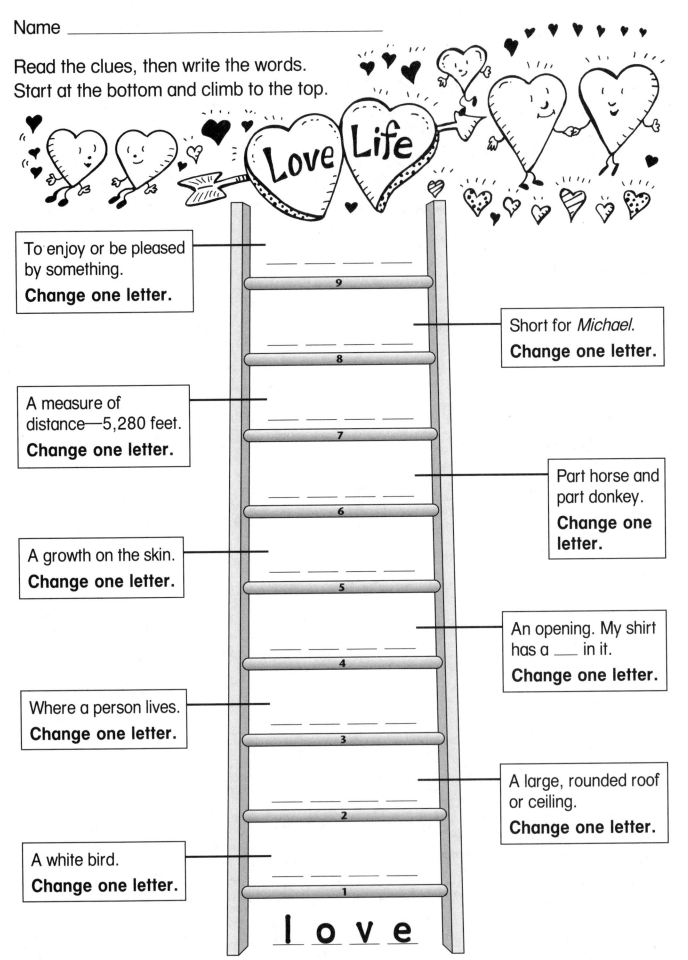

Love Life

To enjoy or be pleased by something.
Change one letter.

9 _ _ _ _

Short for *Michael.*
Change one letter.

8 _ _ _ _

A measure of distance—5,280 feet.
Change one letter.

7 _ _ _ _

Part horse and part donkey.
Change one letter.

6 _ _ _ _

A growth on the skin.
Change one letter.

5 _ _ _ _

An opening. My shirt has a ___ in it.
Change one letter.

4 _ _ _ _

Where a person lives.
Change one letter.

3 _ _ _ _

A large, rounded roof or ceiling.
Change one letter.

2 _ _ _ _

A white bird.
Change one letter.

1 _ _ _ _

l o v e

Daily Word Ladders Grades 2–3 Scholastic Teaching Resources

Read the clues, then write the words.
Start at the bottom and climb to the top.

Underfoot

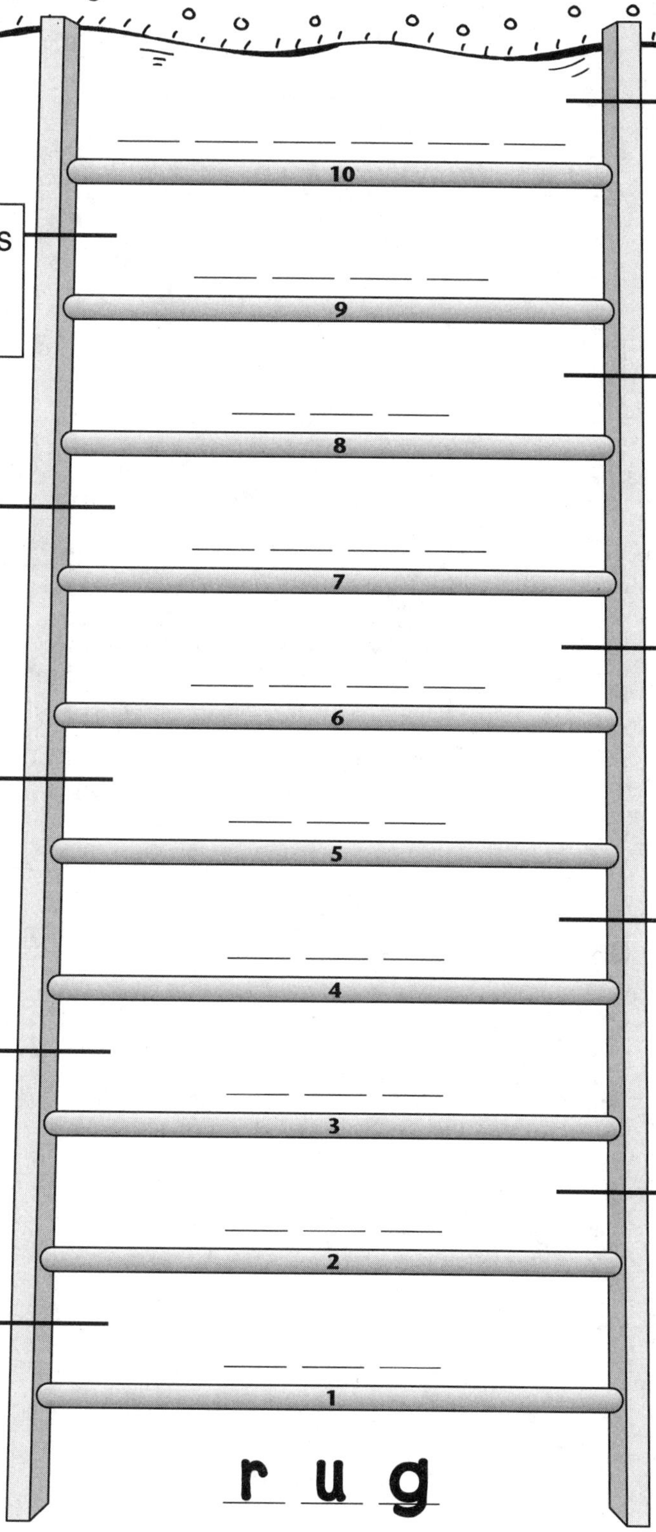

It covers the floor.
Add two letters.

A large fish that looks like a goldfish.
Add one letter.

An automobile.
Take away one letter.

What might be left on your skin after a wound heals.
Change one letter.

Something in the night sky.
Add one letter.

Something black and gooey for fixing roads.
Rearrange the letters.

A rodent.
Change one letter.

An old cloth for cleaning.
Change one letter.

A container for carrying things, like groceries.
Change one letter.

An insect.
Change one letter.

10
9
8
7
6
5
4
3
2
1

r u g

Name _____

Read the clues, then write the words.
Start at the bottom and climb to the top.

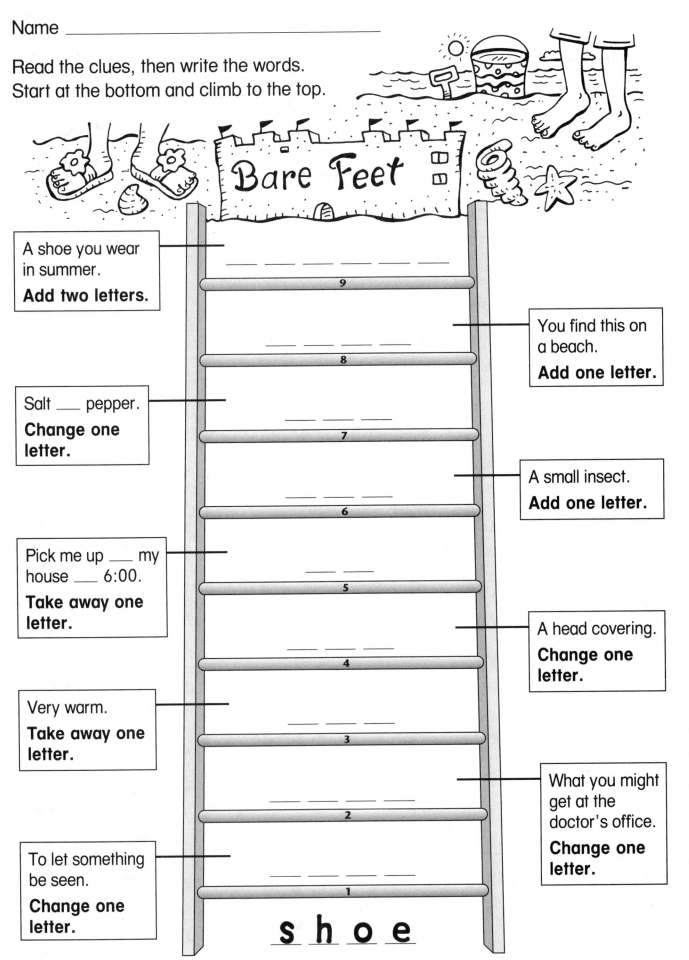

Bare Feet

A shoe you wear in summer.
Add two letters.

9 _____

You find this on a beach.
Add one letter.

8 _____

Salt __ pepper.
Change one letter.

7 _____

A small insect.
Add one letter.

6 _____

Pick me up __ my house __ 6:00.
Take away one letter.

5 _____

A head covering.
Change one letter.

4 _____

Very warm.
Take away one letter.

3 _____

What you might get at the doctor's office.
Change one letter.

2 _____

To let something be seen.
Change one letter.

1 _____

s h o e

54

Name _____

Read the clues, then write the words.
Start at the bottom and climb to the top.

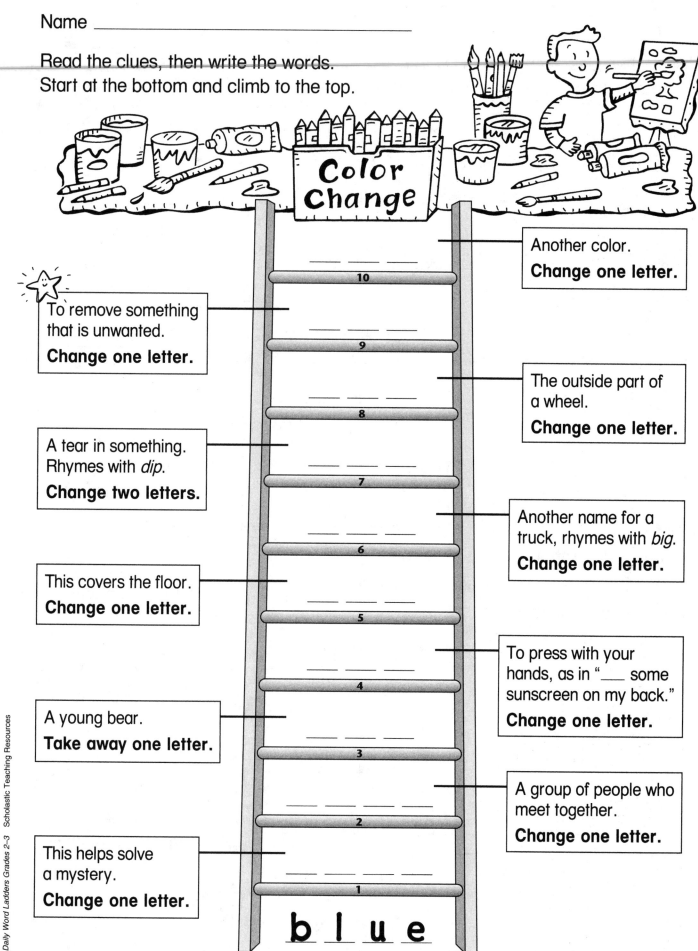

Color Change

Another color.
Change one letter.

To remove something that is unwanted.
Change one letter.

The outside part of a wheel.
Change one letter.

A tear in something. Rhymes with *dip*.
Change two letters.

Another name for a truck, rhymes with *big*.
Change one letter.

This covers the floor.
Change one letter.

To press with your hands, as in "___ some sunscreen on my back."
Change one letter.

A young bear.
Take away one letter.

A group of people who meet together.
Change one letter.

This helps solve a mystery.
Change one letter.

b l u e

Name _____

Read the clues, then write the words.
Start at the bottom and climb to the top.

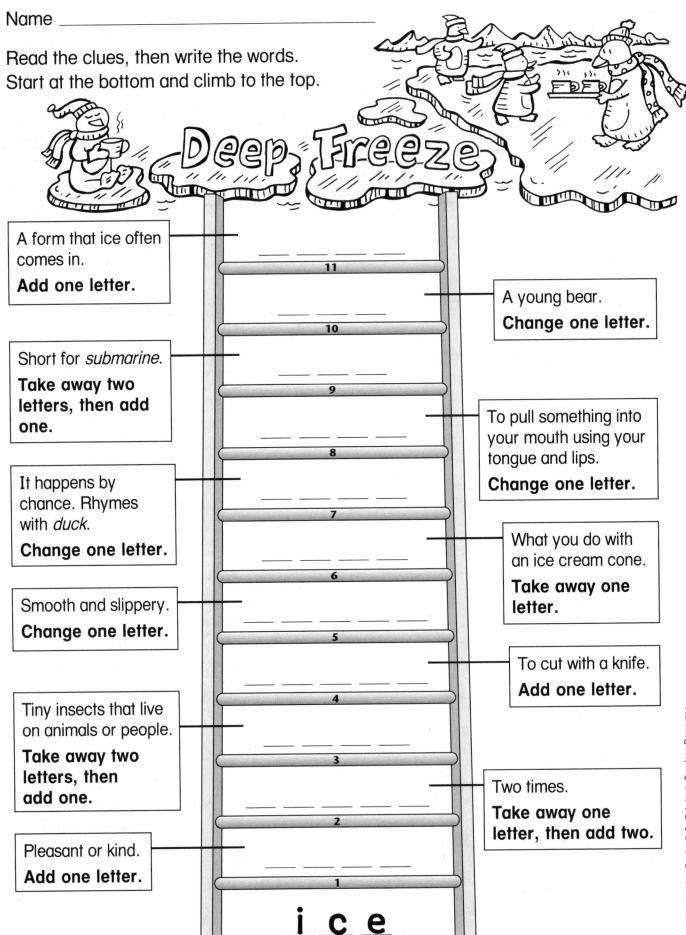

Deep Freeze

A form that ice often comes in.
Add one letter.

A young bear.
Change one letter.

Short for *submarine*.
Take away two letters, then add one.

To pull something into your mouth using your tongue and lips.
Change one letter.

It happens by chance. Rhymes with *duck*.
Change one letter.

What you do with an ice cream cone.
Take away one letter.

Smooth and slippery.
Change one letter.

To cut with a knife.
Add one letter.

Tiny insects that live on animals or people.
Take away two letters, then add one.

Two times.
Take away one letter, then add two.

Pleasant or kind.
Add one letter.

11
10
9
8
7
6
5
4
3
2
1

i c e

Daily Word Ladders Grades 2–3 Scholastic Teaching Resources

Name _____

Read the clues, then write the words.
Start at the bottom and climb to the top.

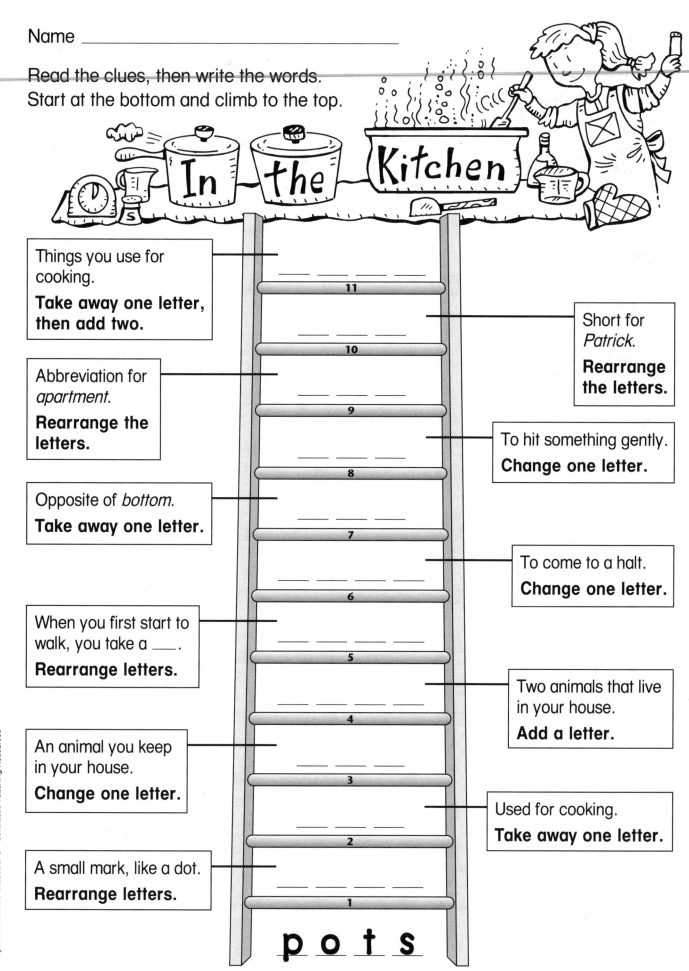

In the Kitchen

Things you use for cooking.
Take away one letter, then add two.

Short for *Patrick*.
Rearrange the letters.

Abbreviation for *apartment*.
Rearrange the letters.

To hit something gently.
Change one letter.

Opposite of *bottom*.
Take away one letter.

To come to a halt.
Change one letter.

When you first start to walk, you take a ___.
Rearrange letters.

Two animals that live in your house.
Add a letter.

An animal you keep in your house.
Change one letter.

Used for cooking.
Take away one letter.

A small mark, like a dot.
Rearrange letters.

11
10
9
8
7
6
5
4
3
2
1

p o t s

Name _____

Read the clues, then write the words.
Start at the bottom and climb to the top.

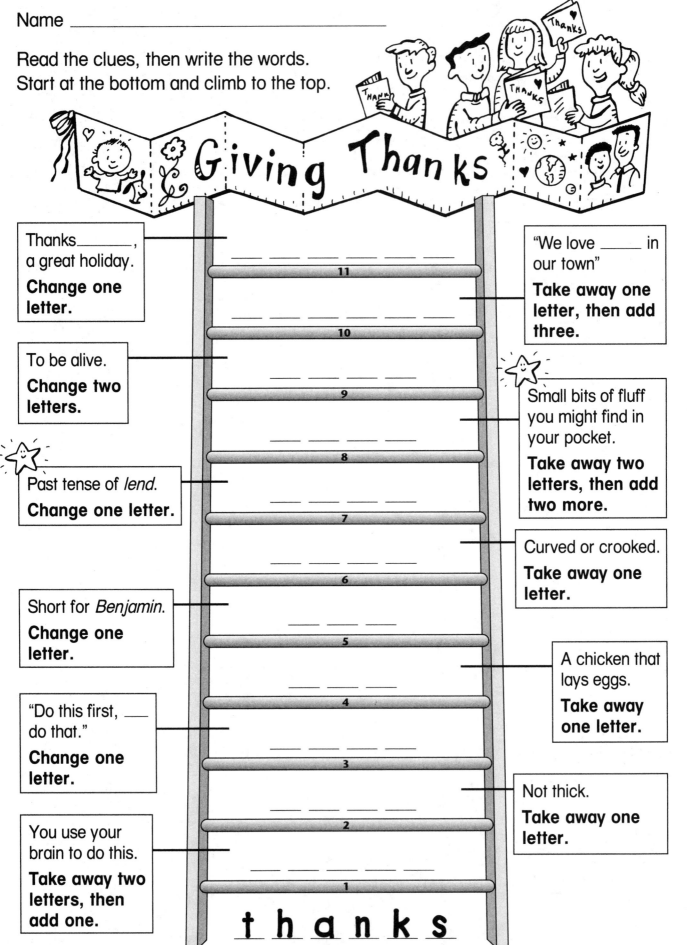

Giving Thanks

Thanks_____,
a great holiday.
**Change one
letter.**

— — — — — — —
11

To be alive.
**Change two
letters.**

— — — — — — —
10

Past tense of *lend*.
Change one letter.

— — — — —
9

Short for *Benjamin*.
**Change one
letter.**

— — — — —
8

"Do this first, ___
do that."
**Change one
letter.**

— — — —
7

You use your
brain to do this.
**Take away two
letters, then
add one.**

— — — —
6

— — — — —
5

— — — —
4

— — — — —
3

— — — —
2

— — — —
1

"We love ___ in
our town"
**Take away one
letter, then add
three.**

Small bits of fluff
you might find in
your pocket.
**Take away two
letters, then add
two more.**

Curved or crooked.
**Take away one
letter.**

A chicken that
lays eggs.
**Take away
one letter.**

Not thick.
**Take away one
letter.**

t h a n k s

58

Daily Word Ladders Grades 2–3   Scholastic Teaching Resources

Name _____

Read the clues, then write the words.
Start at the bottom and climb to the top.

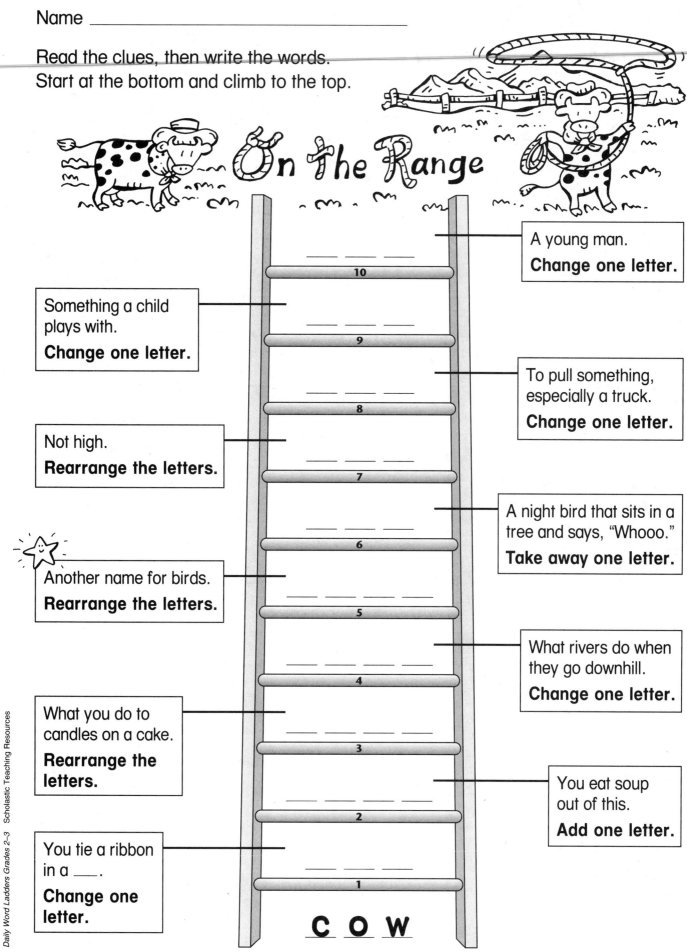

On the Range

A young man.
Change one letter.

Something a child plays with.
Change one letter.

To pull something, especially a truck.
Change one letter.

Not high.
Rearrange the letters.

A night bird that sits in a tree and says, "Whooo."
Take away one letter.

Another name for birds.
Rearrange the letters.

What you do to candles on a cake.
Rearrange the letters.

What rivers do when they go downhill.
Change one letter.

You eat soup out of this.
Add one letter.

You tie a ribbon in a ___.
Change one letter.

10
9
8
7
6
5
4
3
2
1

C   O   W

Name _____

Read the clues, then write the words.
Start at the bottom and climb to the top.

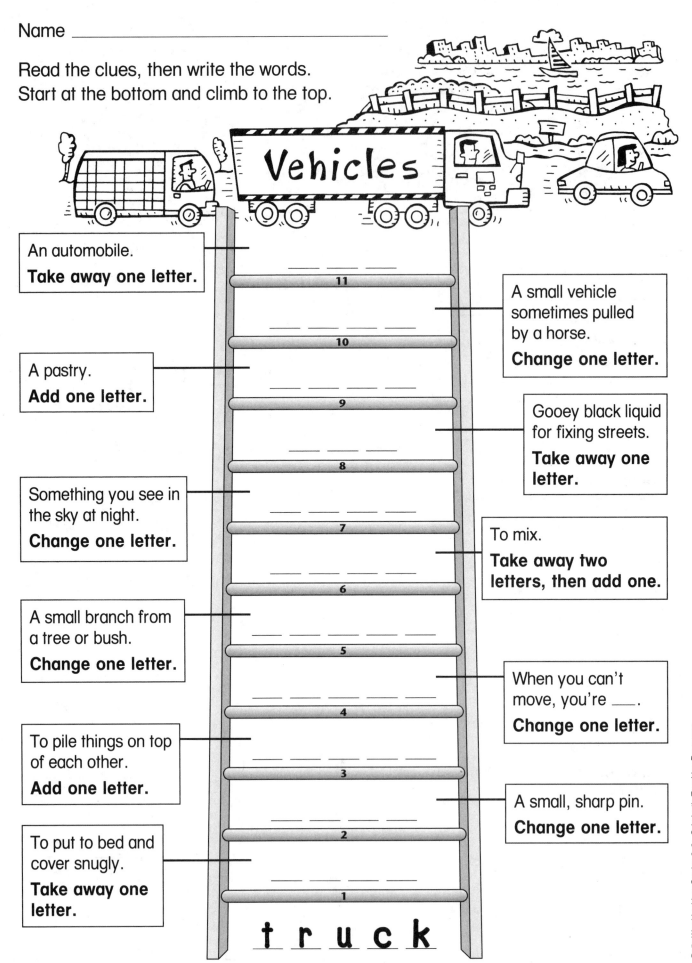

Vehicles

An automobile.
Take away one letter.

A pastry.
Add one letter.

Something you see in the sky at night.
Change one letter.

A small branch from a tree or bush.
Change one letter.

To pile things on top of each other.
Add one letter.

To put to bed and cover snugly.
Take away one letter.

A small vehicle sometimes pulled by a horse.
Change one letter.

Gooey black liquid for fixing streets.
Take away one letter.

To mix.
Take away two letters, then add one.

When you can't move, you're ___.
Change one letter.

A small, sharp pin.
Change one letter.

11

10

9

8

7

6

5

4

3

2

1

t r u c k

Daily Word Ladders Grades 2–3   Scholastic Teaching Resources

Name _____

Read the clues, then write the words.
Start at the bottom and climb to the top.

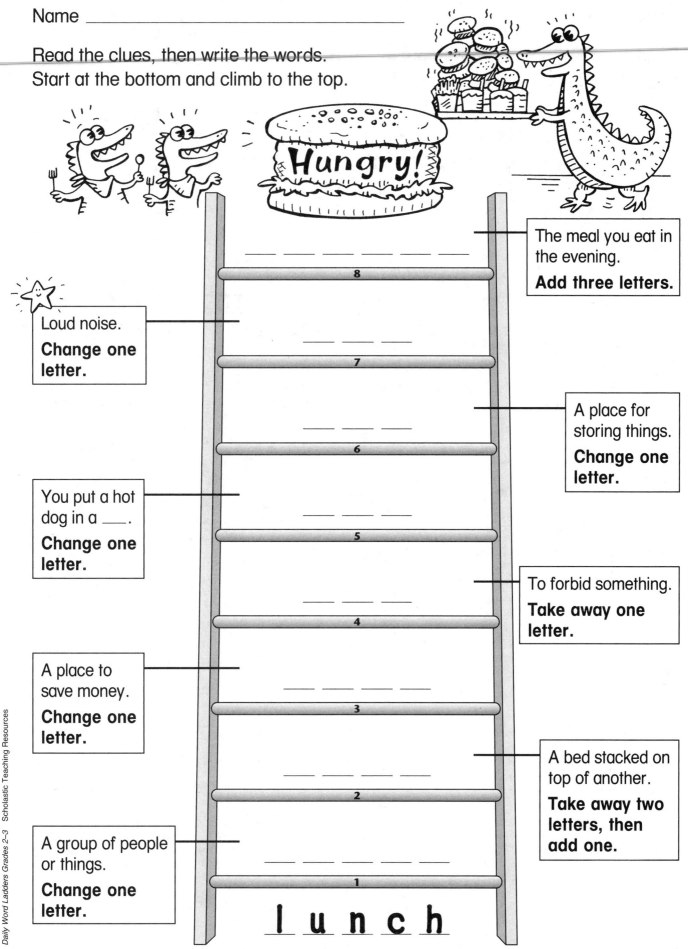

Hungry!

The meal you eat in the evening.
Add three letters.

Loud noise.
Change one letter.

A place for storing things.
Change one letter.

You put a hot dog in a ___.
Change one letter.

To forbid something.
Take away one letter.

A place to save money.
Change one letter.

A bed stacked on top of another.
Take away two letters, then add one.

A group of people or things.
Change one letter.

8

7

6

5

4

3

2

1

l u n c h

Name _____

Read the clues, then write the words.
Start at the bottom and climb to the top.

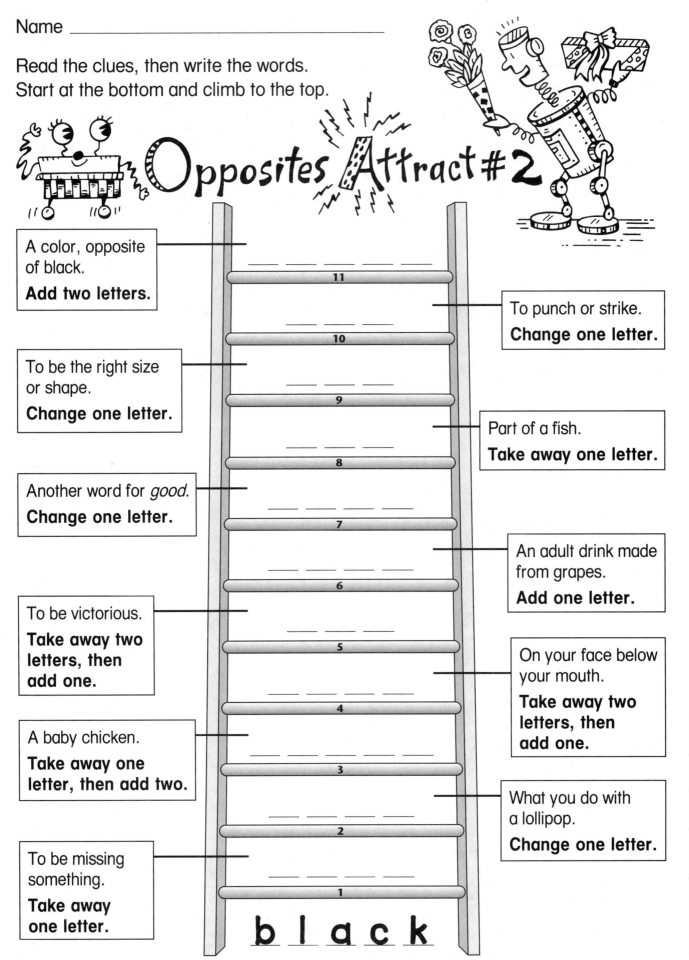

Opposites Attract #2

A color, opposite of black.
Add two letters.

To punch or strike.
Change one letter.

To be the right size or shape.
Change one letter.

Part of a fish.
Take away one letter.

Another word for *good*.
Change one letter.

An adult drink made from grapes.
Add one letter.

To be victorious.
Take away two letters, then add one.

On your face below your mouth.
Take away two letters, then add one.

A baby chicken.
Take away one letter, then add two.

What you do with a lollipop.
Change one letter.

To be missing something.
Take away one letter.

b l a c k

Daily Word Ladders Grades 2–3   Scholastic Teaching Resources

Name _____

Read the clues, then write the words.
Start at the bottom and climb to the top.

Nap Time

What you do when you nap.
Add one letter.

To flow slowly.
Add one letter.

With your eyes, you ___.
Change one letter.

A large body of water.
Change one letter.

A hot drink.
Change one letter.

Golfers put a golf ball on this.
Take away a letter.

Gives us shade outdoors.
Take away one letter, then add two.

To attempt.
Take away three letters.

A place to keep food.
Add three letters.

A flat pot for cooking.
Rearrange the letters.

10

9

8

7

6

5

4

3

2

1

n a p

Name _____

Read the clues, then write the words.
Start at the bottom and climb to the top.

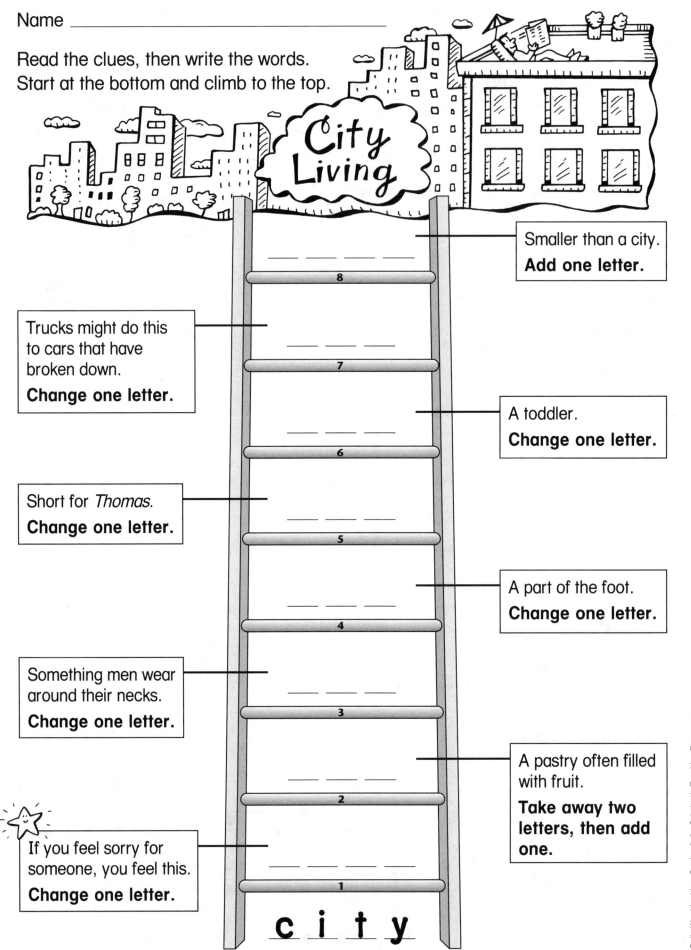

City Living

Smaller than a city.
Add one letter.

Trucks might do this
to cars that have
broken down.
Change one letter.

A toddler.
Change one letter.

Short for *Thomas*.
Change one letter.

A part of the foot.
Change one letter.

Something men wear
around their necks.
Change one letter.

A pastry often filled
with fruit.
**Take away two
letters, then add
one.**

If you feel sorry for
someone, you feel this.
Change one letter.

8

7

6

5

4

3

2

1

c i t y

Daily Word Ladders Grades 2–3 Scholastic Teaching Resources

Name _____

Read the clues, then write the words.
Start at the bottom and climb to the top.

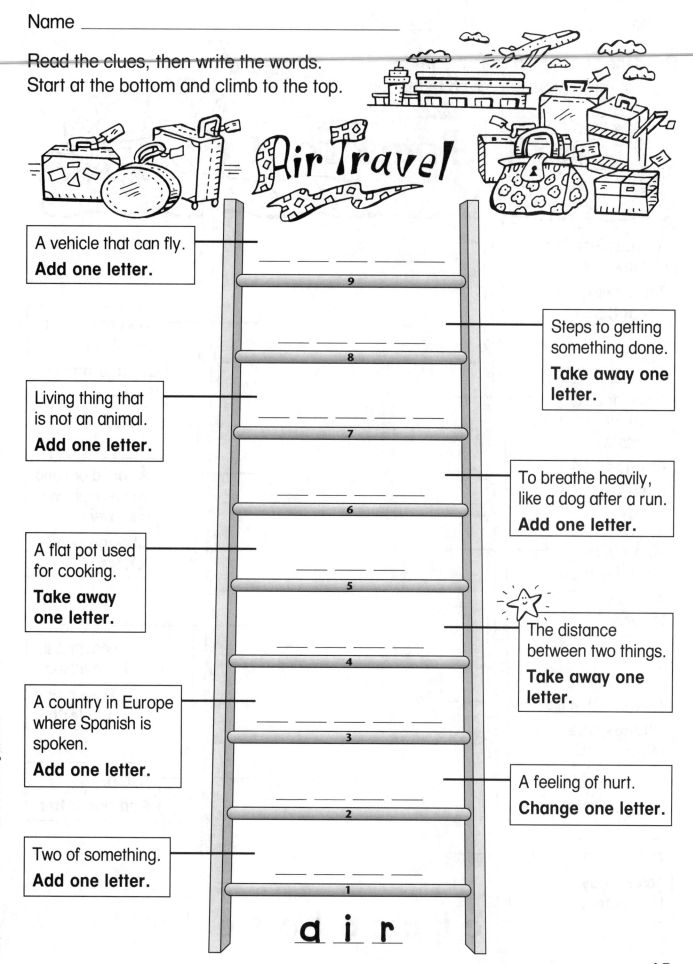

Air Travel

A vehicle that can fly.
Add one letter.

_ _ _ _ _ _ _   9

Steps to getting
something done.
**Take away one
letter.**

_ _ _ _ _ _   8

Living thing that
is not an animal.
Add one letter.

_ _ _ _ _   7

To breathe heavily,
like a dog after a run.
Add one letter.

_ _ _ _   6

A flat pot used
for cooking.
**Take away
one letter.**

_ _ _ _   5

The distance
between two things.
**Take away one
letter.**

_ _ _ _   4

A country in Europe
where Spanish is
spoken.
Add one letter.

_ _ _ _   3

A feeling of hurt.
Change one letter.

_ _ _ _   2

Two of something.
Add one letter.

_ _ _   1

a i r

Name _____

Read the clues, then write the words.
Start at the bottom and climb to the top.

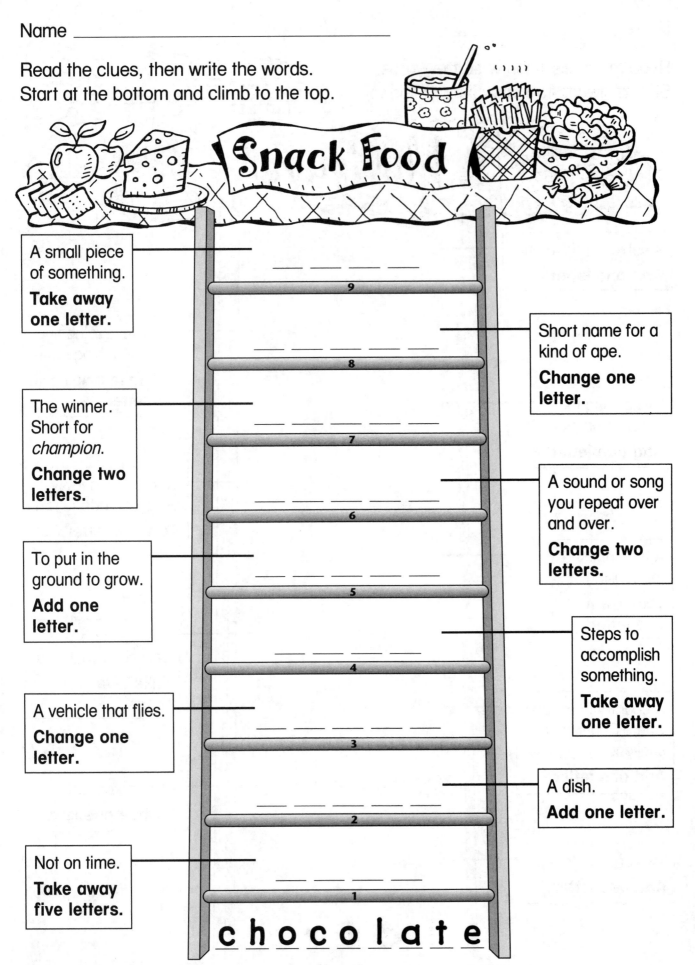

Snack Food

A small piece of something.
Take away one letter.

_ _ _ _ _ _ _ _   9

Short name for a kind of ape.
Change one letter.

_ _ _ _ _ _ _ _   8

The winner. Short for *champion*.
Change two letters.

_ _ _ _ _ _   7

A sound or song you repeat over and over.
Change two letters.

_ _ _ _ _ _   6

To put in the ground to grow.
Add one letter.

_ _ _ _ _   5

Steps to accomplish something.
Take away one letter.

_ _ _ _ _   4

A vehicle that flies.
Change one letter.

_ _ _ _   3

A dish.
Add one letter.

_ _ _ _   2

Not on time.
Take away five letters.

_ _ _ _   1

c h o c o l a t e

66

Name _____

Read the clues, then write the words.
Start at the bottom and climb to the top.

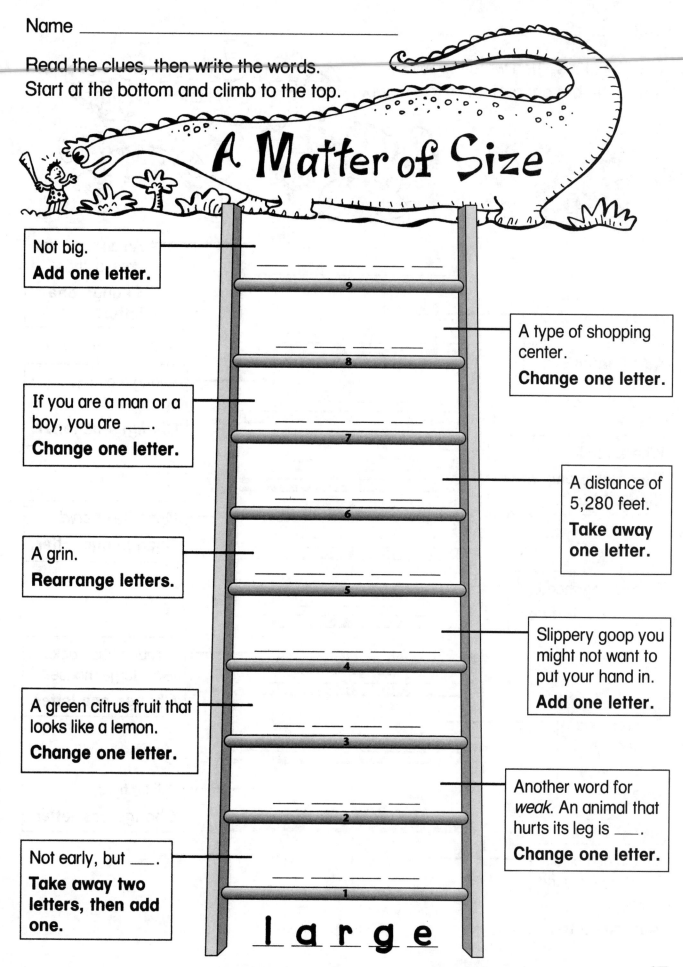

A Matter of Size

Not big.
Add one letter.

_ _ _ _ 9

A type of shopping center.
Change one letter.

_ _ _ _ 8

If you are a man or a boy, you are ___.
Change one letter.

_ _ _ _ 7

A distance of 5,280 feet.
Take away one letter.

_ _ _ _ 6

A grin.
Rearrange letters.

_ _ _ _ _ 5

Slippery goop you might not want to put your hand in.
Add one letter.

_ _ _ _ 4

A green citrus fruit that looks like a lemon.
Change one letter.

_ _ _ _ 3

Another word for *weak*. An animal that hurts its leg is ___.
Change one letter.

_ _ _ _ 2

Not early, but ___.
Take away two letters, then add one.

_ _ _ _ 1

l a r g e

Name _____

Read the clues, then write the words.
Start at the bottom and climb to the top.

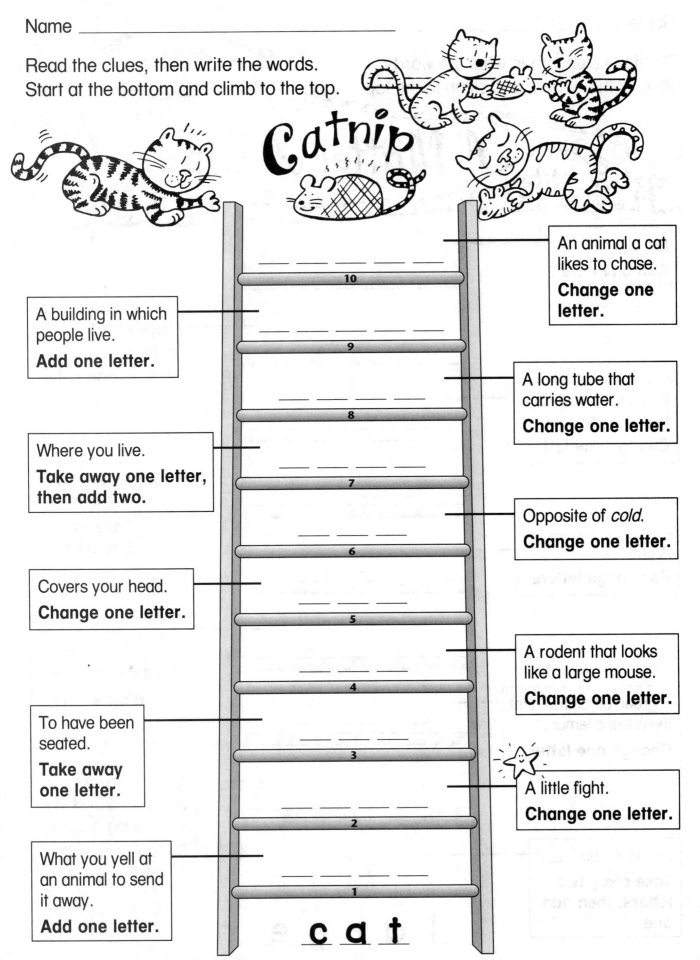

Catnip

An animal a cat likes to chase.
Change one letter.

A building in which people live.
Add one letter.

10

A long tube that carries water.
Change one letter.

9

Where you live.
Take away one letter, then add two.

8

Opposite of *cold*.
Change one letter.

7

Covers your head.
Change one letter.

6

A rodent that looks like a large mouse.
Change one letter.

5

To have been seated.
Take away one letter.

4

A little fight.
Change one letter.

3

What you yell at an animal to send it away.
Add one letter.

2

1

c a t

Daily Word Ladders Grades 2–3 Scholastic Teaching Resources

Name _____

Read the clues, then write the words.
Start at the bottom and climb to the top.

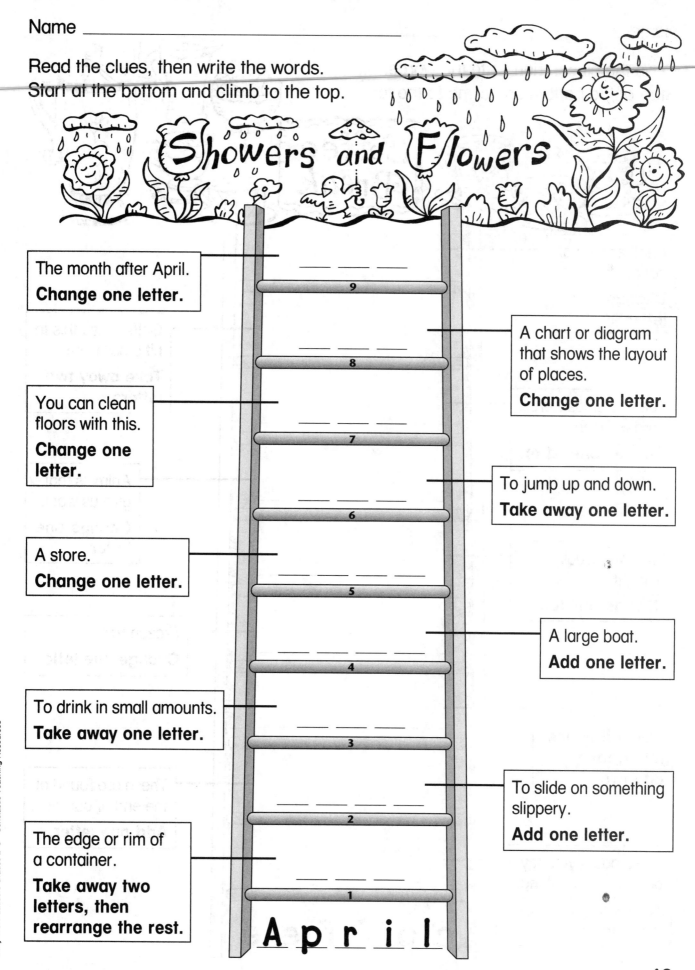

Showers and Flowers

The month after April.
Change one letter.

9 ___ ___ ___

A chart or diagram that shows the layout of places.
Change one letter.

8 ___ ___ ___

You can clean floors with this.
Change one letter.

7 ___ ___ ___

To jump up and down.
Take away one letter.

6 ___ ___ ___

A store.
Change one letter.

5 ___ ___ ___ ___

A large boat.
Add one letter.

4 ___ ___ ___ ___

To drink in small amounts.
Take away one letter.

3 ___ ___ ___

To slide on something slippery.
Add one letter.

2 ___ ___ ___ ___

The edge or rim of a container.
Take away two letters, then rearrange the rest.

1 ___ ___ ___

A p r i l

Name _____

Read the clues, then write the words.
Start at the bottom and climb to the top.

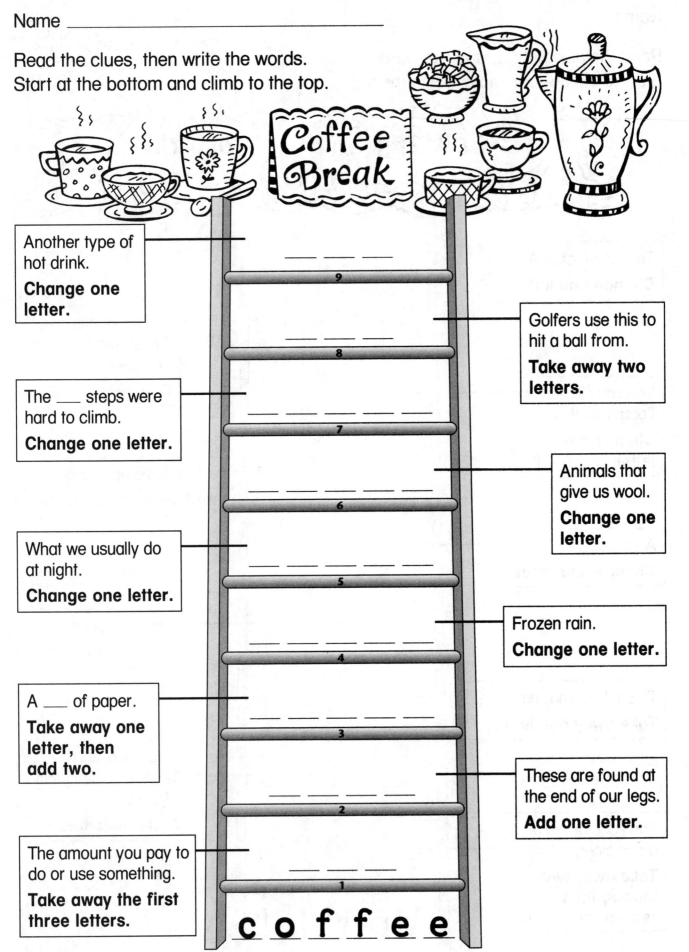

Coffee Break

Another type of hot drink.
Change one letter.

Golfers use this to hit a ball from.
Take away two letters.

The ___ steps were hard to climb.
Change one letter.

Animals that give us wool.
Change one letter.

What we usually do at night.
Change one letter.

Frozen rain.
Change one letter.

A ___ of paper.
Take away one letter, then add two.

These are found at the end of our legs.
Add one letter.

The amount you pay to do or use something.
Take away the first three letters.

9
8
7
6
5
4
3
2
1

c o f f e e

Daily Word Ladders Grades 2–3   Scholastic Teaching Resources

Read the clues, then write the words.
Start at the bottom and climb to the top.

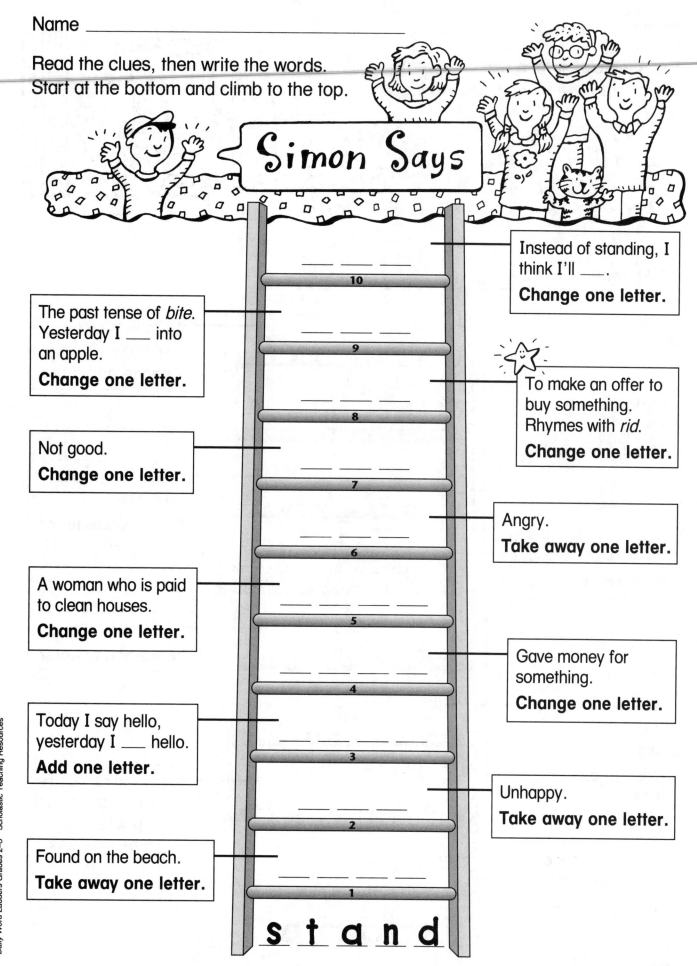

Simon Says

Instead of standing, I think I'll ___.
Change one letter.

The past tense of *bite*. Yesterday I ___ into an apple.
Change one letter.

To make an offer to buy something. Rhymes with *rid*.
Change one letter.

Not good.
Change one letter.

Angry.
Take away one letter.

A woman who is paid to clean houses.
Change one letter.

Gave money for something.
Change one letter.

Today I say hello, yesterday I ___ hello.
Add one letter.

Unhappy.
Take away one letter.

Found on the beach.
Take away one letter.

10
9
8
7
6
5
4
3
2
1

s t a n d

Name _____

Read the clues, then write the words.
Start at the bottom and climb to the top.

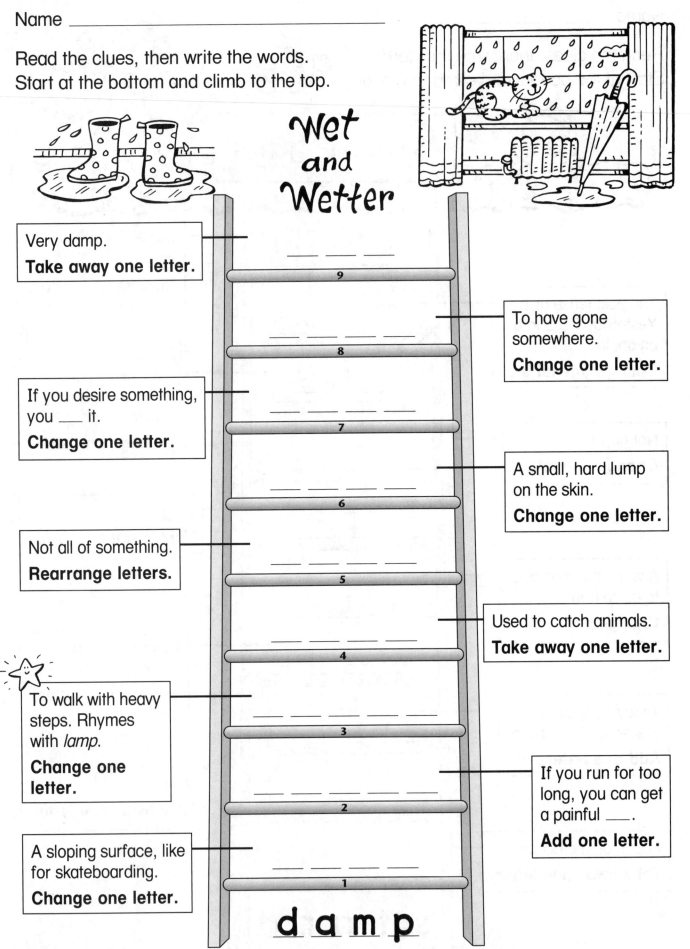

Wet and Wetter

Very damp.
Take away one letter.

— — — — 9

To have gone somewhere.
Change one letter.

— — — — 8

If you desire something, you ___ it.
Change one letter.

— — — — 7

A small, hard lump on the skin.
Change one letter.

— — — — 6

Not all of something.
Rearrange letters.

— — — — 5

Used to catch animals.
Take away one letter.

— — — — 4

To walk with heavy steps. Rhymes with *lamp*.
Change one letter.

— — — — 3

If you run for too long, you can get a painful ___.
Add one letter.

— — — — 2

A sloping surface, like for skateboarding.
Change one letter.

— — — — 1

d a m p

Daily Word Ladders Grades 2–3   Scholastic Teaching Resources

Read the clues, then write the words.
Start at the bottom and climb to the top.

After Dinner

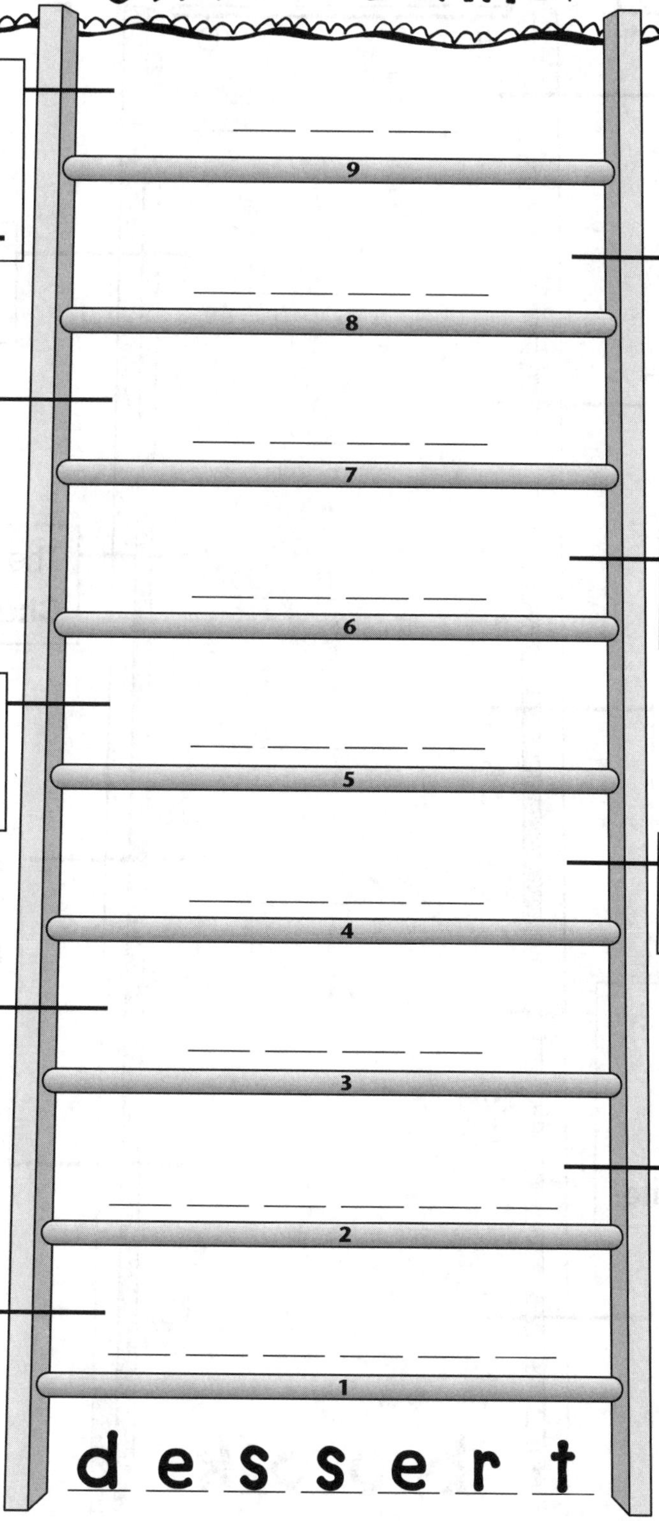

A pastry dessert.
Take away one letter, then rearrange the rest.

_ _ _ _ _ _ _ _ _   9

_ _ _ _ _ _ _ _   8

When fruit is ready to eat, it's ___.
Change one letter.

A strong string for tying things.
Change one letter.

_ _ _ _ _ _ _   7

_ _ _ _ _ _   6

A large city in Italy.
Change one letter.

A flower that grows on a thorny bush.
Rearrange letters.

_ _ _ _ _   5

_ _ _ _   4

Painful.
Change one letter.

To put things into categories.
Take away two letters.

_ _ _ _ _   3

_ _ _ _ _ _   2

A place to go for vacation.
Change two letters.

A place with little rain.
Take away one letter.

_ _ _ _ _ _ _   1

d e s s e r t

Name _____

Read the clues, then write the words.
Start at the bottom and climb to the top.

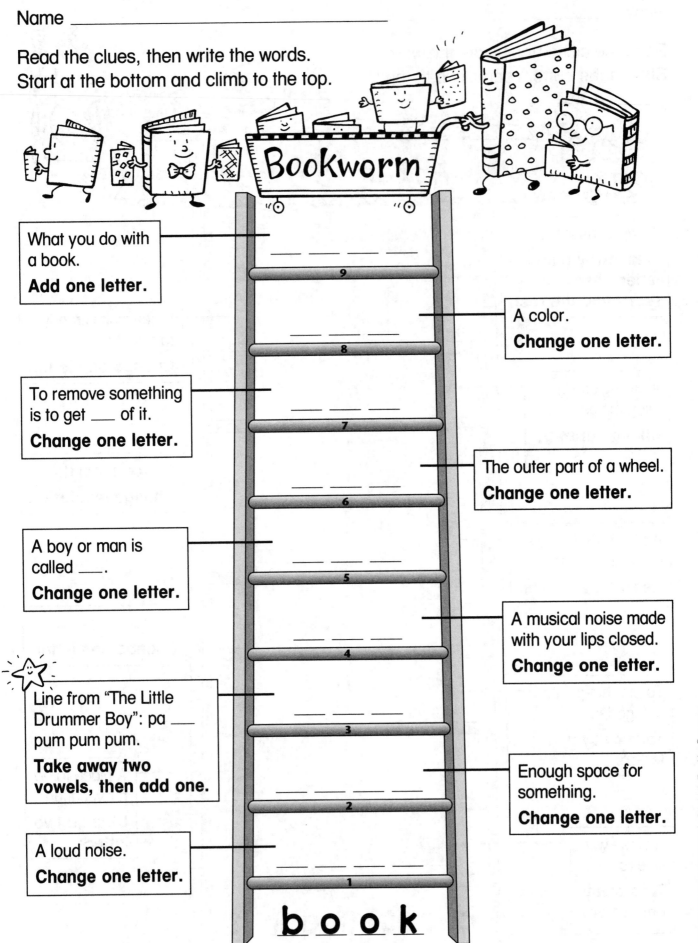

Bookworm

What you do with a book.
Add one letter.

_ _ _ _ _   9

A color.
Change one letter.

_ _ _ _   8

To remove something is to get ___ of it.
Change one letter.

_ _ _   7

The outer part of a wheel.
Change one letter.

_ _ _   6

A boy or man is called ___.
Change one letter.

_ _ _   5

A musical noise made with your lips closed.
Change one letter.

_ _ _ _   4

Line from "The Little Drummer Boy": pa ___ pum pum pum.
Take away two vowels, then add one.

_ _ _   3

Enough space for something.
Change one letter.

_ _ _ _   2

A loud noise.
Change one letter.

_ _ _ _   1

b o o k

Daily Word Ladders Grades 2–3   Scholastic Teaching Resources

Name _____

Read the clues, then write the words.
Start at the bottom and climb to the top.

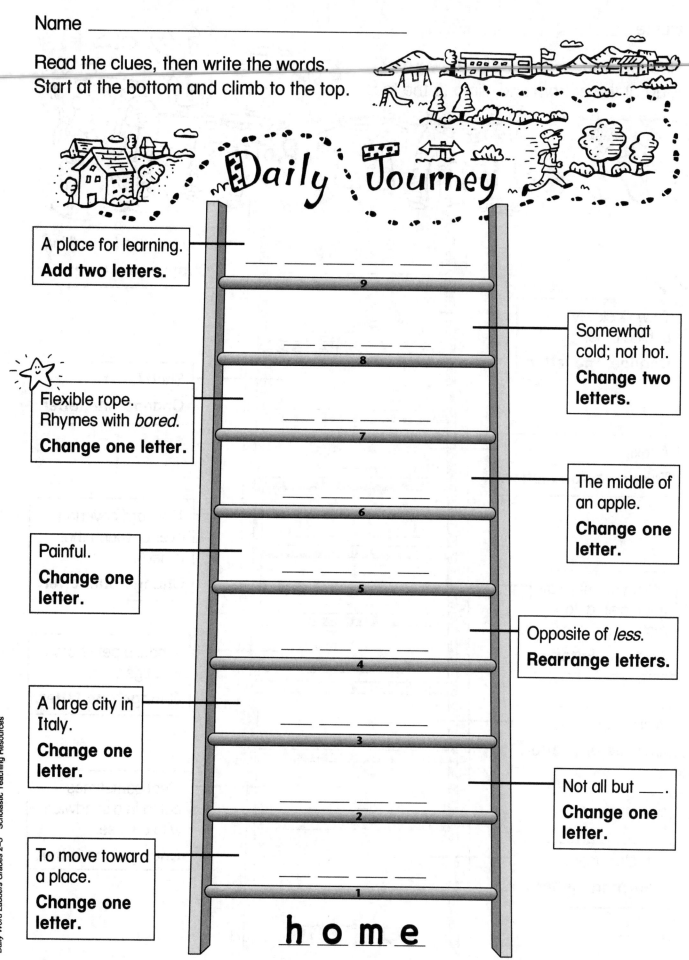

Daily Journey

A place for learning.
Add two letters.

Somewhat cold; not hot.
Change two letters.

Flexible rope.
Rhymes with *bored*.
Change one letter.

The middle of an apple.
Change one letter.

Painful.
Change one letter.

Opposite of *less*.
Rearrange letters.

A large city in Italy.
Change one letter.

Not all but ___.
Change one letter.

To move toward a place.
Change one letter.

h o m e

Name _____

Read the clues, then write the words.
Start at the bottom and climb to the top.

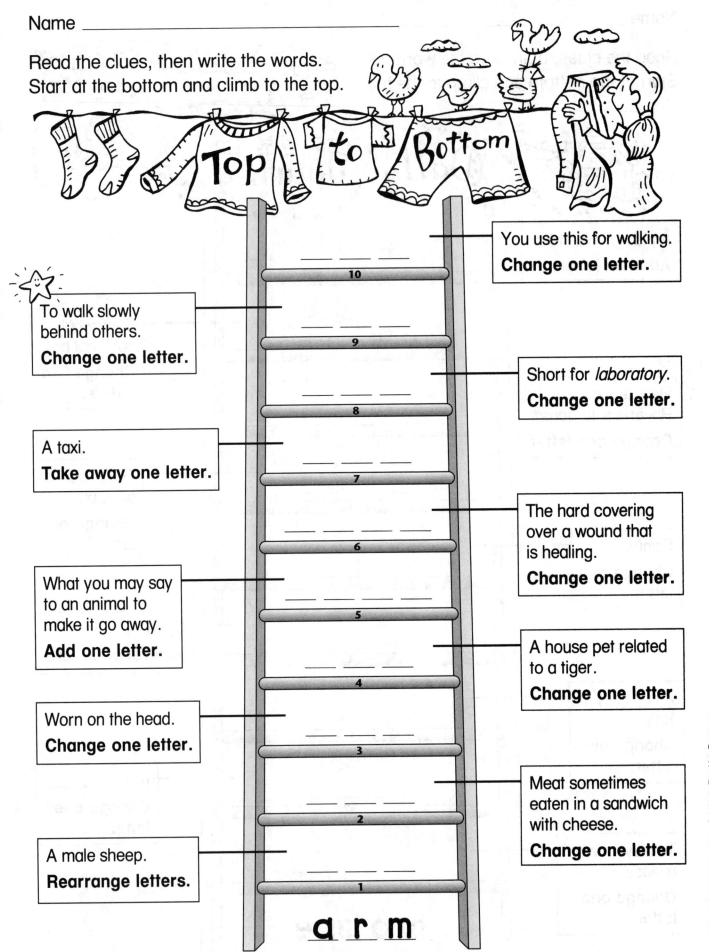

You use this for walking.
Change one letter.

To walk slowly
behind others.
Change one letter.

Short for *laboratory*.
Change one letter.

A taxi.
Take away one letter.

The hard covering
over a wound that
is healing.
Change one letter.

What you may say
to an animal to
make it go away.
Add one letter.

A house pet related
to a tiger.
Change one letter.

Worn on the head.
Change one letter.

Meat sometimes
eaten in a sandwich
with cheese.
Change one letter.

A male sheep.
Rearrange letters.

10
9
8
7
6
5
4
3
2
1

a r m

Daily Word Ladders Grades 2–3   Scholastic Teaching Resources

Read the clues, then write the words.
Start at the bottom and climb to the top.

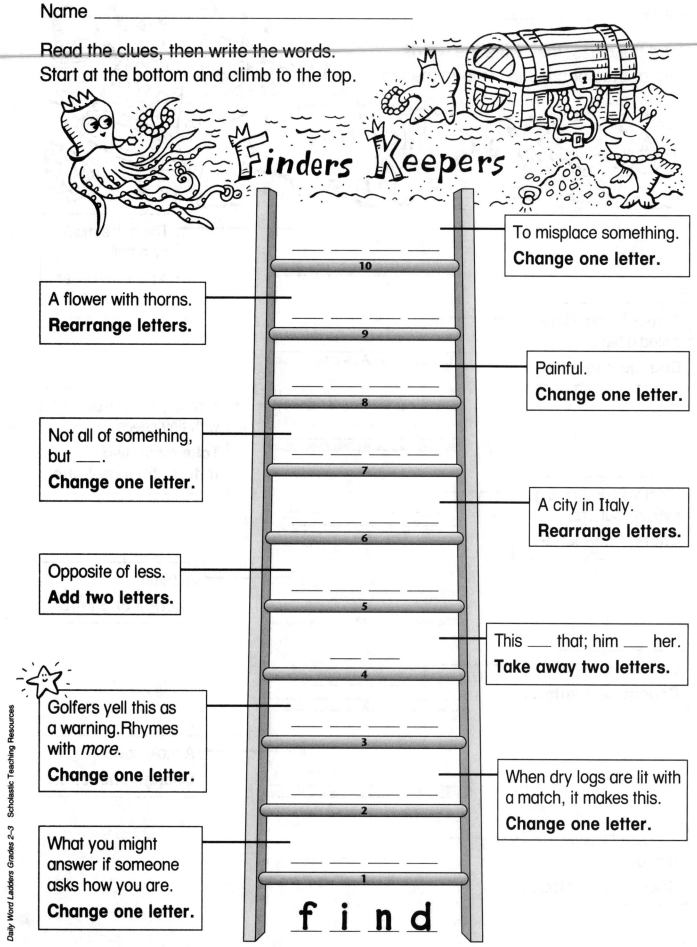

Finders Keepers

To misplace something.
Change one letter.

A flower with thorns.
Rearrange letters.

Painful.
Change one letter.

Not all of something,
but ___.
Change one letter.

A city in Italy.
Rearrange letters.

Opposite of less.
Add two letters.

This ___ that; him ___ her.
Take away two letters.

Golfers yell this as
a warning. Rhymes
with *more*.
Change one letter.

When dry logs are lit with
a match, it makes this.
Change one letter.

What you might
answer if someone
asks how you are.
Change one letter.

f i n d

Name _____

Read the clues, then write the words.
Start at the bottom and climb to the top.

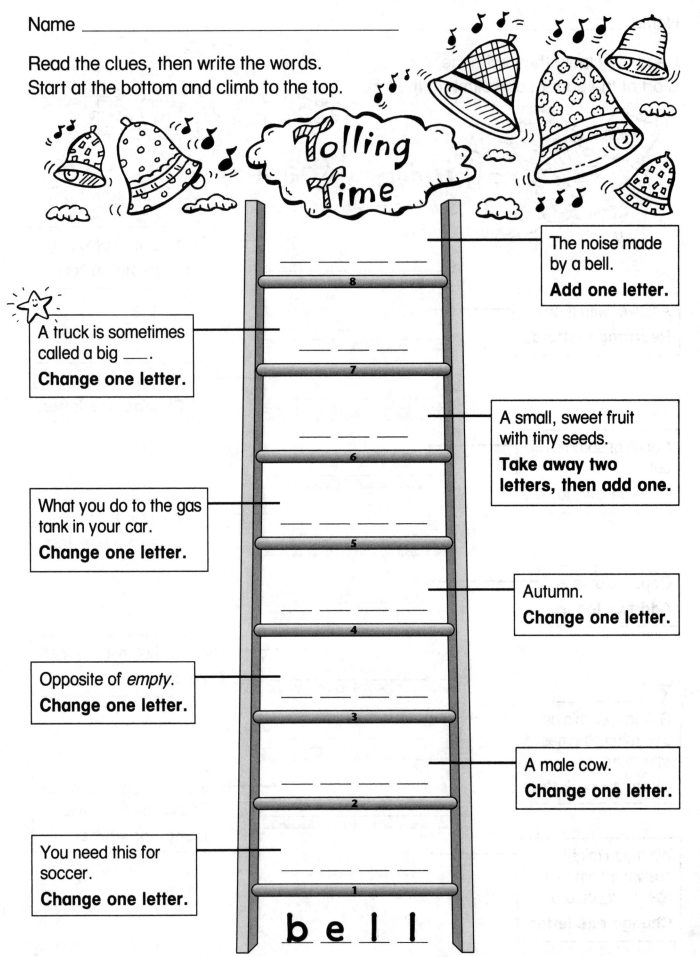

Tolling Time

The noise made by a bell.
Add one letter.

A truck is sometimes called a big ____.
Change one letter.

A small, sweet fruit with tiny seeds.
Take away two letters, then add one.

What you do to the gas tank in your car.
Change one letter.

Autumn.
Change one letter.

Opposite of *empty*.
Change one letter.

A male cow.
Change one letter.

You need this for soccer.
Change one letter.

8

7

6

5

4

3

2

1

b e l l

Daily Word Ladders Grades 2–3 Scholastic Teaching Resources

Name _____

Read the clues, then write the words.
Start at the bottom and climb to the top.

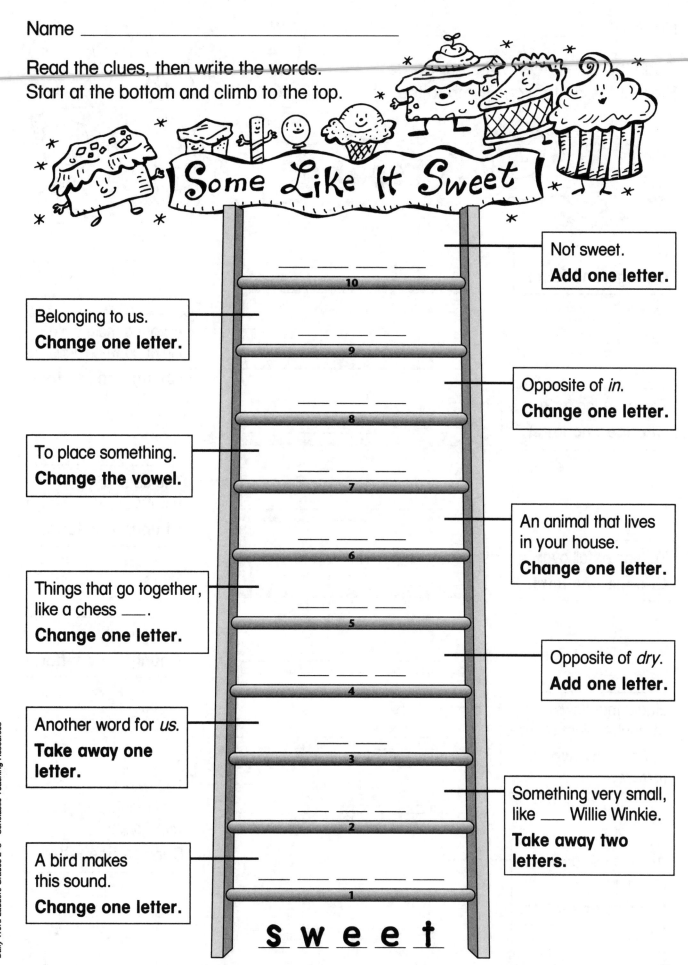

Some Like It Sweet

10 _ _ _ _ _ _ _ — Not sweet.
Add one letter.

Belonging to us.
Change one letter. — **9** _ _ _ _

Opposite of *in*.
Change one letter. — **8** _ _ _

To place something.
Change the vowel. — **7** _ _ _

An animal that lives
in your house.
Change one letter. — **6** _ _ _

Things that go together,
like a chess ___.
Change one letter. — **5** _ _ _

Opposite of *dry*.
Add one letter. — **4** _ _ _

Another word for *us*.
**Take away one
letter.** — **3** _ _

Something very small,
like ___ Willie Winkie.
**Take away two
letters.** — **2** _ _ _

A bird makes
this sound.
Change one letter. — **1** _ _ _ _ _

s w e e t

Name _____

Read the clues, then write the words.
Start at the bottom and climb to the top.

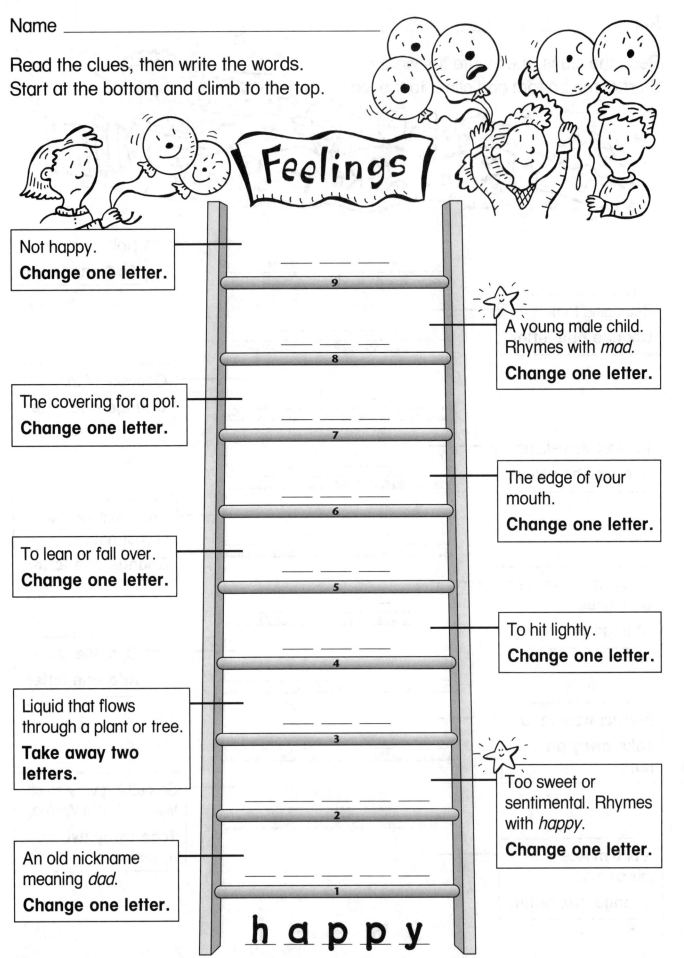

Feelings

Not happy.
Change one letter.

__ __ __ __ __   9

A young male child.
Rhymes with *mad*.
Change one letter.

__ __ __ __ __   8

The covering for a pot.
Change one letter.

__ __ __ __ __   7

The edge of your mouth.
Change one letter.

__ __ __ __ __   6

To lean or fall over.
Change one letter.

__ __ __ __ __   5

To hit lightly.
Change one letter.

__ __ __ __ __   4

Liquid that flows through a plant or tree.
Take away two letters.

__ __ __ __ __   3

Too sweet or sentimental. Rhymes with *happy*.
Change one letter.

__ __ __ __ __   2

An old nickname meaning *dad*.
Change one letter.

__ __ __ __ __   1

h a p p y

Daily Word Ladders Grades 2–3 Scholastic Teaching Resources

Name _____

Read the clues, then write the words.
Start at the bottom and climb to the top.

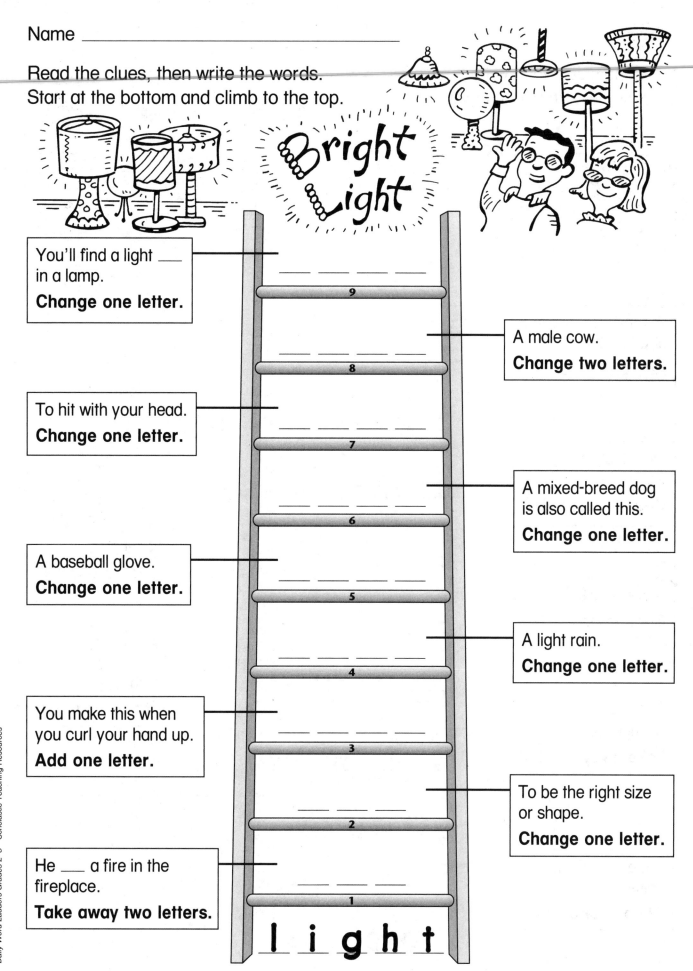

Bright Light

You'll find a light ___ in a lamp.
Change one letter.

A male cow.
Change two letters.

To hit with your head.
Change one letter.

A mixed-breed dog is also called this.
Change one letter.

A baseball glove.
Change one letter.

A light rain.
Change one letter.

You make this when you curl your hand up.
Add one letter.

To be the right size or shape.
Change one letter.

He ___ a fire in the fireplace.
Take away two letters.

9
8
7
6
5
4
3
2
1

l i g h t

Name _____

Read the clues, then write the words.
Start at the bottom and climb to the top.

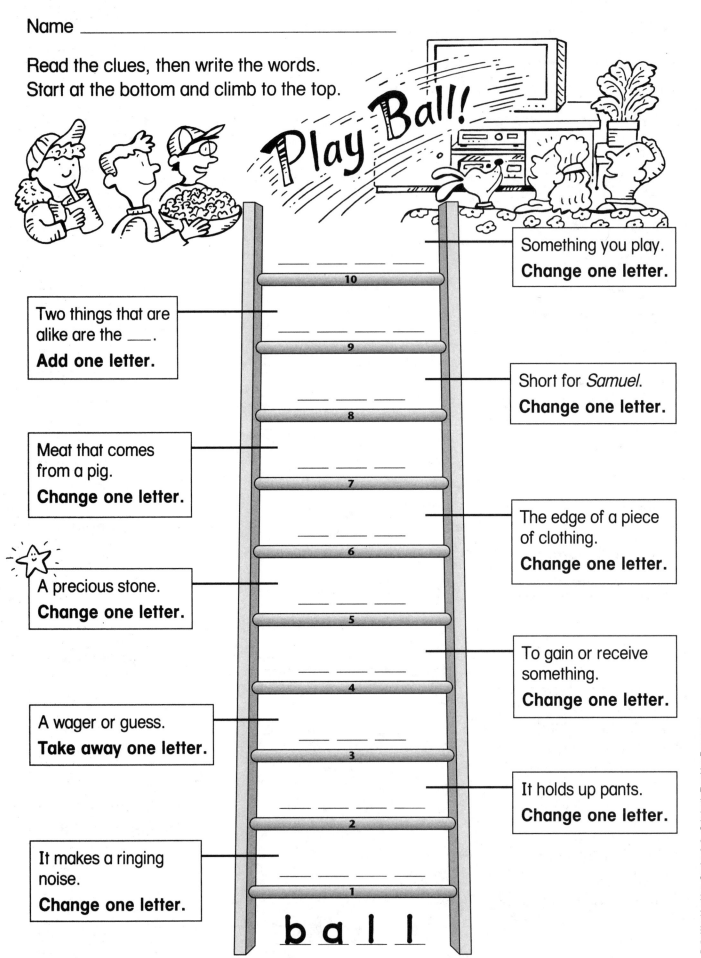

Play Ball!

Something you play.
Change one letter.

Two things that are alike are the ___.
Add one letter.

Short for *Samuel*.
Change one letter.

Meat that comes from a pig.
Change one letter.

The edge of a piece of clothing.
Change one letter.

A precious stone.
Change one letter.

To gain or receive something.
Change one letter.

A wager or guess.
Take away one letter.

It holds up pants.
Change one letter.

It makes a ringing noise.
Change one letter.

b a l l

10
9
8
7
6
5
4
3
2
1

Daily Word Ladders Grades 2–3 Scholastic Teaching Resources

Name _____

Read the clues, then write the words.
Start at the bottom and climb to the top.

Good, Clean Fun

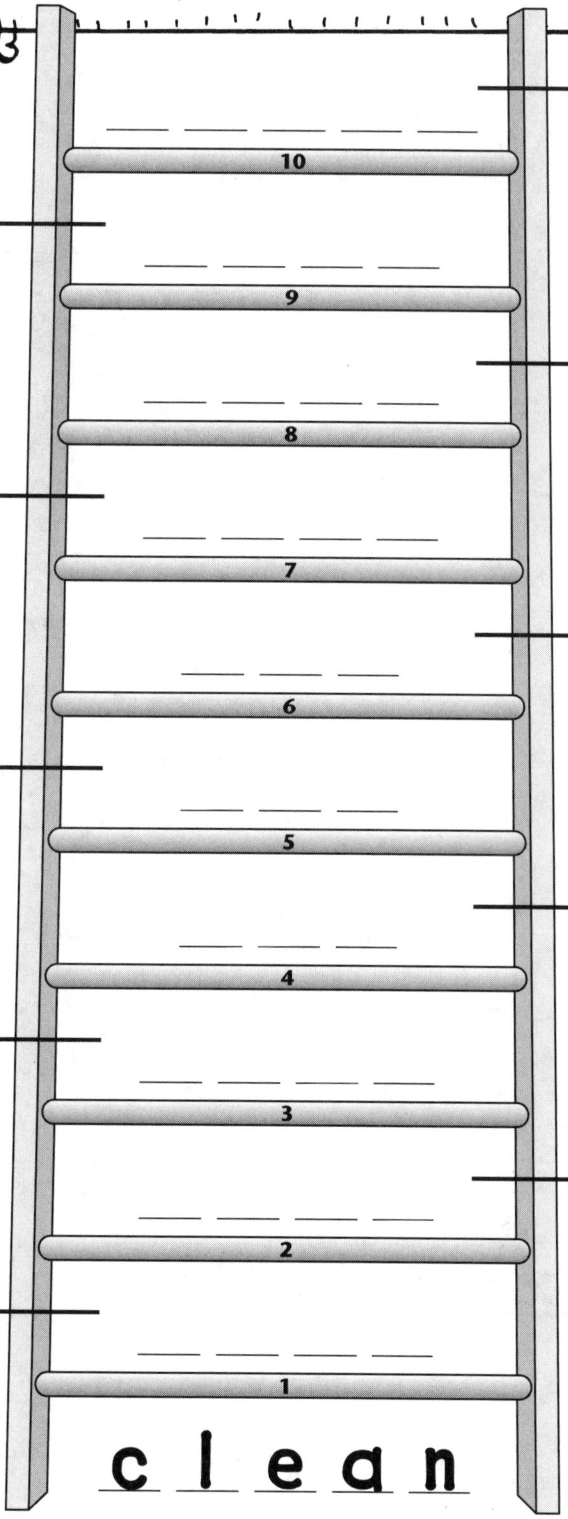

Not clean.
Add one letter.

Soil.
Change one letter.

You throw this pointed object at a board.
Change one letter.

A vehicle that carries things.
Add one letter.

An animal that is often a pet.
Change one letter.

A winged animal that flies at night.
Change one letter.

A wager or guess.
Take away one letter.

To win or be victorious.
Change one letter.

Something you eat, maybe with rice.
Change one letter.

Not fat.
Take away one letter.

10

9

8

7

6

5

4

3

2

1

c l e a n

Name _____

Read the clues, then write the words.
Start at the bottom and climb to the top.

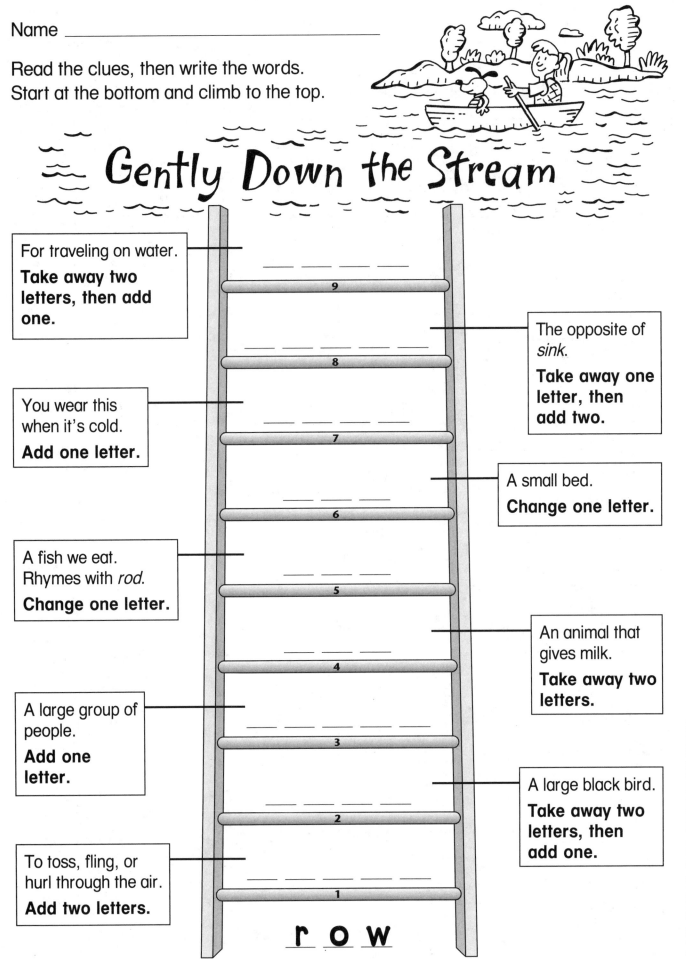

Gently Down the Stream

For traveling on water.
Take away two letters, then add one.

The opposite of *sink*.
Take away one letter, then add two.

You wear this when it's cold.
Add one letter.

A small bed.
Change one letter.

A fish we eat. Rhymes with *rod*.
Change one letter.

An animal that gives milk.
Take away two letters.

A large group of people.
Add one letter.

A large black bird.
Take away two letters, then add one.

To toss, fling, or hurl through the air.
Add two letters.

9
8
7
6
5
4
3
2
1

r o w

84

Daily Word Ladders Grades 2–3 Scholastic Teaching Resources

Name _____

Read the clues, then write the words.
Start at the bottom and climb to the top.

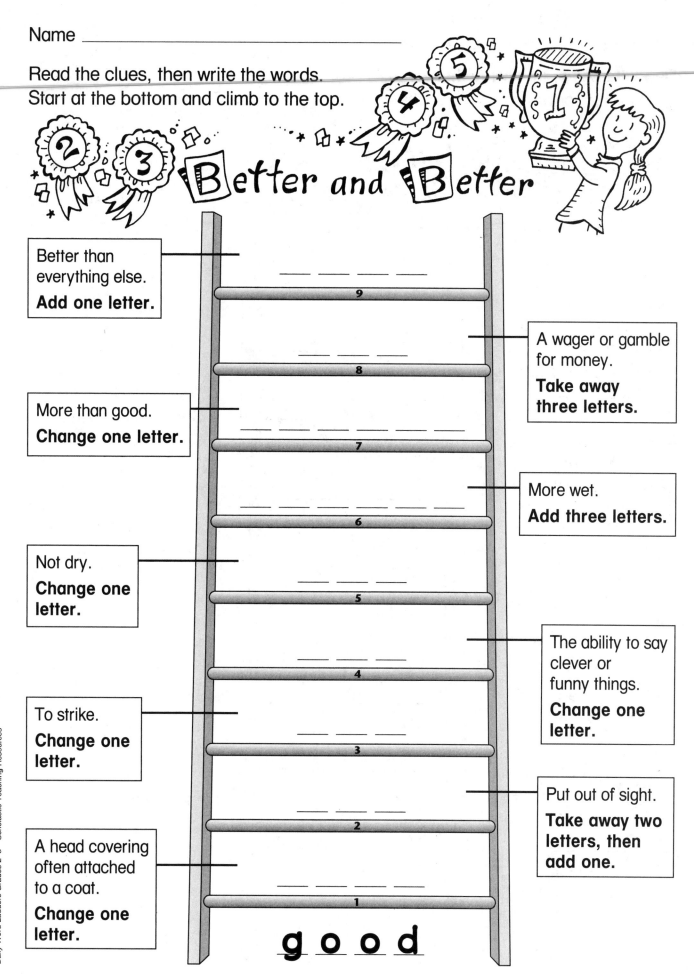

Better and Better

Better than everything else.
Add one letter.

9 ___ ___ ___ ___ ___

A wager or gamble for money.
Take away three letters.

8 ___ ___ ___ ___

More than good.
Change one letter.

7 ___ ___ ___ ___ ___ ___

More wet.
Add three letters.

6 ___ ___ ___

Not dry.
Change one letter.

5 ___ ___ ___

The ability to say clever or funny things.
Change one letter.

4 ___ ___ ___

To strike.
Change one letter.

3 ___ ___ ___

Put out of sight.
Take away two letters, then add one.

2 ___ ___ ___ ___

A head covering often attached to a coat.
Change one letter.

1 ___ ___ ___ ___

g o o d

Name _____

Read the clues, then write the words.
Start at the bottom and climb to the top.

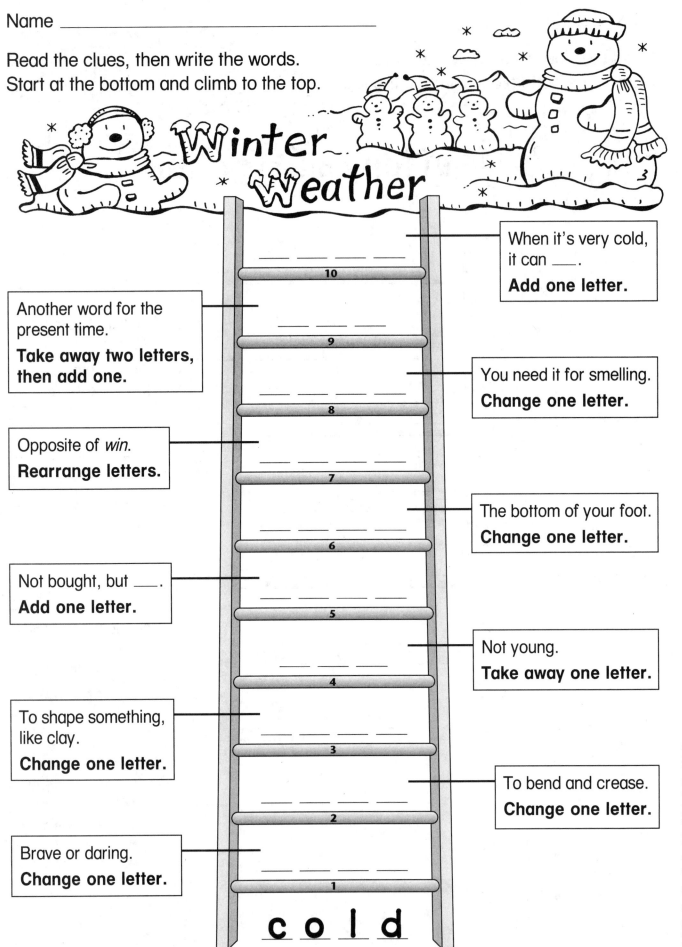

Winter Weather

When it's very cold, it can ___.
Add one letter.

Another word for the present time.
Take away two letters, then add one.

You need it for smelling.
Change one letter.

Opposite of *win*.
Rearrange letters.

The bottom of your foot.
Change one letter.

Not bought, but ___.
Add one letter.

Not young.
Take away one letter.

To shape something, like clay.
Change one letter.

To bend and crease.
Change one letter.

Brave or daring.
Change one letter.

10

9

8

7

6

5

4

3

2

1

c o l d

Daily Word Ladders Grades 2–3   Scholastic Teaching Resources

Name _____

Read the clues, then write the words.
Start at the bottom and climb to the top.

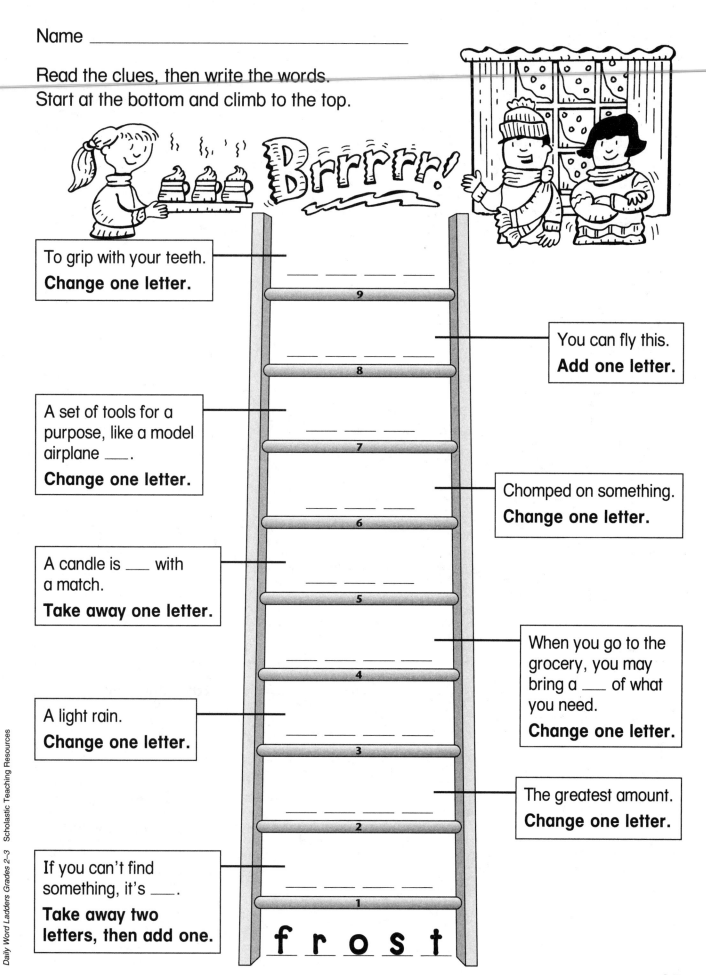

To grip with your teeth.
Change one letter.

9

You can fly this.
Add one letter.

8

A set of tools for a
purpose, like a model
airplane ___.
Change one letter.

7

Chomped on something.
Change one letter.

6

A candle is ___ with
a match.
Take away one letter.

5

When you go to the
grocery, you may
bring a ___ of what
you need.
Change one letter.

4

A light rain.
Change one letter.

3

The greatest amount.
Change one letter.

2

If you can't find
something, it's ___.
**Take away two
letters, then add one.**

1

f r o s t

Name _____

Read the clues, then write the words.
Start at the bottom and climb to the top.

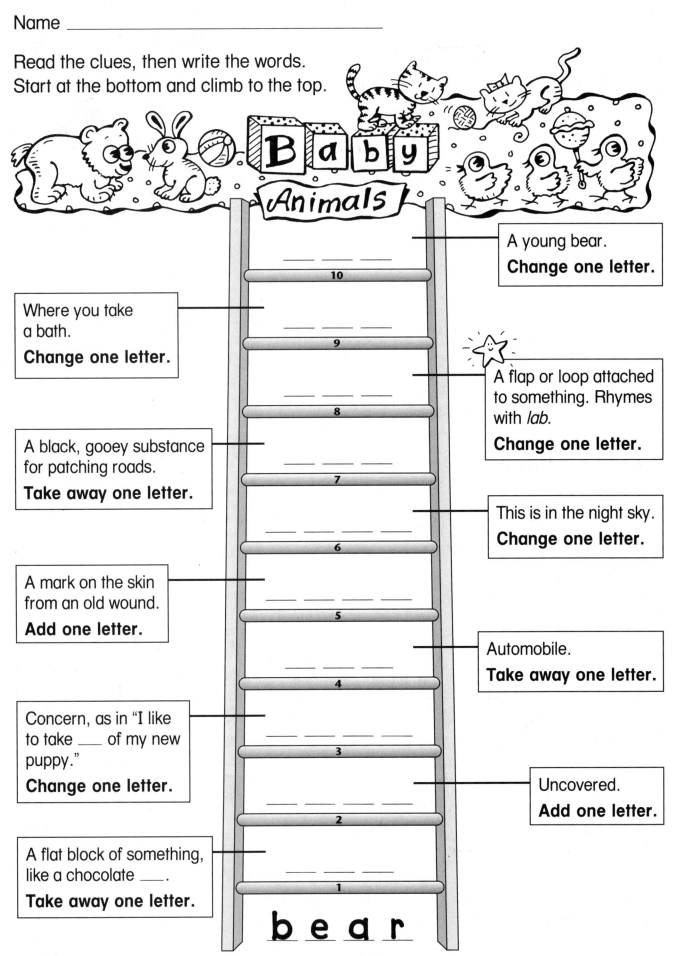

Baby Animals

A young bear.
Change one letter.

— — — — 10

Where you take
a bath.
Change one letter.

— — — 9

A flap or loop attached
to something. Rhymes
with *lab*.
Change one letter.

— — — 8

A black, gooey substance
for patching roads.
Take away one letter.

— — — 7

This is in the night sky.
Change one letter.

— — — 6

A mark on the skin
from an old wound.
Add one letter.

— — — 5

Automobile.
Take away one letter.

— — — 4

Concern, as in "I like
to take ___ of my new
puppy."
Change one letter.

— — — 3

Uncovered.
Add one letter.

— — 2

A flat block of something,
like a chocolate ___.
Take away one letter.

— — — 1

b e a r

Daily Word Ladders Grades 2–3   Scholastic Teaching Resources

Name _____

Read the clues, then write the words.
Start at the bottom and climb to the top.

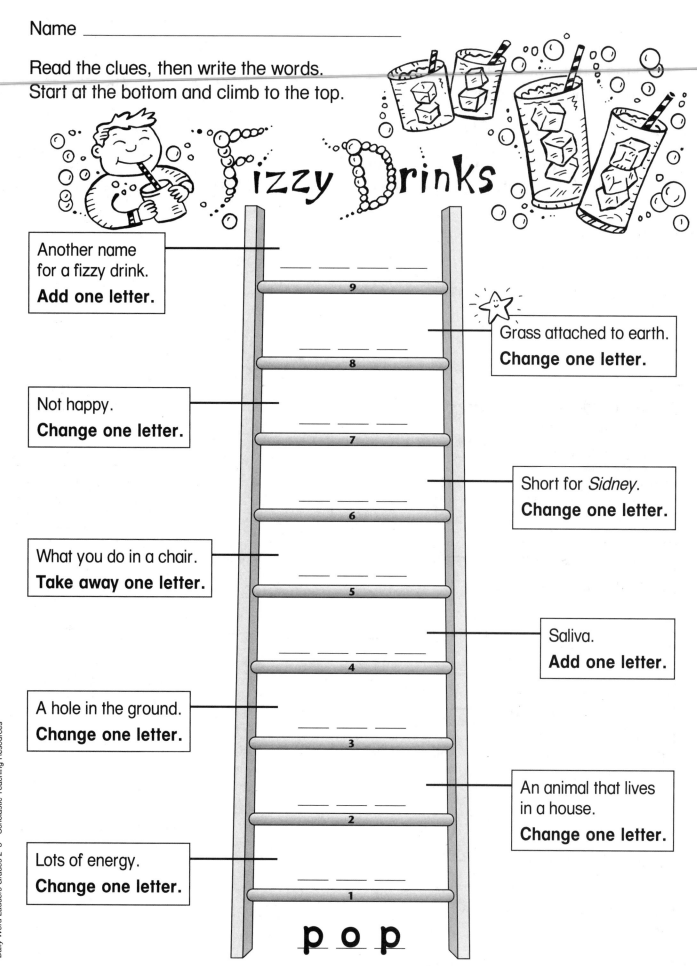

Fizzy Drinks

Another name for a fizzy drink.
Add one letter.

— — — — (9)

Grass attached to earth.
Change one letter.

— — — — (8)

Not happy.
Change one letter.

— — — — (7)

Short for *Sidney*.
Change one letter.

— — — — (6)

What you do in a chair.
Take away one letter.

— — — — (5)

Saliva.
Add one letter.

— — — — (4)

A hole in the ground.
Change one letter.

— — — — (3)

An animal that lives in a house.
Change one letter.

— — — — (2)

Lots of energy.
Change one letter.

— — — — (1)

p o p

Name _____

Read the clues, then write the words.
Start at the bottom and climb to the top.

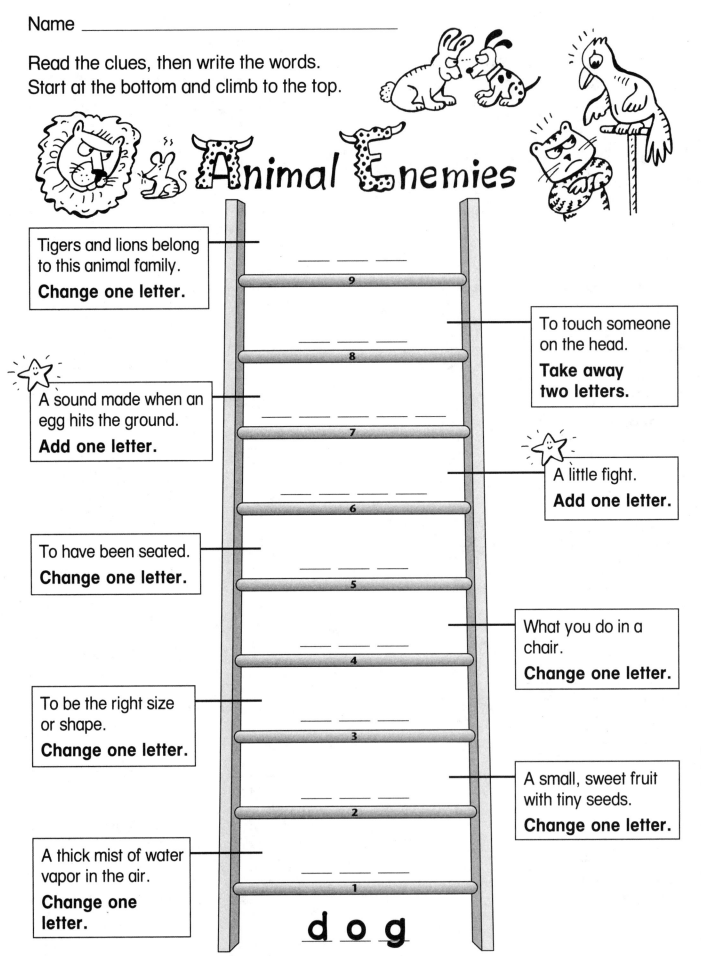

Animal Enemies

Tigers and lions belong to this animal family.
Change one letter.

9 ___ ___ ___ ___

To touch someone on the head.
Take away two letters.

8 ___ ___ ___ ___

A sound made when an egg hits the ground.
Add one letter.

7 ___ ___ ___ ___

A little fight.
Add one letter.

6 ___ ___ ___ ___

To have been seated.
Change one letter.

5 ___ ___ ___ ___

What you do in a chair.
Change one letter.

4 ___ ___ ___ ___

To be the right size or shape.
Change one letter.

3 ___ ___ ___ ___

A small, sweet fruit with tiny seeds.
Change one letter.

2 ___ ___ ___ ___

A thick mist of water vapor in the air.
Change one letter.

1 ___ ___ ___ ___

d o g

Daily Word Ladders Grades 2–3   Scholastic Teaching Resources

Read the clues, then write the words.
Start at the bottom and climb to the top.

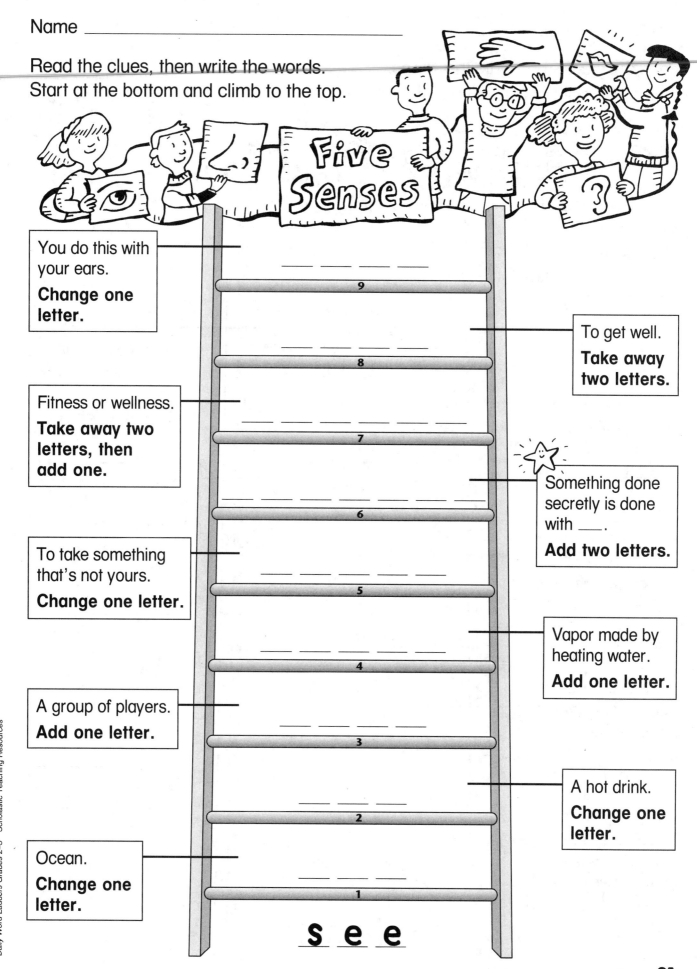

Five Senses

You do this with your ears.
Change one letter.

To get well.
Take away two letters.

Fitness or wellness.
Take away two letters, then add one.

Something done secretly is done with ___.
Add two letters.

To take something that's not yours.
Change one letter.

Vapor made by heating water.
Add one letter.

A group of players.
Add one letter.

A hot drink.
Change one letter.

Ocean.
Change one letter.

9
8
7
6
5
4
3
2
1

S e e

Name _____

Read the clues, then write the words.
Start at the bottom and climb to the top.

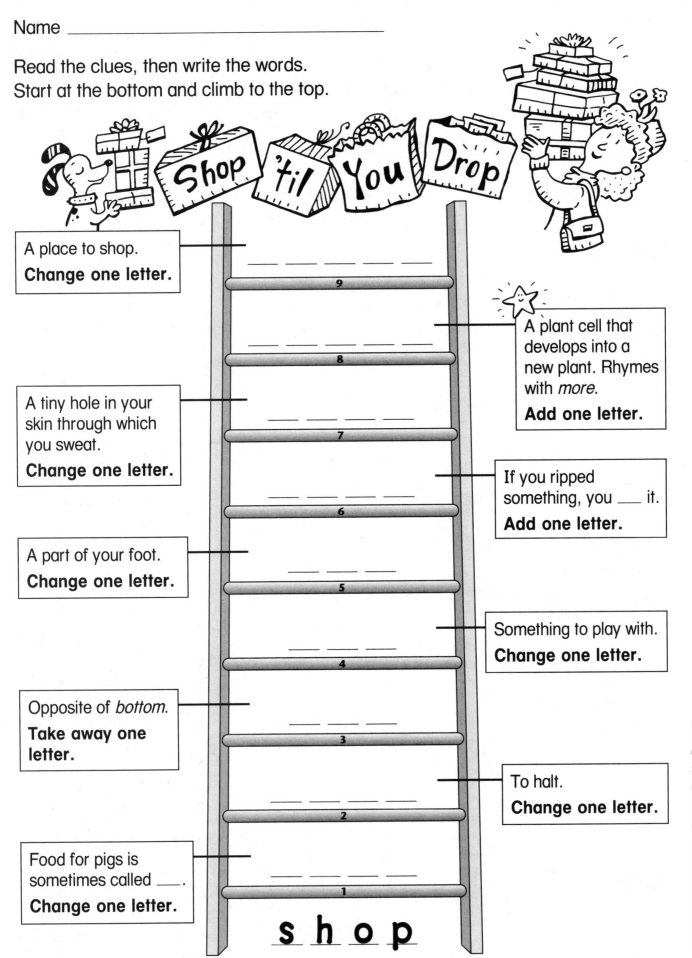

Shop 'til You Drop

A place to shop.
Change one letter.

9 _ _ _ _

8 _ _ _ _

A tiny hole in your
skin through which
you sweat.
Change one letter.

7 _ _ _ _

A plant cell that
develops into a
new plant. Rhymes
with *more*.
Add one letter.

6 _ _ _ _

If you ripped
something, you __ it.
Add one letter.

A part of your foot.
Change one letter.

5 _ _ _

4 _ _ _ _

Something to play with.
Change one letter.

Opposite of *bottom*.
**Take away one
letter.**

3 _ _ _

To halt.
Change one letter.

2 _ _ _ _

Food for pigs is
sometimes called ___.
Change one letter.

1 _ _ _ _

s h o p

Daily Word Ladders Grades 2–3 Scholastic Teaching Resources

Read the clues, then write the words.
Start at the bottom and climb to the top.

Cross-Country

Where the sun sets.
Add one letter.

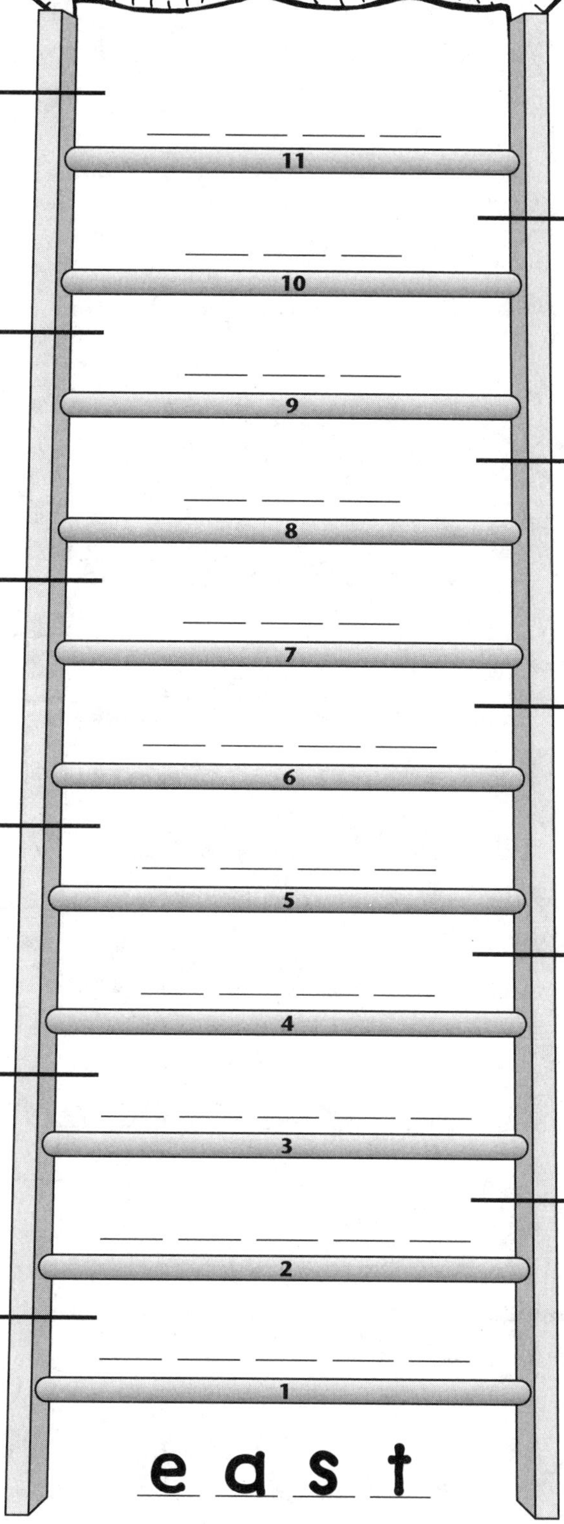

_____ 11

Not dry.
Change one letter.

_____ 10

An animal you keep
in your home.
Change one letter.

_____ 9

Short for *Patricia*.
Change one letter.

_____ 8

A rodent that looks
like a large mouse.
**Take away one
letter.**

_____ 7

To judge or grade
something.
Change one letter.

_____ 6

Opposite of *early*.
Rearrange letters.

_____ 5

A story.
**Take away
one letter.**

_____ 4

Not fresh.
Rearrange letters.

_____ 3

The fewest or lowest.
Change one letter.

_____ 2

Beauty and the ___.
Add one letter.

_____ 1

e a s t

Name _____

Read the clues, then write the words.
Start at the bottom and climb to the top.

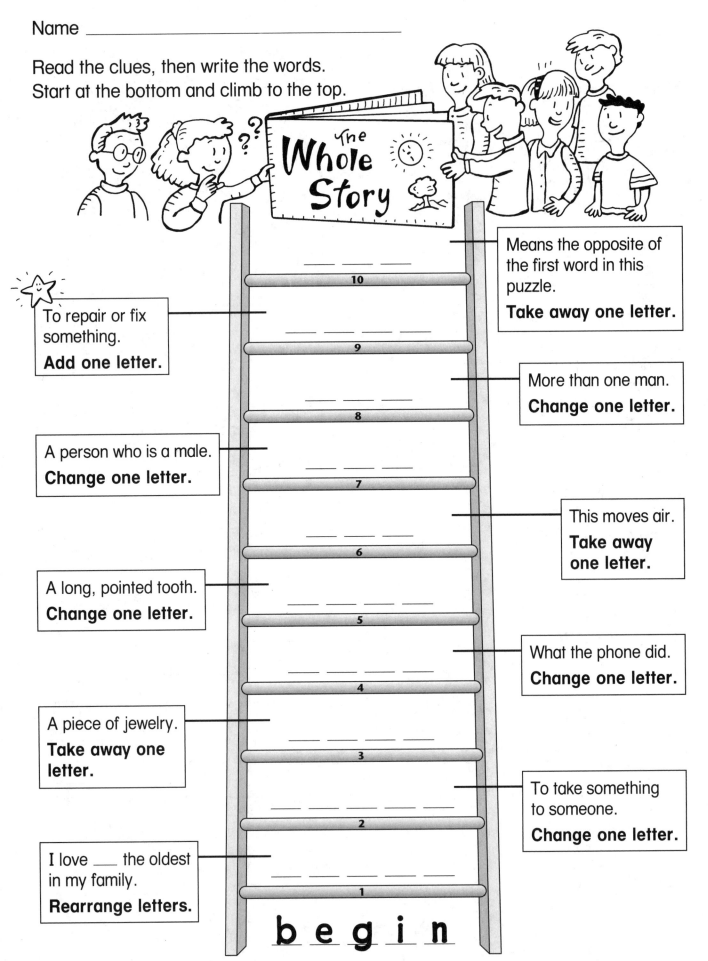

The Whole Story

Means the opposite of the first word in this puzzle.
Take away one letter.

To repair or fix something.
Add one letter.

More than one man.
Change one letter.

A person who is a male.
Change one letter.

This moves air.
Take away one letter.

A long, pointed tooth.
Change one letter.

What the phone did.
Change one letter.

A piece of jewelry.
Take away one letter.

To take something to someone.
Change one letter.

I love ___ the oldest in my family.
Rearrange letters.

10

9

8

7

6

5

4

3

2

1

b e g i n

Daily Word Ladders Grades 2–3   Scholastic Teaching Resources

Name _____

Read the clues, then write the words.
Start at the bottom and climb to the top.

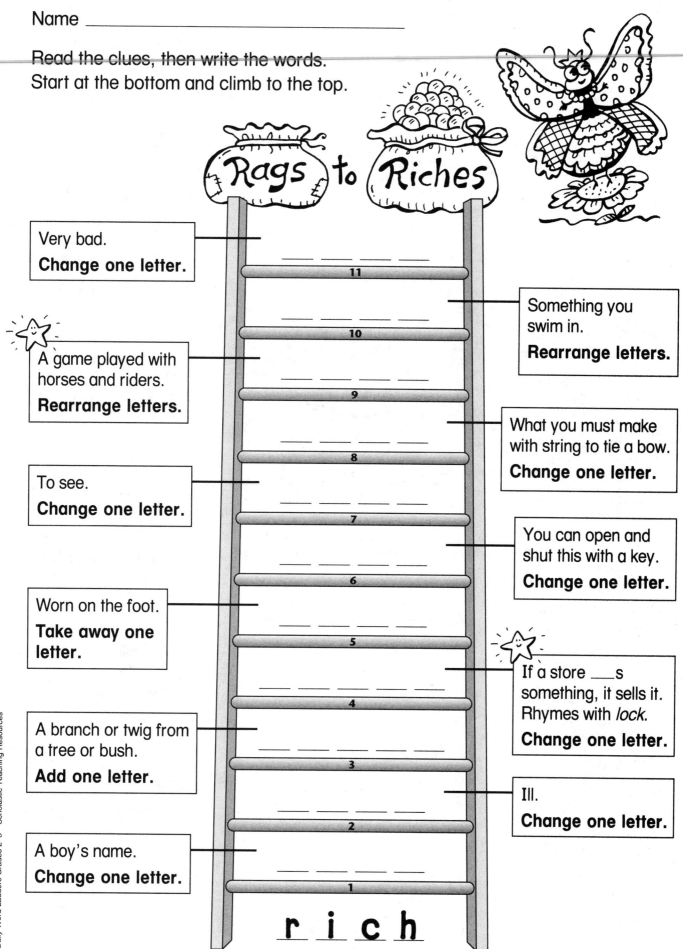

Rags to Riches

Very bad.
Change one letter.

Something you swim in.
Rearrange letters.

A game played with horses and riders.
Rearrange letters.

What you must make with string to tie a bow.
Change one letter.

To see.
Change one letter.

You can open and shut this with a key.
Change one letter.

Worn on the foot.
Take away one letter.

If a store ___s something, it sells it. Rhymes with *lock*.
Change one letter.

A branch or twig from a tree or bush.
Add one letter.

Ill.
Change one letter.

A boy's name.
Change one letter.

11
10
9
8
7
6
5
4
3
2
1

r i c h

Name _____

Read the clues, then write the words.
Start at the bottom and climb to the top.

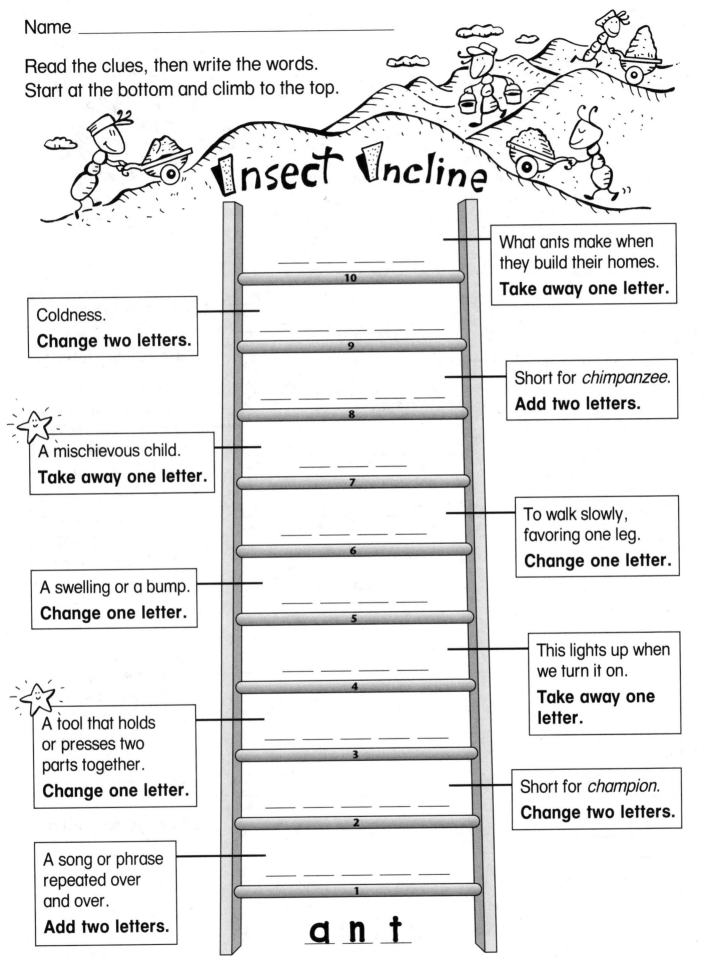

Insect Incline

What ants make when they build their homes.
Take away one letter.

Coldness.
Change two letters.

Short for *chimpanzee*.
Add two letters.

A mischievous child.
Take away one letter.

To walk slowly, favoring one leg.
Change one letter.

A swelling or a bump.
Change one letter.

This lights up when we turn it on.
Take away one letter.

A tool that holds or presses two parts together.
Change one letter.

Short for *champion*.
Change two letters.

A song or phrase repeated over and over.
Add two letters.

10
9
8
7
6
5
4
3
2
1

a n t

Daily Word Ladders Grades 2–3   Scholastic Teaching Resources

Name _____

Read the clues, then write the words.
Start at the bottom and climb to the top.

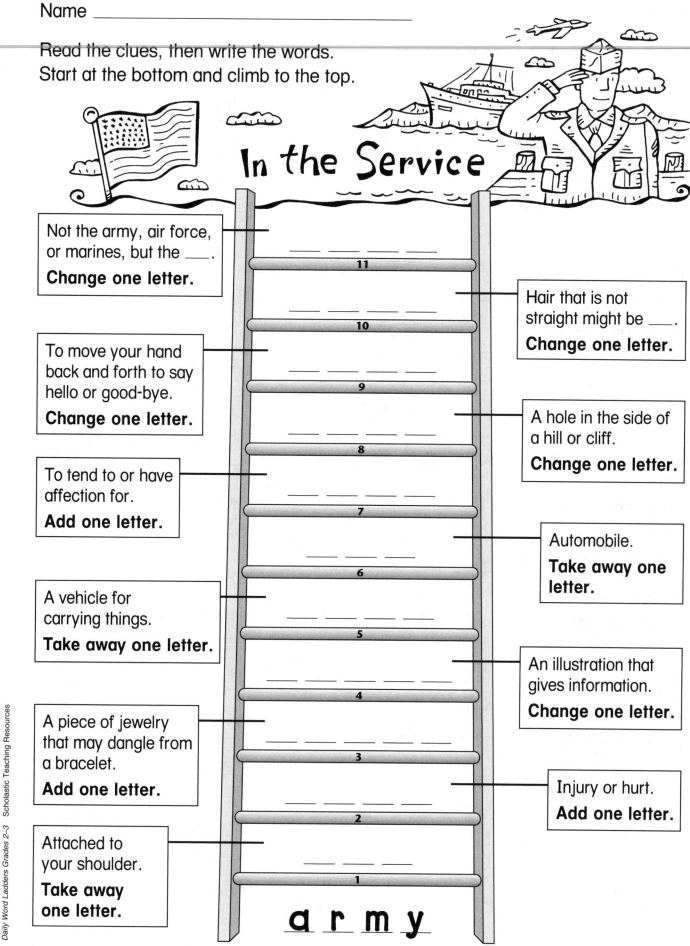

In the Service

Not the army, air force, or marines, but the ___.
Change one letter.

_ _ _ _ 11

Hair that is not straight might be ___.
Change one letter.

_ _ _ _ 10

To move your hand back and forth to say hello or good-bye.
Change one letter.

_ _ _ _ 9

A hole in the side of a hill or cliff.
Change one letter.

_ _ _ _ 8

To tend to or have affection for.
Add one letter.

_ _ _ _ 7

Automobile.
Take away one letter.

_ _ _ 6

A vehicle for carrying things.
Take away one letter.

_ _ _ _ 5

An illustration that gives information.
Change one letter.

_ _ _ _ 4

A piece of jewelry that may dangle from a bracelet.
Add one letter.

_ _ _ _ _ 3

Injury or hurt.
Add one letter.

_ _ _ _ 2

Attached to your shoulder.
Take away one letter.

_ _ _ 1

a r m y

Name _____

Read the clues, then write the words.
Start at the bottom and climb to the top.

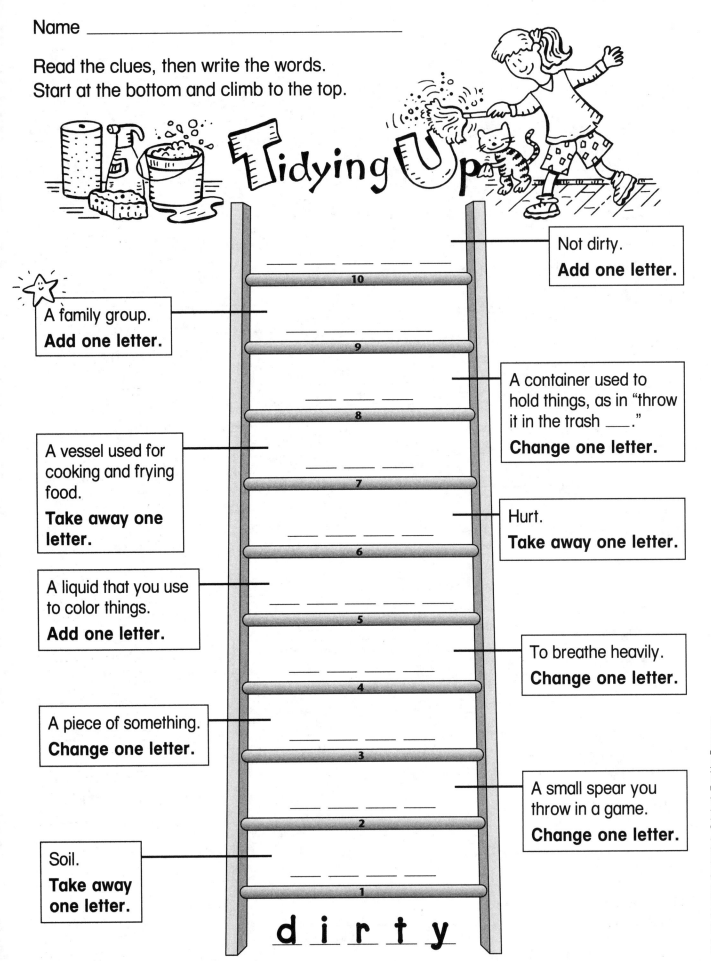

Tidying Up

Not dirty.
Add one letter.

A family group.
Add one letter.

A container used to
hold things, as in "throw
it in the trash ___."
Change one letter.

A vessel used for
cooking and frying
food.
**Take away one
letter.**

Hurt.
Take away one letter.

A liquid that you use
to color things.
Add one letter.

To breathe heavily.
Change one letter.

A piece of something.
Change one letter.

A small spear you
throw in a game.
Change one letter.

Soil.
**Take away
one letter.**

10
9
8
7
6
5
4
3
2
1

d i r t y

Daily Word Ladders Grades 2–3   Scholastic Teaching Resources

Name _____

Read the clues, then write the words.
Start at the bottom and climb to the top.

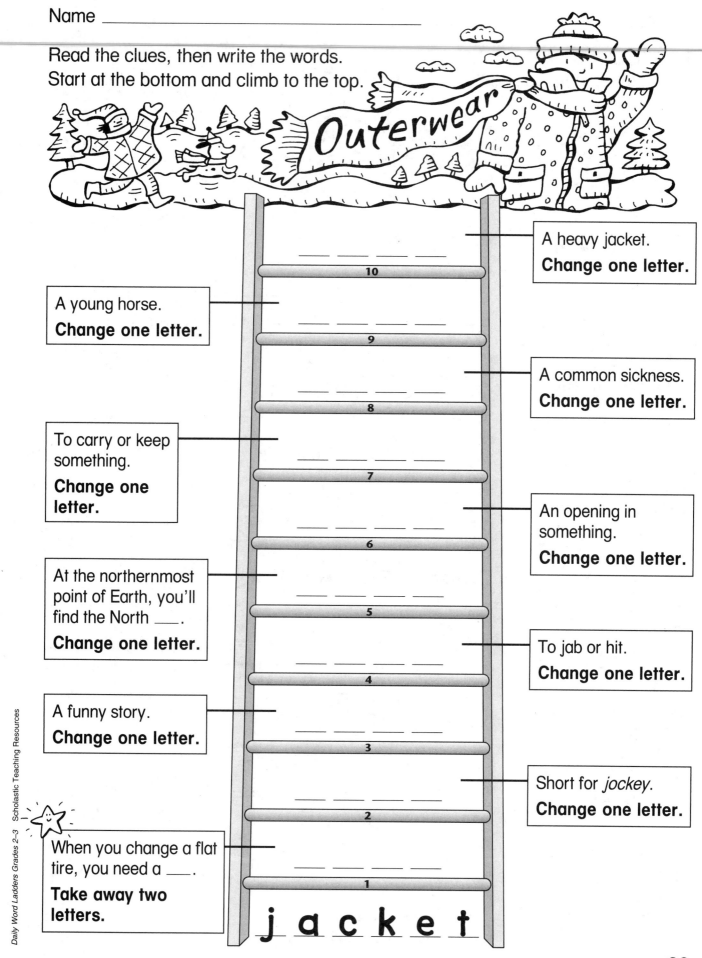

A heavy jacket.
Change one letter.

A young horse.
Change one letter.

A common sickness.
Change one letter.

To carry or keep something.
Change one letter.

An opening in something.
Change one letter.

At the northernmost point of Earth, you'll find the North ___.
Change one letter.

To jab or hit.
Change one letter.

A funny story.
Change one letter.

Short for *jockey*.
Change one letter.

When you change a flat tire, you need a ___.
Take away two letters.

10

9

8

7

6

5

4

3

2

1

j a c k e t

Name _____

Read the clues, then write the words.
Start at the bottom and climb to the top.

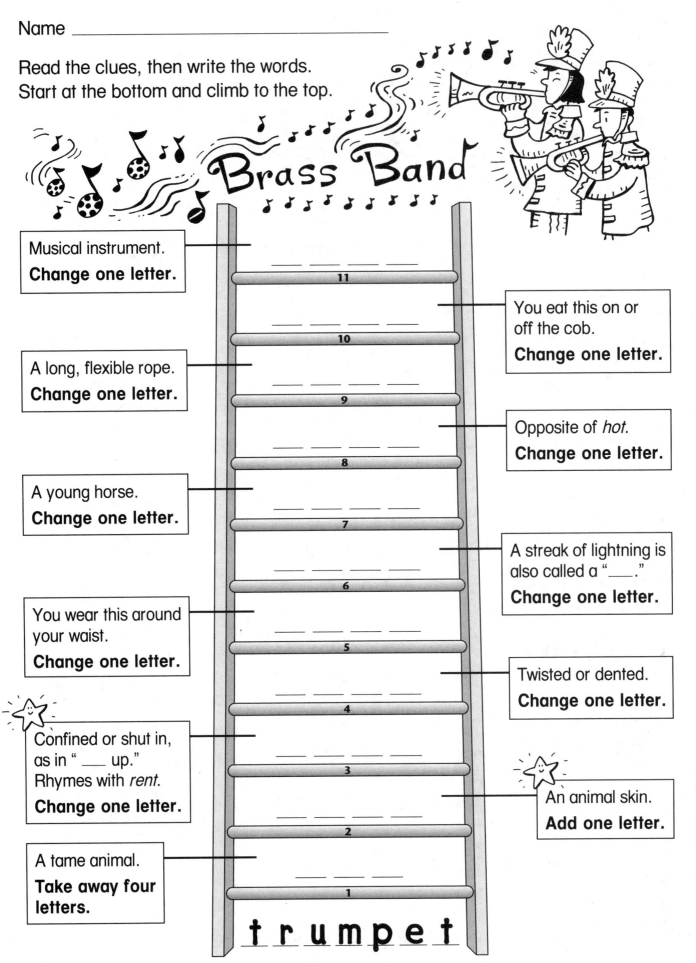

Brass Band

Musical instrument.
Change one letter.

A long, flexible rope.
Change one letter.

A young horse.
Change one letter.

You wear this around
your waist.
Change one letter.

Confined or shut in,
as in " ___ up."
Rhymes with *rent*.
Change one letter.

A tame animal.
**Take away four
letters.**

You eat this on or
off the cob.
Change one letter.

Opposite of *hot*.
Change one letter.

A streak of lightning is
also called a " ___ ."
Change one letter.

Twisted or dented.
Change one letter.

An animal skin.
Add one letter.

11
10
9
8
7
6
5
4
3
2
1

t r u m p e t

Daily Word Ladders Grades 2–3 Scholastic Teaching Resources

Read the clues, then write the words.
Start at the bottom and climb to the top.

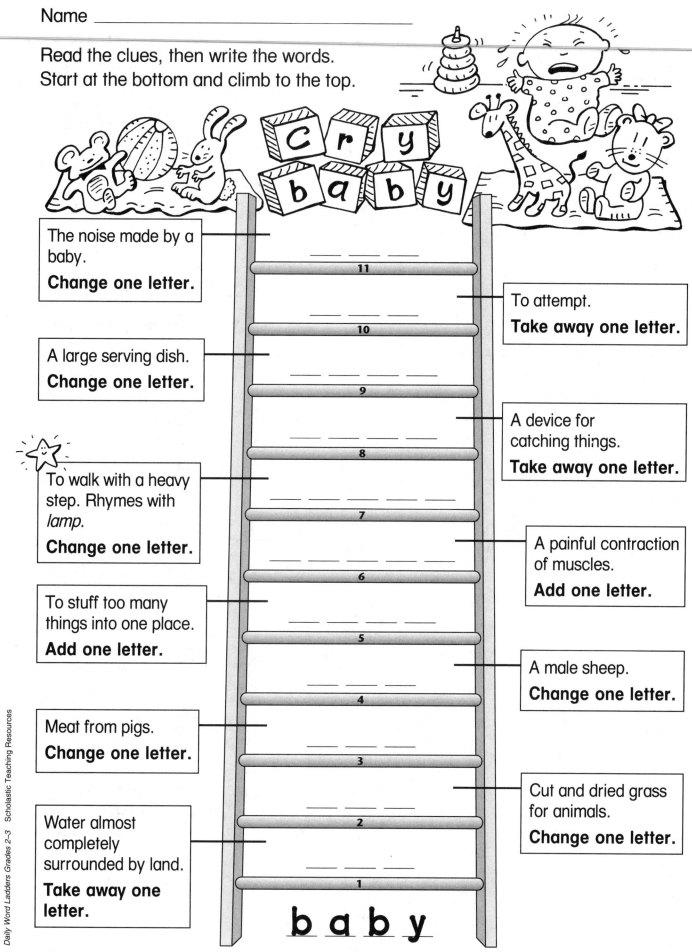

The noise made by a baby.
Change one letter.

A large serving dish.
Change one letter.

To walk with a heavy step. Rhymes with *lamp*.
Change one letter.

To stuff too many things into one place.
Add one letter.

Meat from pigs.
Change one letter.

Water almost completely surrounded by land.
Take away one letter.

To attempt.
Take away one letter.

A device for catching things.
Take away one letter.

A painful contraction of muscles.
Add one letter.

A male sheep.
Change one letter.

Cut and dried grass for animals.
Change one letter.

11

10

9

8

7

6

5

4

3

2

1

b a b y

Name _____

Read the clues, then write the words.
Start at the bottom and climb to the top.

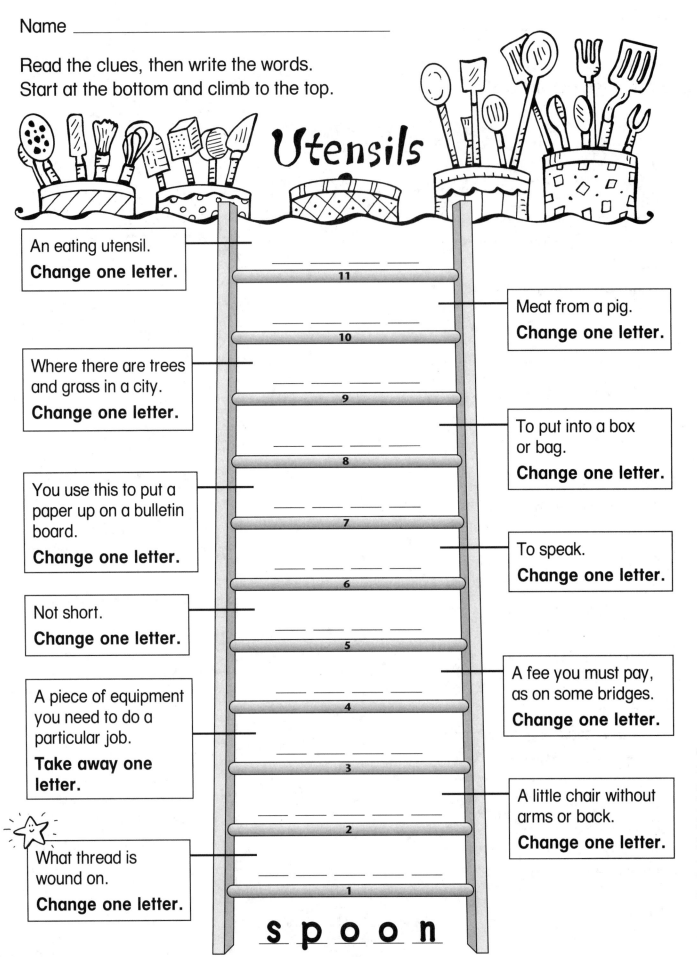

Utensils

An eating utensil.
Change one letter.

Where there are trees
and grass in a city.
Change one letter.

You use this to put a
paper up on a bulletin
board.
Change one letter.

Not short.
Change one letter.

A piece of equipment
you need to do a
particular job.
**Take away one
letter.**

What thread is
wound on.
Change one letter.

Meat from a pig.
Change one letter.

To put into a box
or bag.
Change one letter.

To speak.
Change one letter.

A fee you must pay,
as on some bridges.
Change one letter.

A little chair without
arms or back.
Change one letter.

11

10

9

8

7

6

5

4

3

2

1

s p o o n

Daily Word Ladders Grades 2–3   Scholastic Teaching Resources

Name _____

Read the clues, then write the words.
Start at the bottom and climb to the top.

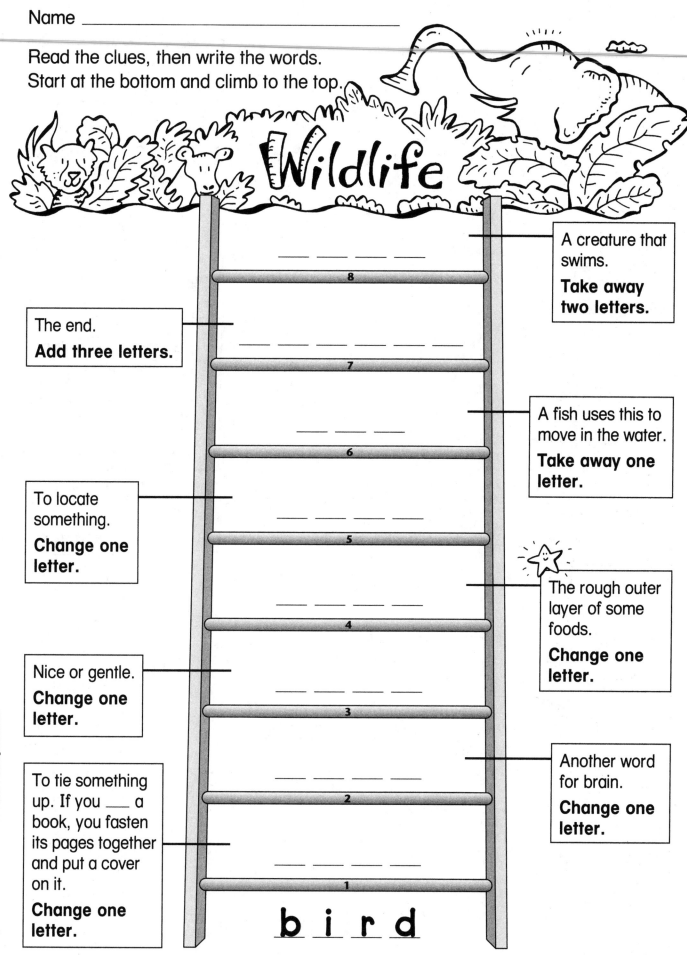

Wildlife

A creature that swims.
Take away two letters.

The end.
Add three letters.

8 _ _ _ _ _

7 _ _ _ _ _ _

A fish uses this to move in the water.
Take away one letter.

6 _ _ _ _

To locate something.
Change one letter.

5 _ _ _ _

The rough outer layer of some foods.
Change one letter.

4 _ _ _ _

Nice or gentle.
Change one letter.

3 _ _ _ _

Another word for brain.
Change one letter.

2 _ _ _ _

To tie something up. If you ___ a book, you fasten its pages together and put a cover on it.
Change one letter.

1 _ _ _ _

b i r d

Name _____

Read the clues, then write the words.
Start at the bottom and climb to the top.

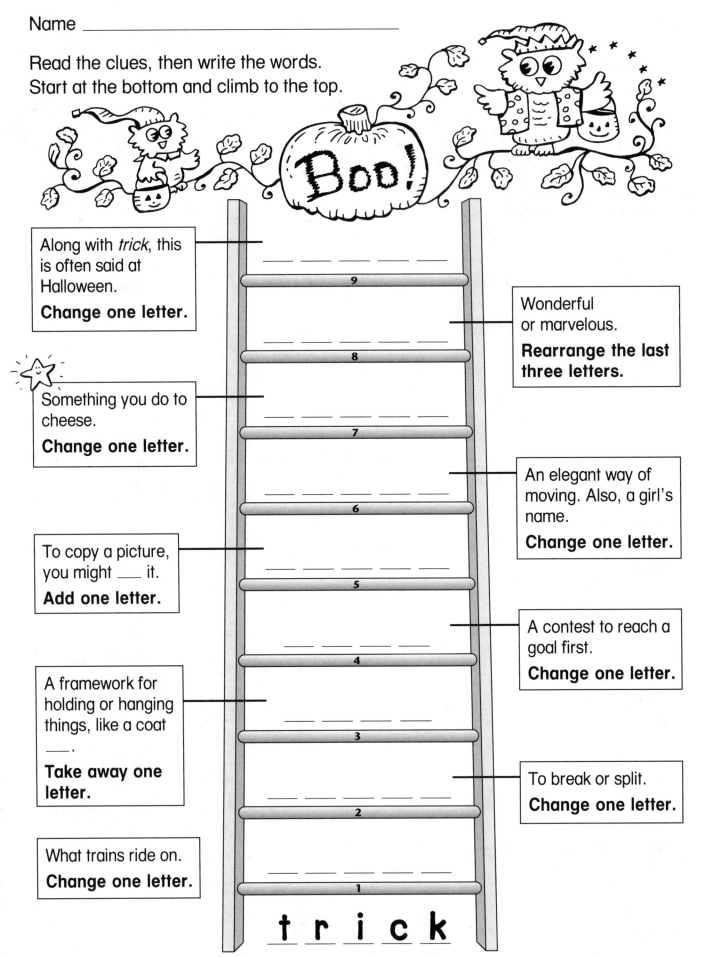

Along with *trick*, this is often said at Halloween.
Change one letter.

Something you do to cheese.
Change one letter.

To copy a picture, you might ___ it.
Add one letter.

A framework for holding or hanging things, like a coat ___.
Take away one letter.

What trains ride on.
Change one letter.

Wonderful or marvelous.
Rearrange the last three letters.

An elegant way of moving. Also, a girl's name.
Change one letter.

A contest to reach a goal first.
Change one letter.

To break or split.
Change one letter.

9

8

7

6

5

4

3

2

1

t r i c k

Daily Word Ladders Grades 2–3   Scholastic Teaching Resources

Name _____

Read the clues, then write the words.
Start at the bottom and climb to the top.

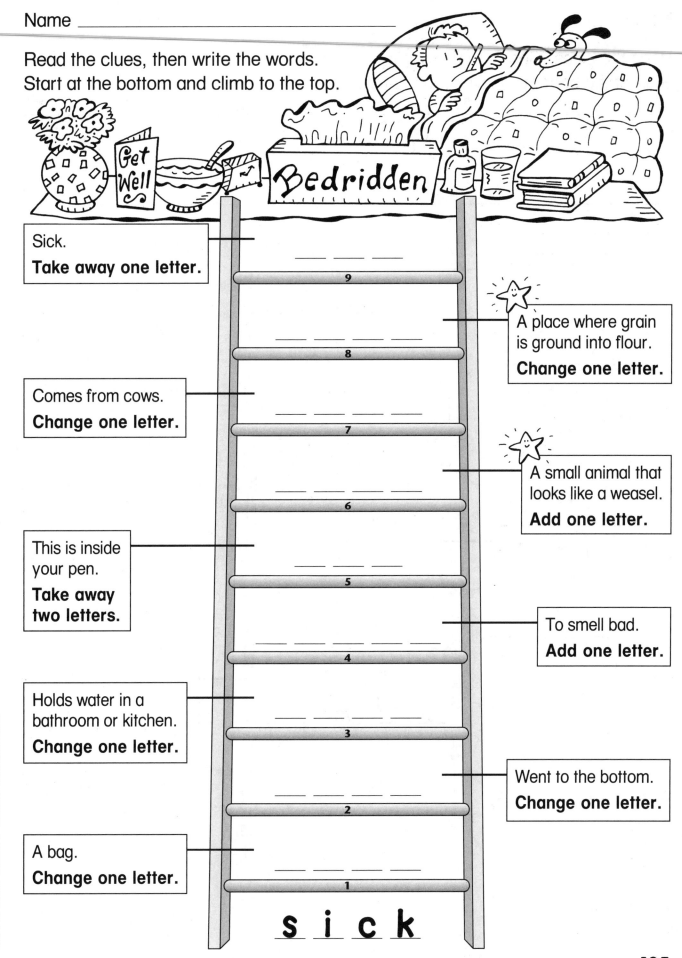

Get Well

Bedridden

Sick.
Take away one letter.

— — — — 9

A place where grain
is ground into flour.
Change one letter.

— — — — 8

Comes from cows.
Change one letter.

— — — — 7

A small animal that
looks like a weasel.
Add one letter.

— — — — 6

This is inside
your pen.
**Take away
two letters.**

— — — 5

Holds water in a
bathroom or kitchen.
Change one letter.

— — — — 4

To smell bad.
Add one letter.

— — — 3

Went to the bottom.
Change one letter.

— — — 2

A bag.
Change one letter.

— — — — 1

s i c k

Name _____

Read the clues, then write the words.
Start at the bottom and climb to the top.

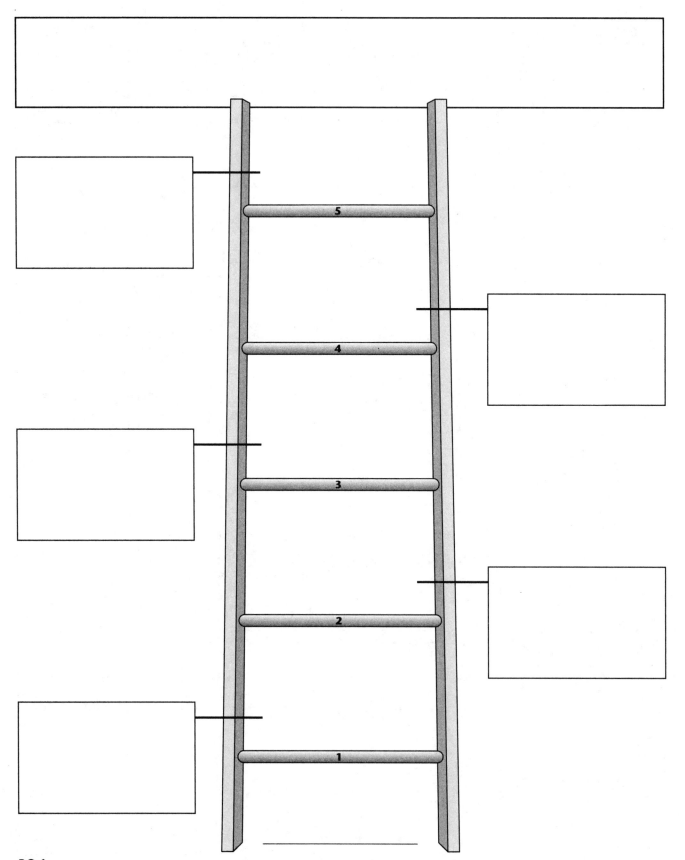

Daily Word Ladders Grades 2–3 Scholastic Teaching Resources

Name _____

Read the clues, then write the words.
Start at the bottom and climb to the top.

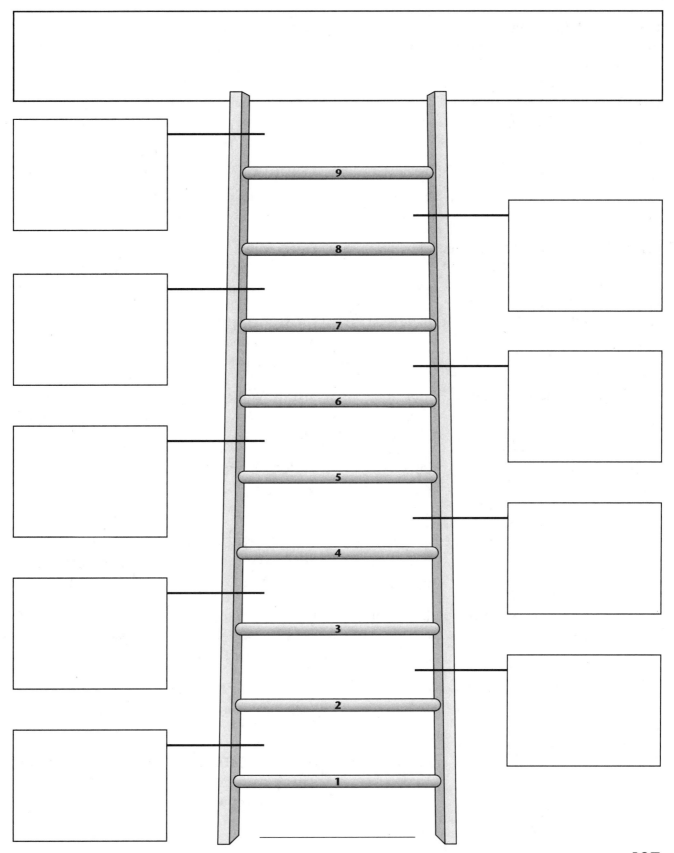

Answer Key

Counting Up, page 7
few, new, now, mow, meow, men, hen, ten, tan, man, many

Home, Sweet Home, page 8
wigwam, wig, wag, sag, Sam, same, some, home

In the Doghouse, page 9
dog, dot, Don, nod, cod, code, coke, cone, bone

Shrinking Sizes, page 10
little, lit, it, ill, mill, mall, small

All Wet, page 11
wet, we, well, bell, belt, bet, bat, bad, Brad, bran, ran, rain

Go, Team!, page 12
team, meat, tame, take, tale, tall, toll, ton, torn, worn, work

Sweet Seasons, page 13
spring, string, sting, sing, sling, slim, slime, slimmer, simmer, summer

Art Smart, page 14
art, cart, car, cat, scat, sat, saw, raw, straw, draw

Sleepytime, page 15
sleep, seep, sheep, peep, peel, pail, pal, lap, clap, slap, sap, nap

More or Less, page 16
most, mast, mask, cask, cast, coast, cost, lost, last, least

Barbershop, page 17
hair, chair, char, chart, charm, harm, arm, ram, rat, rut, cut

Inside Out, page 18
door, floor, flop, lop, lip, lit, wit, win, wind, window

Weighty Matters, page 19
pint, paint, pant, chant, chart, cart, tart, part, art, quart

Restful Vacation, page 20
hotel, hot, hat, hate, haste, host, shot, shoot, sheet, sleep

Gardening, page 21
flower, flow, slow, show, shower, shopper, shop, stop, top, pot

Birdsong, page 22
bird, bid, bed, shed, sled, slid, ship, hip, chip, chirp

Shady Glade, page 23
shade, shape, shame, same, Sam, sap, sip, seep, see, tee, tree

School Days, page 24
school, cool, coal, coat, cat, can't, cent, gent, dent, student

Transportation, page 25
boat, float, moat, mat, pat, pan, plan, plank, plant, plane

Good Cooking, page 26
fried, freed, feed, fed, bed, bad, bar, bark, barn, bare, baked

America's Pastime, page 27
base, bass, mass, miss, mist, mast, melt, mall, small, ball

Sweet Treats, page 28
candy, can, cat, sat, Sam, seem, seam, seat, sweat, sweet